Environmental Law for Biologists

Environmental Law for Biologists

TRISTAN KIMBRELL

The University of Chicago Press

CHICAGO AND LONDON

TRISTAN KIMBRELL is an environmental writer who focuses on the intersection of law and biology. He has a JD from Temple University and a PhD in ecology from the University of Florida and has taught at Southern University, New Orleans.

The University of Chicago Press, Chicago 60637
The University of Chicago Press, Ltd., London
© 2016 by The University of Chicago
All rights reserved. Published 2016.
Printed in the United States of America

25 24 23 22 21 20 19 18 17 16   1 2 3 4 5

ISBN-13: 978-0-226-33368-7 (cloth)
ISBN-13: 978-0-226-33385-4 (paper)
ISBN-13: 978-0-226-33371-7 (e-book)

DOI: 10.7208/chicago/9780226333717.001.0001

Library of Congress Cataloging-in-Publication Data

Kimbrell, Tristan, author.
Environmental law for biologists / Tristan Kimbrell.
pages ; cm
Includes bibliographical references and index.
ISBN 978-0-226-33368-7 (cloth : alk. paper) — ISBN 978-0-226-33385-4
(pbk. : alk. paper) — ISBN 978-0-226-33371-7 (ebook) 1. Environmental law—
United States. 2. Law and biology. I. Title.
KF3775.K56 2016
344.7304′6—dc23   2015019940

♾ This paper meets the requirements of
ANSI/NISO Z39.48-1992 (Permanence of Paper).

Contents

Boxes

Preface

Environmental laws have an enormous influence on species, ecosystems, and landscapes. Environmental laws decide which species are protected from human interference, and which are hunted. Environmental laws control the management of nature reserves, and determine where development is encouraged. Environmental laws limit the release of some pollutants, and create incentives for the release of others.

Despite the importance of environmental law, many ecology and wildlife biology students learn little about the subject before graduating from college or graduate school. This in turn creates research scientists and environmental professionals who spend relatively little time thinking about how environmental law impacts the environment. A search of the Web of Science database illustrates this point. The introduction and spread of nonnative species is one of the primary causes of species extinction in the United States. The Lacey Act (discussed in chapter 2) is the most important federal statute prohibiting the import of nonnative species into the United States, and the transport of those species between the states. Consequently, the Lacey Act is a vital tool for stopping the introduction and spread of nonnative species, and possibly helping to prevent extinctions. Web of Science, however, returns only 14 papers that have the Lacey Act as a topic, and 2 of those papers are law review articles.

To be fair, other environmental laws have received considerably more attention from scientists. For example, searching for "Endangered Species Act" returns 1,146 articles, with roughly 1,000 dealing directly with ecological issues. But even for environmental statutes that do receive attention from ecologists, engagement with the statutes tends to be fairly superficial. The precise language of a statute is important because it draws a sharp line around the subject of the statute, and determines how the statute is meant to influence that subject. For instance, the species listed as threatened or endangered under the Endangered Species Act receive considerable federal protection, while species that are not listed receive zero protection. The precise words of the statute that determine which species are listed and which are not become very important. Just as important are the federal regulations and court decisions interpreting the language of statutes. These regulations and court decisions provide the specific rules determining how statutes are implemented. All these components must be examined together to understand how environmental law affects species and ecosystems.

To provide a more in-depth understanding of environmental law, the chapters in this book explain the details of the most important US environmental statutes, regulations, and court decisions. At the same time, the chapters also discuss ecological research that shows how those details work together to influence the environment. As a result, this book examines the intersection of law and ecology, and will hopefully be useful for ecology students and environmental professionals interested in understanding how the law affects ecological systems.

Admittedly, it was a struggle in several chapters to explain the intersection of law and ecology because there was frequently very little ecological research touching on a particular environmental law or regulation. Consequently, this book attempts to highlight areas where additional scientific research is needed to better understand how environmental law influences ecological systems. It would be wonderful if students find this book useful as a source of ideas for choosing areas for future research. Such research may eventually influence future environmental statutes, regulations, and court decisions, thereby improving the way the law protects the environment.

This book will hopefully also be useful for law students interested in environmental law. The vast majority of books on environmental law spend very little time explaining current ecological research, and include very little discussion of how the law influences species and ecosystems. A greater understanding of the effects of environmental law on ecosystems should lead to a better understanding of why certain statutes are structured the way they are, and why agencies implement those statutes the way they do.

Most of the laws discussed in the book are US federal laws, but state and international law are examined when they have a significant influence on species and ecosystems. The book contains several text boxes that explain basic legal concepts, as well as boxes that explain basic ecological concepts. If a particular concept is familiar to a reader, the text box can safely be skipped.

The book is divided into an introductory chapter and four parts. The first part focuses on laws that directly regulate individual species of wild plants and animals. The second part focuses on laws that regulate how land is used, with the third part examining laws regulating water. The fourth part focuses on the air, with the final chapter devoted to laws that have a direct relation to global climate change. Finally, an appendix lists the most important statutes and treaties discussed in the book, and provides brief descriptions of each.

CHAPTER ONE

Introduction to Environmental Law

The Chequamegon-Nicolet National Forest covers more than 600,000 hectares (1.5 million acres) in northern Wisconsin. As a national forest, Chequamegon-Nicolet is owned and managed by the federal government. Managing such a large area is by no means easy, considering that the forest contains hundreds of lakes and streams, thousands of miles of roads and trails, and considerable recreational and logging activity. Before being combined in 1998, the Chequamegon and Nicolet forests were officially two separate national forests, both managed by the US Forest Service. As part of its management duties, the Forest Service in 1979 began developing management plans for the separate Nicolet and Chequamegon National Forests. By 1986 the Forest Service had issued final drafts for both management plans.

In writing management plans, the Forest Service cannot simply do whatever it pleases. The federal statute that determines how the Forest Service must manage national forests is the National Forest Management Act (NFMA). Among other things, the NFMA requires Forest Service management plans to "provide for diversity of plant and animal communities."[1] As a consequence, the Forest Service is legally required to protect biodiversity on the lands the agency manages. The NFMA, however, does not give any specifics on how the Forest Service is meant to protect biodiversity, leaving that up to the agency to decide. (The NFMA is discussed in greater detail in chapter 5.)

To protect biodiversity in Nicolet and Chequamegon, the Forest Service made an important assumption: biodiversity could be protected by maintaining the diversity of habitats in the forests. In other words, the Forest Service assumed that habitat diversity could act as a proxy for species diversity. To maintain habitat diversity, the Forest Service management plans identified several representative animal species from the two forests, and then calculated the habitat types and patch sizes necessary to maintain minimum viable populations of those species. The plans then divided the two forests into patches of those habitat types, with the patches being just large enough for the maintenance of the representative species. These relatively small patches were interspersed with other areas slated for logging or road building.

In 1990 the Sierra Club sued the Forest Service over the contents of the Nicolet and Chequamegon management plans.[2] The Sierra Club argued that the Forest Service had not used scientific knowledge gained from conservation biology in creating its management plans. Conservation biology research clearly shows that viable populations are best maintained in large patches, ideally extending over an entire landscape. The management plans by the Forest Service instead fragmented the forests "into a patchwork of different habitats."[3] These small patches were bound to suffer from edge effects, limited migration, small population sizes, and other problems that would make the survival of species on those patches difficult. The Sierra Club maintained that the Forest Service had completely ignored ecological research on population dynamics, fragmentation, edge effects, and island biogeography in writing its management plans. The Sierra Club also claimed that by not using the findings of conservation biology, the Forest Service was not fulfilling the requirement of the NFMA to protect biodiversity.

The Forest Service countered that the hypothesis that fragmentation of a forest from timber harvesting and road building may be detrimental to plant and animal diversity had "not been applied to forest management in the Lake States."[4] The Forest Service added that the theories of conservation biology and island biogeography were of interest, but that "there is not sufficient justification at this time to make research of the theory a Forest Service priority."[5]

A federal appellate court, in *Sierra Club v. Marita*, agreed with the Sierra Club that the principles of conservation biology are sound. Nevertheless, the court held that the Forest Service management plans for Nicolet and Chequamegon had met the requirement of the NFMA to protect biodiversity. How could the court come to both of these conclusions? As discussed later in this chapter, courts are required to give considerable deference to the decisions made by a federal agency. The court in *Marita* deferred to the Forest Service in its determination that conservation biology principles were uncertain in their application to Lake States forests in the 1980s. The court further wrote that the Forest Service is "entitled to use its own methodology, unless it is irrational."[6] Consequently, the Forest Service was free to use the management plans it had written, ignoring some of the most widely accepted principles of conservation biology in managing the Nicolet and Chequamegon National Forests.

The *Marita* case illustrates that in environmental law, scientific rigor is not always considered to be the highest value. Here the court believed that the value of the judicial branch's respecting the decisions of the executive branch outweighed the value of adhering to conservation biology principles. As the rest of the chapter explores, this outcome is far from an anomaly.

Environmental law may be broadly defined as the statutes, regulations, and court decisions that manage the effects of human activity on the natural environment. Managing how humans impact the environment is a hugely difficult undertaking, and the length and complexity of many environmental laws is a testament to that difficulty. To manage anything effectively, you of course have to understand what you are managing. The best way to understand the natural environment is through scientific inquiry. The three branches of the federal government certainly understand this, and they rely on scientists and scientific data when writing and implementing environmental laws. As the *Marita* case illustrates, however, the three branches also frequently consider factors other than science when writing and implementing those laws. An understanding of environmental law is incomplete without an understanding of how the three branches of government attempt to balance these other factors along with science when managing how humans impact the natural environment.

As will be seen throughout the book, this balancing often results in conflict between science and law. Ecological principles may point to a certain action to protect species or ecosystems, but the law may allow or even require a completely different action. At a more fundamental level, the law frequently calls on science to make value judgments, even though scientific inquiry is ideally objective and without a value system. Science alone cannot decide how much money should be spent to save a species from extinction, or place a dollar value on a single human life. Finally, science is usually an integrative discipline, interested in understanding the interconnectedness of different systems. Statutes, on the other hand, tend to regulate one discrete area at a time. For instance, a specific statute may regulate only marine mammals, or only migratory birds. Because of this focus on one discrete area, statutes are often oblivious to how regulating that one area will influence other areas outside the focus of the statute. The result is that the law treats nature as discrete units that can be manipulated and swapped for each other without impact to other units of nature. As a consequence, the ways in which science and the law attempt to understand and categorize the world are at odds with each other. Each of these conflicts is discussed in this chapter, and they present themselves throughout the rest of the book.

The chapter first explains that Congress often writes environmental statutes to require federal agencies to use science in their decision making, but the statutes rarely indicate how science is specifically meant to be used. This vagueness gives federal agencies flexibility in applying the law, but also allows Congress to hide behind ambiguity when hard decisions need to be made. Next the chapter discusses that although federal agencies are the most adept branch of the government in using science, they are frequently tempted to use the trappings of science to further agency goals. Finally, the chapter discusses how courts defer to the scientific expertise of federal agencies; in other types of cases not involving agencies, however, judges with no scientific training are left to rule on the validity of scientific findings.

CONGRESS

When Congress writes statutes, it quite often includes requirements that federal agencies rely, at least in part, on science in making decisions. Several environmental statutes specifically state that agencies must make certain decisions based on the "best available scientific information."[7] Other environmental statutes do not use such specific language, but imply that only scientific information should be considered when making certain decisions (see box 1.1 for a discussion of the types of problems environmental statutes often address).[8]

Congress includes requirements for the use of science because science is generally viewed as being objective and nonpartisan. Including a requirement for scientifically based decision making helps make a statute appear legitimate to the public. Congress may also mistrust the politically appointed officials running federal agencies, and statutorily requiring that decisions be scientifically based may be a way to limit the discretion of those officials.[9]

Less charitably, a requirement for decisions based on science may be a way for lawmakers to create cover for themselves to avoid having to take responsibility for unpopular decisions made under the statute.[10] This is problematic because it perpetuates an unrealistic view of what science can accomplish.[11] As mentioned in the introduction to this chapter, science by itself cannot place a value on environmental resources, or determine how

BOX 1.1. EXTERNALITIES AND COMMAND AND CONTROL

Many of the environmental laws Congress passes are attempts to deal with externalities. An externality is a cost that must be borne by those who did not choose to incur that cost. For example, in the absence of environmental laws, a factory that releases pollutants into the air would not have to pay anything to release those pollutants, because the atmosphere is a public good and does not belong to anybody. The public, on the other hand, would have to bear the cost of that pollution in the form of health problems and environmental degradation. This is called a negative externality.

There are also positive externalities. A positive externality imposes a positive effect on those who did not choose to receive that effect. For example, a farmer may have a wetland on her property that helps prevent pollution from reaching a river that is used as a source of drinking water by a downstream city. The people of the city benefit from the wetland's providing them with clean water to drink, but the farmer is not paid for that benefit. As a result, the farmer has no incentive to keep the wetland, and may decide to fill it in for additional cropland.

Environmental laws often try to internalize externalities. For instance, a law may impose a tax on facilities that emit pollutants to internalize negative externalities, or a law may pay landowners not to destroy valuable land to internalize positive externalities.

The most common way environmental laws internalize negative externalities is through command-and-control regulations. These regulations are laws that mandate what an individual or business may or may not do. A law that requires a factory to install a specific type of scrubber on a smokestack to reduce its emission of air pollutants would be a command-and-control regulation. More recently, governments have begun creating environmental laws that reduce negative externalities through the use of market-based approaches. Market-based approaches utilize economic markets to provide incentives to reduce negative externalities. The most common type of market approach gives permits to polluters for the right to emit pollutants, and then allows polluters to trade those permits on an open market. The acid rain program (see chapter 10) is the most famous example of a market-based approach to internalizing negative externalities. Command-and-control regulations and market-based approaches are seen throughout the book.

to balance the protection of those resources against other needs, such as for development or private property rights. By requiring decisions to be scientifically based, environmental statutes imply that science can do that balancing, and that such decisions can be completely objective and not value judgments. As an example, the Endangered Species Act states that when deciding whether to list a species as threatened or endangered, the Fish and Wildlife Service must make that decision "solely on the basis of the best scientific" information (see chapter 3).[12] While science can provide information on the probability of a species going extinct within a certain number of years, science alone cannot decide at what probability public resources should be spent to reduce the likelihood of extinction.[13] The probability of extinction that society is willing to accept is not a strictly scientific determination. Congress, however, sidesteps deciding the acceptable probability when it requires that listing be based solely on science. Instead, Congress forces the Fish and Wildlife Service to make the decision. Requirements for the use of science in environmental statutes may frequently be a way for Congress to tell federal agencies to make the value judgments that are fundamental to the statute because it is too difficult for Congress to do.

An additional problem with environmental statutes requiring federal agencies to use science is that the statutes rarely indicate how the agencies are supposed to use science, or suggest what kinds of science to use. There are several reasons for this vagueness. In order for a bill to pass Congress, the language often must be ambiguous enough to garner a majority of votes. Any language in a bill that is too specific on how science should be used may lead to a loss of votes. Additionally, lawmakers understand that scientific knowledge is constantly changing, and any specific requirements may quickly become outdated. Finally, lawmakers realize that they are not scientific experts, and they may very well make mistakes if they try to put specific scientific requirements into statutes.

The lack of scientific knowledge in Congress is best illustrated by looking at how bills are drafted. Almost no member of Congress actually drafts bills. Some bills are drafted by lobbyists or federal agencies and then presented to members of Congress. Most commonly, members of Congress rely on the Offices of Legislative Counsel to draft bills. Both the Senate and House of Representatives have nonpartisan Offices of Legislative Counsel that focus entirely on writing legislation at the direction of lawmakers. In this arrangement, members of Congress and their staffs focus on policy, and then work with the Offices of Legislative Counsel in translating that policy into a written bill.[14] Although members of Congress and their staffs oversee the drafting process, it is ultimately the lawyers in the Offices of Legislative Counsel that choose most of the words in a bill, and that choice of words may have great importance in how a law is interpreted and implemented. Courts frequently base their interpretation of a provision in a statute on a single word or phrase in that provision (see box 1.2 for a description of how legal materials are cited and how to perform legal research).

BOX 1.2. LEGAL CITATIONS AND RESEARCH

One of the first things to understand before performing legal research is how laws and other legal materials are cited. Federal statutes are officially compiled in the *United States Code* (U.S.C.). The *United States Code* is composed of several titles, and each title has multiple sections (§). For instance, 16 U.S.C. § 703 refers to title 16, section 703 of the *United States Code*. Section 703 happens to be the first section of the Migratory Bird Treaty Act. Most statutes appear in the code in consecutive sections, although there are exceptions. The citation 16 U.S.C. §§ 703–712 refers to all the sections of the Migratory Bird Treaty Act. To get a good grasp on an environmental statute, there is no substitute for sitting down and simply reading all the sections of the statute. The fastest way to access the *United States Code* is through the Legal Information Institute website, operated by Cornell University (www.law.cornell.edu).

Federal agencies in the executive branch write federal regulations. Federal regulations fill in the details of statutes, and can therefore be very long and complex. Federal regulations are compiled in the *Code of Federal Regulations* (C.F.R.). The *Code of Federal Regulations* is also composed of multiple titles and sections, so 50 C.F.R. § 17.1 refers to title 50, section 17.1 of the *Code of Federal Regulations*. The sections of a regulation implementing a particular statute are usually listed consecutively in the *Code of Federal Regulations*. Federal regulations are most easily accessed through the US Government Printing Office website (www.ecfr.gov).

Federal agencies provide announcements and explanations for the regulations they write in the *Federal Register* (Fed. Reg.). *Federal Register* documents are not legally binding, but are important for understanding the reasoning and interpretation of federal regulations. The *Federal Register* can be searched through www.federalregister.gov. Almost all the federal agencies have frequently updated websites that contain numerous documents that help explain environmental laws and regulations.

Opinions written by courts are assembled in case reporters. The decisions are placed in the reporters in chronological order, not based on topic. Federal district court decisions are published in the *Federal Supplement* (F. Supp.), which has been continued into a second series (F. Supp. 2d). A district court decision is cited as *Natural Resources Defense Council v. Kempthorne*, 506 F. Supp. 2d 322

(E.D. Cal. 2007), indicating that the decision can be found in volume 506, page 322, of the second *Federal Supplement* series. The citation also includes the court issuing the decision and the year of the decision, in this case the Eastern District of California in 2007. If a second number follows the page number in the citation, that is a pincite, indicating that the citation is meant to point to that particular page of the decision.

Federal appellate court decisions are reported in the *Federal Reporter* (F.), which is currently in the third series. The federal appellate case *Sierra Club v. United States Fish and Wildlife Service*, 245 F.3d 434 (5th Cir. 2001) refers to a decision published in volume 245, page 434, of the third series *Federal Reporter* by the Fifth Circuit Court of Appeals in 2001.

Supreme Court decisions are reported in the *United States Reports* (U.S.), with a citation such as *Babbitt v. Sweet Home Chapter of Communities for a Great Oregon*, 515 U.S. 687 (1995). As there is only one Supreme Court, there is no need to specify the court in the citation.

Last, law review papers are an excellent way to understand the broader legal and policy implications of environmental laws. Every law school has at least one, and usually several, law review journals that it publishes multiple times a year. Law review papers written by professors or practicing attorneys are usually referred to as articles, while papers written by law students are termed notes or comments.

When doing legal research, the best way to search for information is through the databases Westlaw and LexisNexis. Both Westlaw and LexisNexis archive all manner of legal materials, from statutes to law review articles. The opinions written by the various courts can be difficult to find without using one of these databases. Unfortunately, neither Westlaw nor LexisNexis is free to access. The Web of Knowledge database indexes several law review journals, but again, the database is not free to access. A free but less complete database of legal materials is available at www.justia.com.

The staff of the Offices of Legislative Counsel are lawyers, and do not necessarily have any training in science or the subject of the bill they are drafting. There are only 83 staff members in the two offices, meaning that each must work on drafting bills covering a wide array of subjects.[15] As a result, the scientific language in most statutes is likely to have been written by someone who does not have any training in science, or any special knowledge of the subject of the bill. Furthermore, researchers conducting a survey of congressional staff found that members of Congress and their staffs do not bother to read most of the bills the offices draft, and even when they do, they have difficulty understanding the statutory language in the bills and how well it reflects the policy intentions of the members of Congress.[16] Consequently, the scientific language in a bill may not be particularly well understood by the people drafting the bill or the people voting on the bill.

FEDERAL AGENCIES

The task of deciding how to use science in implementing environmental statutes falls most squarely on the federal agencies. Many agencies have trained scientists who conduct research and regularly publish in peer-reviewed journals. They also frequently rely on research by, and consultations with, academic scientists.

Being the branch of government with the greatest scientific expertise, the federal agencies can take advantage of the other branches of government. Federal agencies occasionally use, or ignore, scientific research as a means to justify decisions that they have already made. The *Marita* case at the beginning of this chapter shows that the Forest Service was willing to ignore conservation biology research because it conflicted with how the agency wanted to manage two national forests.

A different example also comes from the Forest Service. By the early twentieth century, the Forest Service had begun to realize that grazing livestock on rangeland managed by the Forest Service was having a detrimental impact on rangeland ecosystems.[17] In response, the Forest Service decided to develop measures to estimate the maximum amount of livestock that could graze a given rangeland without degrading it. This maximum grazing level was termed a carrying capacity, and was considered fixed for a particular rangeland. By setting a fixed carrying capacity, the Forest Service could reduce livestock grazing on a rangeland, and could point to the carrying capacity when pressured by ranchers to increase grazing. Forest Service scientists quickly realized, however, that rangeland conditions vary over time and after disturbances, and that a carrying capacity for grazing that is fixed at one level every year makes little ecological sense. Nonetheless, the Forest Service maintained the fixed carrying-capacity concept for many years because of its usefulness in justifying the decision to reduce grazing pressure on Forest Service rangelands.[18]

Regulations

If Congress passes a statute that makes vague pronouncements on the usage of science, the details are usually filled in by regulations promulgated by a federal agency. Box 1.3 explains the process of promulgating regulations, as

well as the difference between regulations and guidance documents. By filling in the details left out of statutes, regulations are often where the balancing of social values takes place. The way in which proposed regulations are
approved by the White House, however, tends to put a greater emphasis on
economic considerations than on other values.[19]

BOX 1.3. AGENCIES, REGULATIONS, AND GUIDANCE DOCUMENTS

Most federal agencies are created by Congress through statutes. One prominent
exception is the Environmental Protection Agency, which was created through
an order signed by President Nixon. Once a federal agency has sprung into
existence, its authority to act comes from Congress. Congress tells the agencies
what to do through legislation, and then appropriates funds for the agencies
to operate.

Although Congress sets an agency's duties, agencies have a great deal of
discretion in how they go about undertaking those duties. Most employees in
federal agencies are career employees, but the top officials are appointed by the
president. These political appointees set the priorities of the agencies.

When Congress passes a statute that an agency must implement, the details
of how to implement the statute are often sketchy at best. There are several reasons for this: to get a bill through Congress that will receive a majority of votes,
the bill must be vague on details; members of Congress are not experts in the
fields in which they pass legislation, so they leave how to implement statutes
to the agencies that do have that expertise; and Congress could not possibly
anticipate all the details that are necessary in implementing a statute, so they
leave those details to be worked out by the agencies. The way federal agencies
fill in the details of statutes is through promulgating regulations and writing
guidance documents.

Regulations are rules written by federal agencies. A proposed regulation
must go through a notice-and-comment process, meaning that an agency publishes the proposed regulation in the *Federal Register* and then interested parties have an opportunity to send comments to the agency. After considering the
comments, the agency may issue a final regulation. Once final, the regulation
has the force of law.

Promulgating a regulation is a formal process, and takes considerable time.
Agencies often issue guidance documents to more quickly set down rules for
implementing statutes. There is no requirement for a notice-and-comment process to issue a guidance document. As a result, guidance documents do not have
the force of law. Because of this, agencies may not always strictly follow their
own guidance documents, and citizens are not compelled to do so.

It may seem odd that unelected officials in the executive branch can write regulations that have the force of law. There are limits to this power, however. Federal agencies may not go beyond the authority they are given by a statute in promulgating a regulation. For example, FWS could not promulgate a regulation to stop the spread of an invasive species just because the agency thought it would be useful. The regulation would have to fit within the authority given to FWS from a specific statute or set of statutes.

Additionally, if a person believes a federal agency has incorrectly interpreted a statute in its regulations and has standing to do so, the person may sue the agency. In deciding whether a federal agency has correctly interpreted a statute in promulgating a regulation, courts follow the test created by the Supreme Court in *Chevron U.S.A. v. Natural Resources Defense Council.*

*467 U.S. 837 (1984).

Under Executive Order 12,866, when a federal agency is drafting a new regulation, the agency is required to "assess all costs and benefits of available regulatory alternatives, including the alternative of not regulating." Agencies are then directed to choose regulatory approaches that "maximize net benefits" unless a statute specifically requires a different regulatory approach. The executive order also requires the Office of Information and Regulatory Affairs (OIRA) within the White House Office of Management and Budget to review significant proposed regulations from almost all federal agencies. As part of the review, OIRA also looks at the cost-benefit analyses that agencies perform on their proposed regulations. During the review process, OIRA may require an agency to change its proposed regulation, or to withdraw the regulation completely. OIRA is staffed primarily by economists, and therefore tends to view regulations through an economic lens.[20] A report by the US Government Accountability Office found that OIRA reviews of proposed regulations are primarily concerned with reducing the costs or improving the cost-effectiveness of regulations.[21] The report also found that the emphasis on cost did not necessarily result in an increased net benefit to society. Some scholars have concluded that the OIRA review is less of an exercise in overseeing cost-benefit analyses and more of an opportunity for the White House to insert political considerations into agency regulations.[22]

The OIRA review of proposed regulations is particularly intense on regulations touching the environment. Between 1998 and 2000, OIRA changed almost 90% of Environmental Protection Agency (EPA) proposed regulations.[23] For instance, in 2001 the EPA proposed a regulation under the Clean

Water Act that would have required power plants that withdrew at least 189 million liters (50 million gallons) of cooling water every day from estuaries or tidal rivers to meet a new uniform national standard for intake structures as a way to reduce deaths to aquatic organisms.[24] The proposed regulation would have cost $610 million per year, with benefits of $890 million per year, resulting in a net benefit to humans and the environment of $280 million per year. The OIRA then undertook its review of the regulation. Instead of new uniform standards for intake structures, OIRA suggested site-specific standards for intake structures. More importantly, OIRA suggested allowing power plants to restore waterways as a means of substituting for the installation of improved intake structures. The regulation as revised by OIRA would cost $280 million per year, have benefits of $735 million, resulting in a net benefit of $455 million. With the intervention of OIRA, the net benefit of the revised regulation was larger, but the absolute benefit to humans and the environment was $155 million smaller. The revised regulation was published in 2002. In balancing the protection of environmental resources with economic costs, OIRA helps keep the scales tipped toward economic considerations, even if the absolute benefit to the environment is considerably smaller.

Adaptive Management

Once a federal agency has promulgated a set of regulations, the agency must then implement those regulations. For environmental statutes and regulations, that means undertaking the management of environmental resources. Ecological systems are complex, and even with considerable information about a particular ecosystem, how it will respond to an agency's management actions may be difficult to predict. One can think of ways of dealing with this uncertainty as being on a spectrum.[25] On one end is the precautionary principle; in this context, the precautionary principle requires that any management action that could harm the environment, such as allowing oil or gas exploration, requires scientific evidence that the activity will not cause harm before the agency allows the action to proceed. On the other end is the "sound science" approach. The "sound science" approach is the opposite of the precautionary principle, in that it requires scientific evidence that an action will harm the environment before the action is prohibited. Finally, in the middle is adaptive management, and it comes in several guises.

Active adaptive management means implementing a variety of management practices in multiple patches of the management area at the same time.[26] Differences in the results of these various management practices are observed, and then used to create a new set of management practices to implement and observe. This is essentially performing replicated experiments on the outcomes of management practices, learning from those outcomes, and then performing new experiments. While replicated patches are the ideal, active adaptive management can also be done on a single patch as long as an experimental approach is taken to making management decisions. Passive adaptive management on the other hand means creating mathematical or computer models of the ecosystems in a management area. Managers use the models to suggest management actions for the entire management area, then collect data on the outcomes of that management. That data is then fed back into the model, suggesting new management actions. Finally, haphazard adaptive management means undertaking management actions randomly, or based on instinct, and then collecting data on the outcome of that action. This is more akin to trial and error than true adaptive management.

Many federal agencies claim to use active or passive adaptive management in managing environmental resources. Three problems tend to arise, however, when federal agencies actually try to implement adaptive management.[27] First, the individuals responsible for making management decisions in federal agencies are often risk averse, and do not want to approve management experiments that may be considered failures. For instance, the Northwest Forest Plan was adopted in 1994 as the guidelines for managing federal lands in the Pacific Northwest. The plan specifically created 10 adaptive management areas (AMAs) covering 607,000 hectares (1.5 million acres) to experiment with different management actions. The Forest Service and Bureau of Land Management (BLM) jointly oversaw the AMAs. Subsequent interviews with employees at the Forest Service and BLM found, though, that there was very little effort to engage in adaptive management in the AMAs.[28] Employees indicated that experimentation and risk taking were not rewarded by supervisors in these agencies, making the experimentation required by adaptive management professionally unrewarding. To be fair, part of this aversion to risk arose because of the fear of litigation if any experimental management action harmed a threatened or endangered species.

Second, federal agencies may use the process of adaptive management to justify management actions the agency has already decided to pursue. For example, if an agency wants to allow oil or gas drilling on the lands it manages, allowing that drilling in certain areas could be justified as experimentation under active adaptive management. If an agency instead employs passive adaptive management, the agency may be able to subtly alter management models to make the management action that the agency prefers appear to be the best option.

Third, there is a systemic lack of environmental monitoring by federal agencies. Adaptive management is pointless if there is no attempt to observe and learn from the outcomes of different management actions. The lack of environmental monitoring by federal agencies deserves its own section and will be discussed in greater detail next.

While adaptive management continues to be recommended by many scientists and legal scholars, less flexible management methods may occasionally do more to protect environmental resources.[29] A plan that sets out well-defined management actions may be more difficult for an agency to use as cover to implement preferred actions. Additionally, less flexible management plans may be easier for the public to understand and criticize (and possibly litigate). Adaptive management will likely remain the best management option when there is a clear management goal, and when it is respected by an agency's hierarchy as a tool to reach that goal.[30] If the management goal is unclear and the agency hierarchy is very risk averse, however, a less flexible management plan that forces the agency to clearly indicate its management actions may be more beneficial to the environment.

Environmental Monitoring

A federal agency will have difficulty determining the effects of its management actions if there is inadequate monitoring of the area being managed. An agency may monitor the wrong things, or may simply not do any environmental monitoring in the first place.

There has been considerable scientific research on the best ways to monitor ecosystems, but federal agencies have historically been slow to adopt the lessons of this research. The finest example of this is the use of management indicator species (MISs) by the Forest Service (see chapter 5). An MIS is a species that the Forest Service believed could act as a proxy for all the other vertebrate species in a given area. By monitoring the condition of the MIS, the Forest Service thought it would know the condition of all the other vertebrate species in the ecosystem. The Forest Service regulations of 1982 required management plans to include MISs as indicators of the functioning of forest ecosystems. However, the idea that a handful of indicator species could act as proxies for the functioning of an entire ecosystem was widely rejected by ecologists soon after the 1982 regulations were promulgated.[31] It was not for another 30 years, however, that new Forest Service regulations eliminated the requirement for MISs, instead requiring the use of focal species as a way to monitor ecosystem functioning.

Since at least the early 1980s, ecological research has pointed to the use of more holistic ecosystem-based monitoring.[32] By the late 1990s, the EPA decided to reevaluate how it conducted environmental monitoring, and asked the National Research Council (NRC) to evaluate biological indicators.[33] The NRC noted that indicators monitored by federal agencies are of great significance because they will be used in setting public policy. As a consequence, these indicators must be "understandable, quantifiable, and broadly applicable." The NRC argued that indicators are more influential if there are fewer of them, and if the indicators are easily understood by the public. The NRC also reasoned that the rules for calculating an indicator should be objective and clear, so that the public has confidence that the indicators are not easily influenced by outside interests. The NRC recommended indicators in three categories: (1) indicators for the extent and status of ecosystems should be land cover and land use; (2) indicators for ecological structure should be total species diversity, native species diversity, nutrient runoff, and soil organic matter; and (3) indicators for ecological function should be carbon storage, production capacity, net primary productivity, lake trophic status, stream oxygen, nutrient-use efficiency, and nutrient balance. Despite the NRC report, the use of all these indicators by a federal agency in its management areas is extremely rare.

While monitoring the wrong things is problematic, not doing any monitoring is even more so. There are several reasons an agency may decide to make environmental monitoring a very low priority. Long-term ecological monitoring is expensive, and federal agencies would rather spend that money on other activities. For instance, a Government Accountability Office report from 2005 found that BLM had moved funds for wildlife monitoring toward permitting for oil and gas development. Additionally, outside interests may push for less monitoring, so that there is less of a scientific basis for new regulations. Federal agencies may also prefer to collect only the minimum amount of monitoring data necessary to justify maintaining the current management plan—additional data may be used against the agency by politicians or the courts. For that reason, agencies often prefer to use models to justify management actions because the assumptions in a model are easier to subtly alter to produce the preferred results than are empirical monitoring data. Finally, scientists in federal agencies view ecological monitoring as rather boring and unlikely to lead to advancement within an agency.[34]

One legal scholar suggests the creation of an independent federal agency that is solely concerned with conducting environmental monitoring and overseeing the monitoring done by other agencies. He argues that an independent agency would likely be perceived by the public and lawmakers as unbiased and less influenced by outside interests. The US Geological Survey is a federal agency that already has as its primary mission conducting research on environmental issues. The scholar suggests that the US Geological Survey would be well positioned as an independent agency that conducts widespread environmental monitoring, while also overseeing monitoring by other federal agencies.[35]

Regardless of whether an independent agency is created, the current lack of high-quality environmental monitoring makes adaptive management of federal lands difficult at best. It also lets agencies claim their management practices are protecting environmental resources, because the agency is not collecting evidence that would suggest the contrary. This may become especially important if an agency's decisions are reviewed by a court.

COURTS

The last branch of the federal government that must consider the place of science in the law is the courts. Unlike federal agencies that employ hundreds of scientists, judges and the lawyers arguing in front of them do not necessarily have any scientific training. For that reason, the courts have several rules that they use to make sure the science they consider is legitimate and applies to the case at hand.

Expert Witnesses

Before considering how courts judge the use of science by federal agencies, it is worth noting how courts deal with science in other types of cases. There are of course many types of cases where scientific evidence may be important. For example, a jury might rely on the testimony of a scientist that the pollutants released into a river by an industrial facility were the direct cause of a massive fish kill. Scientists testifying in a court case are called expert witnesses. Although courts recognize that expert witnesses are important in helping a jury understand the facts of a case, courts are also wary of the power of expert witnesses to sway juries and judges. Occasionally an expert witness is not actually an expert, or is testifying about scientific ideas that are not considered reliable by the scientific community. A jury or judge, however, may not have the scientific knowledge to know that the expert witness is unreliable. This threat is ideally overcome by the *Daubert* standard and the Federal Rules of Evidence.

In the past, judges would decide whether to allow expert testimony by looking at whether an expert was testifying on a method or theory that had gained general acceptance in the relevant scientific field.[36] Judges played a limited role, allowing in science as long it was generally accepted by scientists as valid. The Supreme Court overturned that way of deciding on expert testimony in *Daubert v. Merrell Dow Pharmaceuticals, Inc.* The court in that case wrote that trial judges must act as gatekeepers to prevent the testimony of unreliable expert witnesses.[37]

In performing the gatekeeping function required by *Daubert,* a trial judge must ensure that an expert's testimony is based on reasoning or methodology that is scientifically valid, and can properly be applied to the facts at issue. The Supreme Court listed five factors that a judge can consider: (1) whether a theory or technique used by the expert can be and has been tested; (2) whether the theory or technique has been subjected to peer review and publication (lack of publication does not automatically eliminate an expert's testimony, but it is relevant for a judge to consider); (3) the known or potential rate of error of a scientific technique; (4) the existence and maintenance of standards controlling a technique; and (5) general acceptance of a theory or technique in the scientific community. These factors are nonexclusive, and courts may consider other factors in deciding whether to allow expert testimony. In a subsequent case, *Kumho Tire Co. v. Carmichael,* the Supreme Court wrote that this gatekeeping function also applies to technical experts, not just scientific experts.[38]

The *Daubert* standard is now affirmed in rule 702 of the Federal Rules of Evidence. For federal courts, the Federal Rules of Evidence govern what evidence may be presented at a trial. Rule 702 states that a person is qualified as an expert by "knowledge, skill, experience, training, or education." Such an expert may testify in a case in the form of an opinion if (1) the expert's scientific knowledge will help the judge or jury understand the evidence or facts in a case; (2) testimony is based on "sufficient facts or data"; (3) the testimony is the product of "reliable principles and methods"; and (4) the expert has "reliably applied" those principles and methods to the facts in the case.

There are critics of the *Daubert* standard, chief among them being former Chief Justice Rehnquist. In his dissent to the *Daubert* decision, Justice Rehnquist wrote that the *Daubert* standard turns trial judges into "amateur scientists" who must determine the scientific validity of an expert's testimony even though most judges do not have a background in science. Other scholars have argued that the lack of a requirement in *Daubert* that theories or techniques be peer-reviewed or have gained general acceptance may result in junk science being presented in court. An empirical study from 2001 found, though, that judges were actually applying stricter standards to expert evidence under the *Daubert* standard, and were more likely to prevent an expert from testifying, than before the advent of the standard.[39] The study also found that even general acceptance of a theory or technique in a scientific field was occasionally insufficient for a judge to consider it reliable. As a consequence, the *Daubert* standard may not only help keep out junk science, but may also tend to prevent new theories and techniques from making it into courtrooms.

Chevron *Deference*

As discussed in a previous section, one of the most important ways federal agencies use science is incorporating it into regulations. When examining regulations, courts do not follow the *Daubert* standard; they are instead much more deferential. When a federal agency makes a rule, such as a regulation, courts usually defer to the agency's rule making. Courts consider this reasonable because federal agencies have much greater expertise in implementing statutes than do judges.

To determine when a court must defer to a rule made by a federal agency, the Supreme Court established a two-step test in *Chevron U.S.A. v. Natural Resources Defense Council*.[40] In the first step of the test, a court determines whether the statute directly addresses the question at issue. If so, then the statute trumps the rule, and the court stops its analysis there. If the statute does not directly address the question, the court moves to the second step. In the second step, the court determines whether the agency's interpretation of the statute in the rule is a "permissible construction of the statute." The agency's interpretation need not be the best interpretation, or even a particularly good one. As long as the interpretation is reasonable, then the agency's rule will stand. As a consequence, the Supreme Court has held that courts must give considerable deference to the rules agencies create.

The Supreme Court curbed the requirement for deference a bit in a subsequent case, *United States v. Mead Corp.*[41] In *Mead*, the court held that another step must come before the two-step *Chevron* test. A court must first ask if Congress delegated authority to a federal agency to make a rule carrying the force of law, and if the rule promulgated by the agency was in the exercise of that authority. If this first step is passed, then the court may move to the two-step *Chevron* test. If, however, an agency created a rule that is not based on a delegation of authority from Congress, then *Chevron* does not apply. Instead, a court must decide whether the agency's rule stands based on several factors. These factors are the thoroughness evident in the agency's consideration; the validity of its reasoning; its consistency with earlier and later decisions; and "all those factors which give it power to persuade, if lacking power to control."

In the facts of the *Mead* case, the US Customs Service classified day planners as diaries, making them subject to a 4% tariff. This tariff classification required no notice-and-comment period, such as for a regulation, and the Supreme Court found that the classification was not intended by Congress to carry the force of law. As a result, the tariff classification did not receive *Chevron* deference. The Supreme Court did write, though, that such decisions by federal agencies must receive respect from courts according to their persuasiveness.

The results of *Chevron* and *Mead* together describe the level of deference courts must give federal agencies. The regulations promulgated by agencies, and the science in those regulations, receive *Chevron* deference. Many specifics of how science is used by an agency, though, appear in the manuals or guidelines written by that agency. As these manuals and guidelines are not rules created under authority delegated by Congress, they do not receive *Chevron* deference; they instead fall under the *Mead* standard. Courts, however, still give deference to such documents based on their persuasiveness. Empirical studies find that when applying *Chevron*, courts affirm agency decisions 73% of the time.[42] When applying *Mead*, courts affirm agency decisions 60% of the time.

Chevron and *Mead* deference apply when an agency is interpreting an ambiguous statute. When an agency is making a specific decision, a different standard applies. Under the Administrative Procedure Act, a court may set aside a decision made by an agency if the court finds the decision to be "arbitrary, capricious, an abuse of discretion, or otherwise not in accordance with law."[43] When a court determines whether an agency's decision is arbitrary and capricious, it often performs a "hard look" review. During a hard look review, a court examines an agency's decision-making process to make sure an agency determination is based on relevant factors and is not a clear error of judgment.[44] Lack of an administrative record documenting how an agency came to a decision often leads to a finding that the decision is arbitrary and capricious. Similar to *Chevron* deference, though, a court may not simply substitute its judgment for the judgment of the agency.

The Supreme Court endorsed hard look review in *Motor Vehicle Manufacturers Association of the United States, Inc. v. State Farm Mutual Auto Insurance Co.*[45] The Supreme Court wrote in *State Farm* that a decision by a federal agency is arbitrary and capricious "if the agency has relied on factors which Congress has not intended it to consider." The court also wrote that a decision will be found arbitrary and capricious if the agency "entirely failed to consider an important aspect of the problem, offered an explanation for its decision that runs counter to the evidence before the agency, or is so implausible that it could not be ascribed to a difference in view or the product of agency expertise." Consequently, an agency decision must be the product of "reasoned decisionmaking," and contain a "rational connection between the facts found and the choice made."

When it comes to agency decision making that involves science, however, hard look review tends to be considerably softer. In *Baltimore Gas & Electric Co. v. Natural Resources Defense Council, Inc.*, the Supreme Court stated that when reviewing a scientific determination made by a federal agency, "a reviewing court must generally be at its most deferential."[46] In the *Marita* case discussed earlier, the court cited the *Baltimore Gas* case in explaining why it was deferring to the scientific determination made by the Forest Service, that conservation biology principles did not necessarily apply to forests in the Lake States. Although it may seem reasonable for a court to defer to the scientific expertise of an agency, such deference may cause a problem discussed earlier in the chapter. When agencies know that courts are particularly deferential to science, the agencies are incentivized to make every decision appear to be a scientific decision, even ones that are mainly policy decisions.[47] Consequently, both Congress and the courts have indirectly encouraged federal agencies to make policy decisions under the guise of science. More recently, there seems to be evidence that courts have begun to move away from giving extreme deference to scientific decisions.[48] Courts certainly still give considerable deference to scientific decisions, but a move away from extreme deference and toward hard look review will likely make it more difficult for federal agencies to continue hiding policy decisions behind science.

ECOLOGY AND THE LAW

Ecologists attempt to understand nature as an interdependent web of connections.[49] There is a realization that what happens at one level of organization, from genes to ecosystem to landscape, and at one time scale, from seconds to centuries, affects all other levels of organization and time scales. Environmental law, conversely, attempts to resolve specific problems occurring in the present, with little thought given to how those problems influence the wider interdependent system in the present or over long time scales.[50]

The narrow focus of the law can be seen in the organization of environmental statutes. Environmental statutes tend to concentrate on one medium, such as water in the Clean Water Act, or level of organization, such as species in the Endangered Species Act. Environmental statutes also tend to require the management of ecosystems for one or a handful of species, such as marine mammals under the Marine Mammal Protection Act, or commercial fish stocks under the Magnuson-Stevens Act.[51] Environmental statutes implicitly assume that regulation of one medium or management for one species can be separated from the surrounding environment in which that medium or species exists. Of course, the interconnectedness of ecosystems makes this impossible. Consequently, the effects of environmental statutes tend to ripple out from the species or medium that is the focus of the law. As will be discussed frequently in the rest of this book, environmental statutes are bad at anticipating or dealing with the consequences of those ripples.

Because of the fundamentally different way in which ecology and environmental law understand and deal with the complexity of the natural world, there will always be friction between the two disciplines. Hopefully, continuing ecological research will help identify when and how environmental laws fall short of protecting the environment, and suggest ways to improve them. Research that shows how the effects of environmental statutes spread through the complex connections within and between ecosystems is particularly important. While environmental statutes tend to have a narrow focus, ecological research is one of the most important drivers in opening up that focus.

PART I

Species

CHAPTER TWO

Wild Plants and Animals

When European settlers first arrived in America, the population of passenger pigeons (*Ectopistes migratorius*) numbered at least three billion.[1] Some scholars estimate that the pigeons constituted 25 to 40% of the total land-bird population in North America. Because of the species' abundance, passenger pigeon meat soon became a staple in growing towns and cities. To capture passenger pigeons, hunters would clear an area of grass and debris, bait the ground with salt, and wait for a flock of passenger pigeons to congregate. When enough birds had arrived, a spring-loaded net would be deployed, capturing the flock. Over the course of 40 days in 1869, almost 12 million passenger pigeons were sent to market from one town in Michigan. By the mid-1800s, overhunting and habitat loss had begun to decimate passenger pigeon numbers. The last wild passenger pigeon was shot in 1908, and the last captive bird died in a zoo on September 1, 1914.

The passenger pigeon was not the only species of bird overhunted in the United States throughout the nineteenth century. Migratory waterfowl and species whose plumages were popular for use on hats were in particular trouble.[2]

Lawmakers knew about the declining bird numbers, but the state legislatures were unwilling to act. The states were especially reluctant to protect migratory birds within their borders. The reason was that doing so would simply provide more birds for hunters in other states to harvest, putting the state that set limits at a competitive disadvantage. The states were willing to risk letting additional bird species follow the demise of the passenger pigeon if it meant the citizens of their states could continue harvesting the birds.

Understanding the problem, Congress finally stepped in and in 1913 passed a federal statute protecting migratory bird species.[3] And then the law was almost immediately struck down. Two federal courts (but not the Supreme Court) ruled the statute to be an unconstitutional violation of the right of the states to harvest the species within their borders.[4]

Undeterred, in 1918 Congress tried again by passing the Migratory Bird Treaty Act. As will be discussed in more detail in the next section, the Migratory Bird Treaty Act actually implements an international treaty. Because the act was not a domestic statute but was implementing a treaty, the Supreme Court in *Missouri v. Holland* upheld the Migratory Bird Treaty Act as constitutional.[5] Overcoming the objections of the states and the possibility of being found unconstitutional, the Migratory Bird Treaty Act is still in force today. The act remains the single most important statute protecting bird species in the United States.

The history of the Migratory Bird Treaty Act illustrates several themes that will weave their way through this chapter. The first theme is suggested by the fate of the passenger pigeon. It may seem legally and economically wasteful to regulate wild species that have large viable populations. Humans, though, have an amazing capacity for altering ecosystems and destroying habitats. The Endangered Species Act, discussed in the next chapter, attempts to protect species that have populations so small they are in danger of going extinct. Species that are close to extinction, though, rarely completely recover. Much more rational is to protect wild species so that they do not reach the brink of extinction.

This points to the second theme. There is no comprehensive federal statute regulating wild animal and plant species in the United States. The Migratory Bird Treaty Act protects many bird species; however, no comparable federal statutes protect other classes of species. Additionally, while the Migratory Bird Treaty Act prohibits the direct killing of bird species, it does little to protect those species' habitats or food resources. There are a great many federal environmental laws, protecting many different aspects of the environment. Cobbled together, the environmental laws of the United States do a fair job of protecting many wild species. But because there is no comprehensive set of laws, many other species receive little or no protection.

The final theme illustrated by the Migratory Bird Treaty Act is the tragedy of the commons. When a resource is open for everyone's use, an individual's best option is to use the resource as quickly as possible before others have time to use it. As a result, no one individual has an incentive to use the resource responsibly so that it remains viable into the future. One way of overcoming the tragedy of the commons is to create private property rights that divide up the resource so that property owners have an incentive to conserve the resources that they own.[6] Another way is for a government to set limits on all users of the resource. In the case of migratory bird species, no individual state wanted to impose limits on hunting those birds because the other states would have more birds for their residents to harvest. It required a law from the federal government to force conservation of migratory birds as a resource. The tragedy of the commons makes frequent appearances in the study of environmental law.

The chapter discusses several federal statutes that regulate the management of wild plant and animal populations, beginning with a more in-depth look at the Migratory Bird Treaty Act. It then explains the federal statutes that help prevent the introduction and spread of invasive species. The second part of this chapter discusses two aspects of plant and wildlife regulation taken on by the states: the public trust and sport hunting.

FEDERAL STATUTES

Because there is no comprehensive federal wildlife law, the federal wildlife laws that do exist tend to be fairly disjointed in what species they regulate and how they go about that regulation. As the second part of this chapter discusses, the individual states own the species within their borders. Consequently, state laws typically determine how populations of individual species are managed. Federal statutes that apply to wild species usually regulate issues of interstate commerce, such as whether organisms may be transported or sold across political borders. Although this may seem like a fairly limited way of influencing the ecology of wild plants and animals, it has the potential to have large effects for many species. The introduction to the chapter suggested that the tragedy of the commons can be particularly pernicious when species cross political borders. Statutes that prevent the overexploitation of species that cross borders may be the best way to protect many migratory species. Even more importantly, it is often species' being brought across political borders by humans that allows the spread of invasive species. As a result, a statute that prevents the transport of nonnative species across political borders may help prevent the spread of invasive species. Before discussing the federal statutes that regulate wild plants and animals, it is worth briefly examining the clauses in the Constitution that provide the authority for Congress to pass such statutes.

Constitutional Clauses

The federal government does not have unlimited power. Any power that the federal government possesses must come directly from the US Constitution. There are three clauses in the Constitution that are often the basis for Congress's authority to pass statutes regulating wild plants and animals. These clauses are known as the commerce clause, treaty clause, and property clause. A brief description of the clauses, and how the interpretations of those clauses have changed through the years, will help illuminate the ways in which federal statutes are written and enforced.

Article I, Section 8 of the US Constitution declares that Congress shall have the power "to regulate Commerce with foreign Nations, and among the several States, and with the Indian Tribes." This provision is known as the commerce clause. The commerce clause gives authority to Congress to make laws that regulate interstate economic activity. The Constitution, however, does not define what constitutes commerce, or when an activity takes place among the several states. This means that the courts have a large role in interpreting exactly how much power the commerce clause gives to Congress.

For almost a century after ratification of the Constitution, courts used the commerce clause mainly to limit state laws that discriminated against interstate commerce. By the late 1800s, though, courts began using the commerce clause to limit the power of Congress to make laws that regulated economic activity. The Supreme Court held that "production," "manufacturing," and "mining" were intrastate activities that could not be regulated by Congress under the commerce clause.[7] After the shock of the Great Depression and the start of the New Deal, the Supreme Court began taking a much more expansive view of the commerce clause. The court started to interpret the commerce clause as providing authority to Congress to regulate any activity that may substantially affect interstate commerce.[8] This interpretation of the commerce clause meant that Congress had authority to regulate a very wide range of activities. More recently, the pendulum has begun to swing back in the other direction. In 1995 the Supreme Court returned to a more restrained interpretation of the commerce clause. The new interpretation began in *United States v. Lopez,* where the Supreme Court held that a federal law forbidding the possession of a firearm in a school zone was unconstitutional because it exceeded Congress's commerce clause authority.[9] The court reasoned that possessing a firearm in a school zone did not have a large enough effect on interstate commerce that Congress could use the commerce clause as authority for the statute.

Although the Supreme Court has begun to restrain Congress's authority under the commerce clause, Congress still has considerable power to regulate intrastate activity that affects interstate commerce. This potentially includes regulating the wildlife within a state. For example, in 1977 the Supreme Court wrote that Congress could regulate fishing within a state as long as there was "some effect" on interstate commerce, even if the fish being regulated did not cross state borders.[10] The court argued that the "movement of vessels from one State to another in search of fish, and back again to processing plants, is certainly activity which Congress could conclude affects interstate commerce."[11] The changing interpretation of the commerce clause since the ratification of the Constitution explains why statutes enacted in the past may make different assumptions about the limits of Congress's authority to regulate wildlife. This will be seen with the Lacey Act, discussed later in the chapter.

The second constitutional provision that acts as a source of authority for regulating wild plants and animals is the treaty clause. The treaty clause gives the president the power, "with the Advice and Consent of the Senate, to make Treaties."[12] After the United States has entered into a treaty, Congress has the power under the "necessary and proper" clause of the Constitution to enact statutes to implement the treaty.[13] As a result, Congress may pass statutes that implement a treaty even if Congress did not have the power to pass those statutes under the other clauses of the Constitution. International treaties and the treaty clause are discussed in greater detail in chapter 4 (see also box 4.1).

Finally, the property clause of the Constitution states that "Congress shall have Power to dispose of and make all needful Rules and Regulations respecting the Territory or other Property belonging to the United States."[14] The property clause gives the federal government authority to control and manage federal lands however it chooses. This is important because the federal government owns 28% of the land in the United States. Originally the property clause was interpreted as giving the federal government the power of a private landowner over the land it owned, subject to the laws of the states where federal land existed.[15] The Supreme Court began to broaden the interpretation of the property clause in the 1800s, giving the federal government greater control over the land it owned. This culminated in the Supreme Court case *Kleppe v. New Mexico* in 1976.[16] In that case, the court held that under the property clause, "the power over the public land thus entrusted to Congress is without limitations."[17] The court also wrote that the power Congress has over the public lands "includes the power to regulate and protect the wildlife living there." The federal government thus has the final say on how wild plants and animals are managed on federal lands, and does not have to abide by the laws of the states where the federal lands exist. The management of federal lands is examined in considerable detail in chapter 5.

The commerce clause, treaty clause, and property clause are the three most important sources of authority for Congress in enacting wildlife laws. Appreciating the relationship between two of these clauses, the commerce clause and treaty clause, is essential to understanding one of the earliest federal wildlife laws, the Migratory Bird Treaty Act.

Migratory Bird Treaty Act

As noted above, in 1914 two federal courts struck down the predecessor to the Migratory Bird Treaty Act as an unconstitutional violation of the right of the states to manage the species within their borders.[18] The courts clearly took a limited view of the power the commerce clause gave to Congress, and did not believe the commerce clause provided authority to Congress to pass legislation regulating wildlife within state borders.

In 1916 the United States entered into a treaty with Great Britain to protect migratory birds. At the time, Great Britain handled the international affairs of Canada, so the treaty, which is still in effect, protects several bird species that migrate between the United States and Canada. Congress passed a statute implementing the treaty in 1918, thereby making the treaty apply to the states. This implementing statute is the Migratory Bird Treaty Act (see chapter 4 for a general discussion of statutes that implement treaties).

The Supreme Court in *Missouri v. Holland* soon took up whether the Migratory Bird Treaty Act also violated the right of the states to manage the species within their borders. The court began by noting that the treaty clause of the Constitution expressly gives the president the power to make treaties. Article I, Section 8 of the Constitution then gives Congress the power to make all "necessary and proper" laws required to execute the powers given to the federal government by the Constitution. The Supreme Court reasoned that because the treaty with Great Britain to protect migratory birds was valid, the Migratory Bird Treaty Act was a necessary and proper means of implementing the treaty. The Supreme Court therefore ruled that the Migratory Bird Treaty Act was constitutional. Consequently, the commerce clause did not give Congress the authority to regulate migratory bird species, but the treaty clause did give Congress that authority. As a result, the federal government could regulate migratory bird species within state borders.

The United States later signed similar treaties with Mexico, Japan, and the former Soviet Union to protect additional migratory bird species. Subsequent amendments to the Migratory Bird Treaty Act by Congress made those treaties also apply to the individual states.

So what does the Migratory Bird Treaty Act do? The act makes it unlawful to pursue, hunt, kill, or capture any migratory bird protected by one of the treaties with Great Britain, Mexico, Japan, or the Soviet Union.[19] The act also makes it unlawful to possess, sell, purchase, import, or export any such migratory bird. The Migratory Bird Treaty Act does not just apply to living birds; it also applies to the eggs, nest, or body part of any migratory bird, including feathers.

The prohibitions of the Migratory Bird Treaty Act do not actually depend on a bird's being migratory. The act applies to any bird species that is included in one of the four migratory bird treaties. The US Fish and Wildlife Service (FWS) regularly updates the list of bird species that fall under the act.[20] Nearly all bird species in the United States are covered by the Migratory Bird Treaty Act—1,026 species in all.

In 2004 Congress amended the Migratory Bird Treaty Act to exclude species that were introduced by humans into the United States.[21] This includes species that are protected in one of the four treaties, but are determined by FWS to be nonnative. Because the amendment to the Migratory Bird Treaty Act occurred after the treaties were signed, the act is considered to be Congress's latest word, and therefore trumps the treaty. The group of species mentioned as protected in a treaty but not protected under the Migratory Bird Act because they are nonnative includes well-known species such as the mute swan (*Cygnus olor*) and the rock pigeon (*Columba livia*).[22]

Prohibition on the killing of migratory birds is not absolute under the Migratory Bird Treaty Act. FWS issues regulations allowing the hunting and killing of several protected bird species.[23] The hunting or possession of a migratory bird requires obtaining a permit, including permits for scientific collecting.[24] There are then several exceptions for the permit requirements. FWS regulations list several migratory bird species that may be killed without a permit if they are eating agricultural crops, or are in flocks of such large numbers that they are a nuisance. Examples of birds that can be killed include American crows (*Corvus brachyrhynchos*) and brown-headed cowbirds (*Molothrus ater*).[25] The Migratory Bird Treaty Act also explicitly states that migratory birds bred on farms for food do not fall under the act.[26] Finally, the act has not historically been applied to feathers and other bird parts used for ornamental purposes by members of recognized American Indian tribes.

While the Migratory Bird Treaty Act clearly prohibits the direct killing of protected species, the act is less clear on whether it prohibits indirect killing of migratory birds. Does cutting down a forest where protected birds roost violate the act? What about destroying the food source for a flock of protected birds? Courts have been very reluctant to extend the Migratory Bird Treaty Act beyond the direct killing of birds. Only under a limited range of circumstances does the indirect killing of a protected bird species violate the act. For example, in 1978 a federal appellate court ruled that a pesticide manufacturer violated the act by discharging pesticide residues into a settlement pond, killing migratory birds visiting the pond.[27] In another case, a power company was found liable for the foreseeable deaths of birds after not installing inexpensive protections on its power lines.[28] The courts appear to have settled on a standard that says that a violation of the Migratory Bird Treaty Act occurs only when harm to birds was proximately caused (caused by the actor without any intervening cause) and reasonably foreseeable.[29] Box 2.1 defines proximate causation.

BOX 2.1. PROXIMATE CAUSATION AND STRICT LIABILITY

The law generally does not penalize a person for undertaking an action that would have been nearly impossible to predict would cause harm. As a result, most statutes require proximate causation before penalizing someone. "Proximate causation" is defined as "a cause that directly produces an event and without which the event would not have occurred."* To prove proximate causation, it must be shown that the harm was reasonably foreseeable.

Some statutes attach strict liability to an activity. Under strict liability, a person can be penalized for an action even if the person did not intend to cause harm or did not foresee that the harm would occur. Strict liability usually applies to very hazardous activities, such as owning a wild animal.

*Bryan A. Garner, ed., *Black's Law Dictionary*, 9th ed. (St. Paul, MN: Thomson/West, 2009).

One area where this standard is still in flux is wind turbines killing protected birds. FWS estimated in 2009 that collisions with wind turbines kill 440,000 birds every year.[30] In late 2013 Duke Energy Corporation pleaded guilty to killing 14 eagles and 149 other birds at wind farms in Wyoming. The company agreed to a settlement of $1 million. This was the first time that the US government had prosecuted a wind energy company for violating the Migratory Bird Treaty Act. Duke Energy admitted that it "constructed these wind projects in a manner it knew beforehand would likely result in avian deaths."[31] That bird deaths must be reasonably foreseeable to be prohibited under the Migratory Bird Treaty Act still seems to hold.

Unlike other environmental laws that will be discussed, the Migratory Bird Treaty Act does not contain a provision allowing citizens to sue other individuals or corporations that are violating the act (although citizens may sue a federal agency for violating the Migratory Bird Treaty Act by invoking the Administrative Procedure Act). As a result, FWS has discretion about when to enforce the act. This is important for wind farm operators because in 2010 FWS created voluntary guidelines for wind farm operators to follow when constructing and operating wind farms.[32] Perhaps the most important suggestion in the guidelines is simply not to construct a wind farm in an area of high ecological value, such as areas used by rare species or areas containing large intact patches of habitat. Wind farm operators that follow the guidelines but still kill birds protected by the Migratory Bird Treaty Act will likely not face prosecution by the federal government.

The greatest ecological criticism that can be leveled at the Migratory Bird Treaty Act is that it does very little to protect bird habitats. Habitat modification and loss that results in decreasing bird population numbers is too indirect a means of bird death to fall under the act. For example, a federal appellate court ruled in 2004 that timber harvesting by the National Park Service that resulted in the cutting of trees with nesting birds did not violate the act.[33]

There has been very little peer-reviewed scientific research done on how the Migratory Bird Treaty Act has affected the ecology of bird species. A report by FWS, however, does show that over the last 40 years, many bird populations in grasslands and arid-land habitats have declined rapidly. Bird populations in forests have also seen steady declines. Conversely, wetland species and hunted waterfowl species have seen population increases during the last 40 years, likely due to the emphasis the federal government places on protecting wetlands (see chapter 8).[34]

Most of the bird species with declining population numbers already receive protection under the Migratory Bird Treaty Act. This suggests that this act alone is not sufficient for protecting many species. While the act protects birds from being directly killed, habitat modification and loss are currently driving the declines in many bird populations. As the Migratory Bird Treaty Act does little to protect bird habitats, it will likely do little to prevent further declines in the population numbers of many bird species. The Endangered Species Act (ESA) does a better job of protecting the habitat of the species it is trying to preserve. Once a species receives protection under the ESA, the federal government is required to designate critical habitat that is essential for the conservation of the species, and federal agencies may not adversely modify or destroy such critical habitat. However, once a species is close enough to extinction that it falls under the ESA, designating critical habitat may not be enough to save the species from extinction. It would be better if wildlife statutes such as the Migratory Bird Treaty Act kept species from needing the protections of the ESA in the first place.

Lacey Act

Although the Migratory Bird Treaty Act is certainly an old statute, the distinction of being the oldest federal statute for protecting wildlife goes to the Lacey Act.[35] Congress passed the Lacey Act in 1900 primarily to enable the federal government to prosecute state game-law violators. Prior to the Lacey Act, a poacher could kill wildlife in one state, violating the laws of that state, and could then transport the dead animal to a different state, thereby evading the jurisdiction of the state where the poaching occurred. To stop this, Congress designed the Lacey Act primarily as a way to enforce state wildlife laws when a violator crossed state lines. As with the Migratory Bird Treaty Act, Congress passed the Lacey Act to prevent a tragedy of the commons, where individuals were able to deplete wildlife resources without punishment. In the case of the Lacey Act, it was people moving wildlife across state borders that created the setting for a tragedy of the commons, not species migrating across state borders.

A violation of the Lacey Act has two requirements: (1) there must be a violation of some other wildlife law—the other law can be a federal law, state law, Indian tribal law, treaty, or even foreign law; and (2) the animal or plant involved in the violation of the other law must be imported, exported, transported, sold, received, acquired, or purchased.[36]

What needs to occur to fulfill the second requirement depends on which wildlife law is violated as part of the first requirement. If the other wildlife law that is violated is a federal law or Indian tribal law, then any action that fulfills the second requirement, such as moving the animal or plant, violates the Lacey Act. For example, if (1) a hunter kills a deer in a portion of a national wildlife refuge closed to hunting, thereby violating federal law, and then (2) transports that deer home, even if the hunter's house is in the same state as the wildlife refuge, prosecution under the Lacey Act would be possible. If the wildlife law that is violated is a state law or foreign law, however, the second requirement must occur in the course of interstate or foreign commerce. For example, if (1) a hunter kills a deer in a closed portion of a state wildlife refuge, thereby violating state law, and then (2) transports that deer on a highway to his house in a different state, prosecution under the Lacey Act would be possible.

In 2003 Congress revised the Lacey Act to do away with the first requirement of the act for big cat species. Under the Lacey Act it is unlawful to import, transport, sell, receive, acquire, or purchase in interstate or foreign commerce any live species of lion, tiger, leopard, cheetah, jaguar, or cougar.[37] For big cats, there is now no requirement to violate a different wildlife law first in order to fall afoul of the Lacey Act.

While the Lacey Act continues to be useful in enforcing other wildlife laws, today the act is more important as the primary tool in protecting US ecosystems from nonnative species. For a species to move outside of its native range and become invasive in a novel ecosystem, the species must often pass through several political borders. Political borders therefore present a potentially good place to stop the movement of nonnative species. The next section discusses how the Lacey Act and other federal laws are used to combat the spread of nonnative species.

Invasive Species and Federal Law

Invasive species present one of the greatest threats to the structure and function of many ecosystems.[38] The National Park Service estimates that invasive species have contributed to 68% of species extinctions in the United States; roughly 40% of species currently threatened with extinction are primarily at risk due to competition or predation from invasive species. Invasive species can also change the characteristics of an ecosystem such that the services provided by that ecosystem are compromised. Invasive species cause huge economic losses, with one estimate suggesting losses of $120 billion each year.[39] As global climate change intensifies, invasive species are expected to become a greater problem as species from more southern latitudes move north, and as weakened ecosystems become more prone to invasion.

There are four steps in the invasion process: transport, introduction, establishment, and spread.[40] A species must be transported from its native range to an area outside of that range, either through natural processes or with the help of humans. The species must then be introduced into the wild in this new area. At this point the species is a nonnative species, but is not yet invasive. The next step in the process is becoming established: the species creates a self-sustaining population outside of its native range. Finally, when the species begins to spread to areas outside of where it first became established, it is an invasive species.

Interrupting any of these steps will stop a species from becoming invasive. Ecological interactions, such as competition with native species, prevent many nonnative introduced species from becoming established, and even more from becoming invasive. For those species that have the ecological potential to become invasive, human intervention may be necessary to prevent an invasion. Interrupting some steps in the invasion process is easier than others, though. Once a potentially invasive species has become established in a new area or has begun to spread to other areas, it is very costly, and frequently impossible, to completely eradicate the species in its new range. Preventing transport and introduction of nonnative species is usually much easier. A species that is never transported outside of its native range will never have the chance to become established in a new area. Moreover, nonnative species must often be introduced several times before they become established in areas outside of their native ranges. As a result, reducing the number of times transported individuals of a nonnative species are released, and the number of individuals released at a time, is also an effective means of reducing the likelihood of a species becoming established. When species are transported outside of their native range, they may cross several political borders during the transportation process. Consequently, stopping nonnative species at political borders may deny species the opportunity to become invasive.

Lacey Act and Invasive Species

The Lacey Act is the primary federal statute for preventing the spread of invasive species.[41] While the Lacey Act can be used to enforce other wildlife laws, its most important use is prohibiting the import into the United States of any animal species listed as "injurious." The Lacey Act also prohibits the transport of any injurious species between the states.

These prohibitions extend to the offspring and eggs of any such species. Injurious species may be imported into the United States or transported between the states only after obtaining a permit from FWS.[42]

The Lacey Act defines injurious species as those that do harm "to human beings, to the interests of agriculture, horticulture, forestry, or to wildlife or the wildlife resources of the United States."[43] FWS considers several factors in deciding whether a species has the potential to become invasive, and therefore whether to list the species as injurious: (1) the likelihood of release or escape; (2) the ability of the species to spread geographically; (3) impact on ecosystems, such as competition, habitat degradation, predation, and hybridization; (4) impact on threatened and endangered species; (5) impact on agriculture, forestry, and horticulture; and (6) ability of managers to control and eradicate the species.[44] FWS also performs an economic analysis to determine the economic impact of listing a species as injurious. The species FWS has decided are injurious are listed in the code of federal regulations.[45] To help enforce the Lacey Act, FWS has wildlife inspectors at 38 US airports, ports, and border crossings.[46]

In several ways the Lacey Act is less than an ideal defender against invasive species. The act prohibits the movement of injurious species between the states, but the Lacey Act has no authority within the states. Each individual state decides whether to prohibit possession or movement of injurious species within that state. Additionally, once a nonnative species is actually present within a state, the Lacey Act gives FWS no authority to manage or eradicate the species, even if becomes invasive. Consequently, the Lacey Act focuses on the transport and introduction stages of the invasion process; the act does little to prevent establishment or spread of invasive species.

Next, the Lacey Act specifies that only species of mammals, birds, fish, mollusks, crustaceans, amphibians, and reptiles may be listed as injurious.[47] FWS may not list plant and invertebrate species under the Lacey Act. The next section of this chapter discusses the Plant Protection Act, which is a statute meant to prevent the spread of potentially invasive plants and plant pest species. The Plant Protection Act, though, emphasizes stopping invasive plants and plant pests that harm agricultural interests. Additionally, the act is enforced by an agency in the Department of Agriculture, not FWS. Neither the Lacey Act nor the Plant Protection Act can be used to prevent the import of infectious wildlife pathogens, such as the chytrid fungus (*Batrachochytrium dendrobatidis*) that has infected many amphibian species.[48] The limited set of taxa that may be listed as injurious under the Lacey Act, along with the agricultural focus of the Plant Protection Act, curtails the combined usefulness of the two statutes in preventing the spread of invasive species.

Lastly, the greatest fault of the Lacey Act is the very structure of the injurious species list. When deciding which species may be imported, a country could potentially create a "dirty list" or a "clean list." A dirty list (or "black list") is a list of species that are not allowed to be imported into the country. Any species not on the dirty list may be freely imported into the country. Conversely, a clean list (or "white list") is a list of species that may be imported into the country, while all other species may not be imported. A dirty-list approach places species on the list only after there is some evidence that they are potentially invasive. A clean-list approach is better at preventing the importation of potentially invasive species because it allows the importation of only those species unlikely to become invasive. All other species are presumed to be potentially invasive and are therefore prohibited from importation.

The injurious species list created under the Lacey Act is a dirty list. Any species may be imported into the United States until there is sufficient evidence to suspect that the species is potentially invasive. Although more difficult to administer, it is possible for a country to adopt a clean-list approach. For instance, New Zealand uses a clean-list approach. Under its Hazardous Safety and New Organism Act of 1996, a species not already present in New Zealand by 1998 may not be imported into the country until a risk assessment is performed for the species. The New Zealand government performs the risk assessment, which must be paid for by the person applying to import the species. The risk assessment determines whether the species has the potential to displace native species or degrade natural habitats.[49] If the species has that potential, it may not be imported into New Zealand.

In the 1970s FWS attempted to move from a dirty list to a clean list. Due to strong opposition, in part from the pet industry, FWS abandoned the attempt.[50] The result is that only the 43 genera listed as injurious by FWS are prohibited from importation into the United States under the Lacey Act. To put this in perspective, in the previous decade, 2,200 nonnative species were involved in the US import trade.[51] The large number of imported species would likely not be a problem if FWS regularly added species to the injurious species list, but FWS on average lists only one taxa every four years.

Given the problems with the Lacey Act, perhaps it is not surprising that the act has not done a stellar job of preventing the spread of invasive species. A study by Fowler et al. showed that at least 56% (9 of 16) taxa listed as injurious were already in the United States at the time of listing, and 44% (7 of 16 taxa) had established populations in the United States at the time of listing.[52] Of the taxa with established populations at the time of listing, 71% have gone on to spread to other states, suggesting that they have become invasive species. Listing a species as injurious after it has already formed an established population in the United States does little to prevent that species from becoming invasive. Conversely, of those taxa that were absent from the

United States at the time FWS listed them as injurious, none have subsequently established populations in the United States. The Lacey Act seems to be doing a good job of preventing the transport stage for those species that make it onto the injurious species list.

The best way to take advantage of this strength of the Lacey Act is to switch from a dirty-list approach to a clean-list approach. By prohibiting the import of species unless they are known not to be invasive, the clean-list approach would stop the transport stage for all other potentially invasive species. However, the likelihood of the United States moving from a dirty- to a clean-list approach is fairly low. The clean-list approach is an implementation of the precautionary principle—a regulatory body stops an action until there is scientific consensus that the action is not harmful. Conversely, most federal environmental laws in the United States, including the dirty-list approach of the Lacey Act, are reactionary—a regulatory body allows an action until the potential for harm has been proven. Switching the Lacey Act from a reactionary to a precautionary approach would be very difficult. Many of the 2,200 nonnative species imported into the United States in the last decade would have to undergo risk assessments to determine if they could be placed on the clean list. This would be a huge undertaking. Additionally, moving from a reactionary to a precautionary approach would likely face considerable opposition from Congress. In 2009 several US representatives cosponsored a bill that would have created a clean list for the importation of nonnative species into the United States.[53] The bill never received a vote.

Plant Protection Act

As mentioned in the last section, a different federal statute attempts to control the spread of potentially invasive plant and plant pest species. Congress passed the Plant Protection Act in 2000, consolidating ten other statutes. The act gives the Animal and Plant Health Inspection Service (APHIS), an agency of the Department of Agriculture, authority to prohibit the import of "noxious weeds" and "plant pests" into the United States, or the movement of such species in interstate commerce.[54]

A "noxious weed" is defined in the Plant Protection Act as any plant or plant product that can directly or indirectly damage crops, livestock, natural resources, human health, or the environment.[55] A "plant pest" is any species that directly or indirectly injures, damages, or causes disease in a plant.[56] Plant pests under the act include protozoans, bacteria, fungi, viruses, and invertebrate animals such as insects, mollusks, or nematodes.[57]

Congress in 2004 amended the Plant Protection Act with the Noxious Weed Control and Eradication Act.[58] The amendments give greater power to APHIS to control invasive plant species already causing damage in the United States. Under the amendments, the Department of Agriculture may make grants to local or state agencies or nongovernmental organizations for the control or eradication of noxious weed species.[59]

In a different section of the Plant Protection Act, the Department of Agriculture has authority in an "extraordinary emergency" to restrict the movement of a noxious weed or plant pest within a state, quarantine part of a state, and destroy the noxious weed or plant pest.[60] Consequently, and unlike the Lacey Act, the Plant Protection Act provides authority for combating the establishment and spread stages of the invasion process. The reason the Plant Protection Act gives a federal agency the authority to take action within a state and the Lacey Act does not reflects the evolving interpretation of the commerce clause discussed earlier in the chapter. Congress passed the Plant Protection Act when the commerce clause was interpreted to give Congress authority over almost any activity within a state as long as the activity affected interstate commerce. The spread of an invasive species within a state certainly has the potential to affect interstate commerce. Congress passed the Lacey Act, however, when the interpretation of the commerce clause was more constrained, and the management of wildlife within a state was considered to be solely a state's responsibility.

Despite the authority granted to the Department of Agriculture under the Plant Protection Act, the statute has the same major flaw as the Lacey Act: both statutes use a dirty-list approach. The list created by APHIS under the Plant Protection Act consists of 19 aquatic and wetland weed species, 5 genera of parasitic weeds, 95 species of terrestrial weeds, and approximately 400 species of plant pests. While this is a relatively large list of species, a dirty list is still a less effective means of stopping the transport and introduction of potentially invasive species than is a clean list. Like the Lacey Act, however, there is very little peer-reviewed research on the effectiveness of the Plant Protection Act.[61]

An additional problem with the Plant Protection Act is that it has not been used by APHIS to its full capacity to fight invasive species. Critics argue that APHIS focuses primarily on protecting crop species from nonnative species, with little concern given to invasive species that affect anything other than agriculture.[62] If APHIS used the statute to protect more than crops, the Plant Protection Act could be a significantly more powerful statute for combating invasive species.

Nonindigenous Aquatic Nuisance Prevention and Control Act

Ballast water discharges are one of the primary means for the introduction of invasive aquatic species. Ballast water is taken in or released from a ship to help maintain the balance of the ship as cargo is loaded or unloaded. Ballast water taken in from a water body also contains many of the aquatic species found in that water body. If that ballast water is then released in a different water body, all the species in the ballast water are introduced into the second water body. One study concluded that of the 59 invasive species established in the Great Lakes since the 1950s, roughly half arrived in ballast tanks.[63] The most damaging of these is the zebra mussel (*Dreissena polymorpha*), which outcompetes native species and causes significant economic damage by growing on water pipes and boat hulls (figure 2.1). The introduction of aquatic species from ballast water can be prevented by using meshes to screen out organisms, and then using chlorine or ultraviolet rays to kill any organisms that make it past the meshes.

Congress passed the Nonindigenous Aquatic Nuisance Prevention and Control Act in 1990 to prevent the introduction of aquatic invasive species.[64] The statute directed the US Coast Guard to issue guidelines for preventing the introduction of aquatic invasive species from ballast water into the Great Lakes.[65] In 1996 Congress amended the act by passing the National Invasive Species Act. Despite its impressive name, the amendments simply extend the Nonindigenous Aquatic Nuisance Prevention and Control Act to all navigable inland waters of the United States, and 5.6 kilometers (3 nautical miles) out into the ocean.[66] One of the main goals of the act and its amendments is to control the spread of zebra mussels.

There have been recent developments in controlling ballast water discharges under the Clean Water Act (CWA). (The CWA will be examined in much greater detail in chapters 7 and 8.) The EPA had long refused to regulate discharges of ballast water under the CWA. It argued that discharges from vessels cause little release of pollutants, and therefore did not need to be regulated. The establishment of invasive species in the Great Lakes and other water bodies from ballast discharges suggested otherwise. A federal appellate court in 2008 ruled that the EPA's refusal to regulate vessel discharges was contrary to the requirements of the CWA.[67] The court reasoned that the CWA includes biological material in its definition of pollutants, that ballast discharges release such biological material, and that ballast discharges should be regulated under the CWA. The EPA now requires permits under the National Pollutant Discharge Elimination System program for

FIGURE 2.1. Zebra mussels attached to a current meter in Lake Michigan. Photograph from NOAA.

ballast discharges in all inland navigable waters of the United States, and 5.6 kilometers out into the ocean. So far, though, the requirements of the permits are generally the same as the requirements under the Nonindigenous Aquatic Nuisance Prevention and Control Act. Permits for vessel discharges other than ballast water are discussed in chapter 9.

Specific Species Laws

The preceding statutes all endeavor to control the spread of a wide range of potentially invasive species. Several federal statutes, though, focus on controlling specific invasive species. For instance, the Brown Tree Snake Control and Eradication Act of 2004 provides authority and funding to the Departments of Interior and Agriculture to control the spread of the invasive brown tree snake (*Boiga irregularis*), and to restore native wildlife damaged by the snake.[68] Other examples include the Nutria Eradication and Control Act and the Salt Cedar and Russian Olive Control Demonstration Act. These statutes attempt to control species that are already invasive and doing considerable ecological damage. While the statutes described in the previous sections concentrate on stopping the transport and introduction stages of the invasion process, the specific species statutes mainly deal with the establishment and spread stages.

The main criticism of these specific species statutes is that it is very inefficient for Congress to pass legislation for one invasive species at a time. A better approach would be a comprehensive invasive-species statute that provides authority and funding for control measures for any species that becomes invasive.

Executive Order 13,112

President Clinton in 1999 signed Executive Order 13,112 in an effort to create a more comprehensive federal system for preventing the spread of invasive species. Although they are not statutes, executive orders have the force of law because the president has the authority to sign such orders under the Constitution, and because executive orders are usually made in pursuance of an act of Congress. For example, Executive Order 13,112 was signed under the authority granted by Congress to the president from the National Environmental Policy Act, the Nonindigenous Aquatic Nuisance Prevention and Control Act, the Lacey Act, the Federal Plant Pest Act, the Federal Noxious Weed Act, and the Endangered Species Act. None of these statutes actually gave the president explicit authority for the language in the executive order, but the authority was considered to arise from provisions in each act. Additionally, authority to issue executive orders comes directly from the US Constitution. There is no constitutional clause specifically giving presidents the authority to issue executive orders, but the grant of "executive power" in Article II, Section 1, Clause 1 of the constitution is taken as authority for executive orders.

Executive Order 13,112 requires that each federal agency prevent the introduction of invasive species, control populations of invasive species, and provide for restoration of native species. The order defines "invasive species" as nonnative species "whose introduction does or is likely to cause economic or environmental harm or harm to human health." The executive order bars federal agencies from authorizing any activity that will promote the spread of invasive species.

The executive order also creates a National Invasive Species Council made up of senior government officials such as the secretaries of state, interior, and agriculture, and the administrator of the Environmental Protection Agency. The council is directed to take actions to achieve the goals of an overarching National Invasive Species Management Plan. This is to be done by overseeing the actions of federal agencies that touch on invasive species, and recommending actions to local and state agencies.

While Executive Order 13,112 seems like a powerful tool in fighting invasive species, in practice the order does not have much muscle. The order bars a federal agency from taking an action that promotes the spread of invasive species, but the order has an exception if the agency determines that the benefit of the action clearly outweighs the potential harm caused by the spread of invasive species. Additionally, the National Invasive Species Council has done little to provide guidance to federal agencies in managing invasive species.[69] Finally, the executive order states that members of the public may not sue the federal government to force it to comply with the order. As a result, the order creates no legal recourse for citizens to force federal agencies to take invasive species into consideration when making decisions. A more comprehensive approach to preventing the spread of invasive species will likely have to await congressional action.

STATE MANAGEMENT OF PLANTS AND WILDLIFE

So far the laws examined in this chapter have all been federal. The individual states also have laws meant to manage and protect the wild species within their borders. Most of these laws exist as statutes passed by the state legislatures. State hunting laws are examples of such statutes.

Courts also have the ability to create law through the decisions they hand down. This law is called common law and is defined in box 2.2. Over the years, courts have created a common law duty that requires states to manage their natural resources, including wild organisms, for the benefit of the people of the state. This is called the public trust duty.

BOX 2.2. COMMON LAW

There are four types of law: (1) constitutional law; (2) statutory law; (3) regulatory law, created by executive agencies; and (4) common law. Common law is the body of law created by judicial decisions. When a court issues a decision in a case, that decision acts as a precedent, meaning that future cases with similar facts must be decided the same way. This ensures fairness because cases with similar facts receive similar outcomes. The precedents from previous decisions make up the common law. The principle that a court should follow the precedents created by previous court decisions is called *stare decisis*. Under *stare decisis*, a court is bound by the precedent of a higher court, and may overturn its own precedents only if there is a compelling reason to do so.

In the past, the common law was a much larger source of law than it is today. Prior to the twentieth century, statutes tended to be limited or vague, which meant that if a court heard a case distinct from any previous case, it would do little good to look to the existing statutes for help in deciding the case. The court would instead have to make up its own rule, and in so doing make new law. Today, however, statutes are the more prevalent source of law. This is in part because over time, legislatures codified many common law principles in statutes. Because statutes are the more prevalent source of law, common law today is most often created when judges interpret statutes to fit specific facts. Statutes have priority over the common law, however, meaning that legislatures can amend or completely change common law rules whenever they desire. There are still some areas of the law, such as tort law, where common law rules built up over many years continue to define a large portion of the law. Although the common law is less important to environmental law than it was in the past, it still influences many aspects of environmental law and makes several appearances throughout the book.

All the states in the United States have common law systems except for Louisiana, which has a civil law system. In civil law systems, judges do not create law and, when interpreting statutes, give less weight to precedent from previous decisions. Louisiana has a civil law system because of its history of French influence, with France also having a civil law system.

Just as state courts make common law, federal courts also make common law, but in a more limited way. The Supreme Court ruled in 1938 in the case of *Erie Railroad Co. v. Tompkins* that federal courts can create common law only if the issue at stake touches on federal or constitutional concerns.* This means that if a federal court is hearing a matter governed by a federal statute or the US Constitution, its decision creates federal common law. Conversely, if a federal court is hearing a matter governed by state law, it must apply the common law rules created by the state courts.

Why, you might ask, would a federal court be hearing a matter governed by state law? The most common occasion is when a citizen of one state sues a citizen of another state. Allowing a state court to hear the case when one of the parties is a citizen of that state could result in bias against the party from the other state. To encourage fairness to both parties, the plaintiff may file the case in federal court, or if the plaintiff files in a state court, the defendant may file a notice to remove the case to a federal court. The federal court would then decide the case under the relevant state law.

*304 U.S. 64 (1938).

State Ownership of Wildlife

Who owns wild animals? The obvious answer seems to be that if an animal is wild, no one owns it. This answer presents certain problems, though. If no one owns wild animals, then does no one have the right to manage them, or to prevent others from hunting them to extinction? The answer to who owns wild animals has a very long history (which is greatly shortened in the next section). The history of this ownership helps define the states' public trust duty for managing wildlife.

History of State Ownership

Under Roman law, wildlife was owned by no one. Only when a wild animal was captured was it considered to be owned, by the captor. Much later in England, a different approach was taken. The courts slowly built up the common law rule that the king owned all the wild animals in the realm. The king, however, owned the species in a sovereign capacity rather than a proprietary capacity. This meant that the king had to manage wild species in the interest of the entire kingdom, instead of for his personal benefit. In practice this often meant rewarding the nobility with hunting rights while withholding hunting rights from commoners. Indirectly, this may have had the benefit of helping to conserve many species from overhunting.

Courts and legislatures in the newly founded United States viewed wildlife as belonging to no one. In the famous 1805 case *Pierson v. Post*, the New York Supreme Court ruled that ownership of a wild animal requires capture or mortal wounding.[70] The doctrine that any person could capture or kill a wild animal because it belonged to no one quickly resulted in overharvesting of many species.

To deal with this overharvesting, state legislatures in the mid-1800s began passing laws limiting the hunting of some wild animals. State courts upheld these laws by arguing that the states were actually the rightful owners of the wildlife within their borders, and therefore could place limits on hunting. The courts made this argument by asserting that when America declared independence from Great Britain, the sovereign ownership of wild species in America that had been vested in the king transferred to the individual states. When America became independent from Britain, the English common law became American common law. As a consequence, the wildlife within the boundaries of a state was owned by that state.

As with the king's being required to manage wildlife for the benefit of the kingdom, not for his own benefit, each state had to manage the wildlife within its borders for the benefit of the people of that state. This duty defines the public trust duty that states have in managing wildlife.

It should be noted that there is also a "public trust doctrine" that has a distinct historical origin from the public trust duty in managing wildlife. The public trust doctrine dates back to Roman law and holds that the government maintains a public trust over waterways for the purpose of navigation, commerce, and fishing. The public trust doctrine has traditionally related only to waterways, and not to the ownership of wildlife or whether states must manage wildlife for the benefit of the public. The public trust doctrine was incorporated into US law in the 1892 Supreme Court case *Illinois Central Railroad v. Illinois*.[71] More recently, some scholars argue that the traditional public trust doctrine as applied to waterways and the public trust duty in managing wildlife should be merged into a comprehensive public trust doctrine for all natural resources.[72]

The first Supreme Court case touching on the states' public trust duty in managing wildlife was *Martin v. Waddell* in 1842.[73] In that case, the court held that the public has a right to fish in navigable waters because the state owns the underlying lands and the fish within the waters in trust for the people of the state. Fifty-four years later, in the influential case *Geer v. Connecticut*, the Supreme Court endorsed the concept of state ownership of wildlife.[74] As an obligation of owning wildlife, the court further confirmed that states have a public trust duty in managing the wildlife they own. The court wrote that it is "the duty of the legislature to enact such laws as will best preserve the subject of the trust, and secure its beneficial use in the future to the people of the state." The states, however, began using this decision to engage in economic protectionism; for instance, states started preventing out-of-state fishermen from fishing in their waters.

The Supreme Court soon began to weaken state ownership of wildlife. In the case described in the introduction to this chapter, *Missouri v. Holland*, the court held that migratory birds are not the possession of the state they happen to be in at the time. The court noted that a migratory bird may not have been in a state yesterday, and in a week may no longer be in the state, so a state may not claim ownership of such birds.[75] Eventually, the Supreme Court explicitly overruled *Geer*, in the 1979 case *Hughes v. Oklahoma*.[76] In the *Hughes* case, an Oklahoma statute prohibited the export of wild minnows out of the state. The Supreme Court invalidated the Oklahoma statute as interfering with interstate commerce. For good measure, the court also wrote that states do not own the wildlife within their borders.

The *Hughes* case seems to have brought ownership of wildlife full circle. Under Roman law and implicitly in early American law, no one owned wild animals. The Supreme Court then endorsed state ownership of wildlife, but under the *Hughes* decision it again appears no one owns wild animals. What does this mean for state management and preservation of wildlife?

Despite the ruling in the *Hughes* case, the Supreme Court did not actually undo the public trust duty that states owe to their citizens. The court in *Hughes* emphasized that while states do not own the wildlife within their borders, they still had the right to manage and preserve wildlife, thereby confirming the public trust duty. As the Alaska Supreme Court wrote, though Alaska may not own its wildlife, "[n]evertheless, the trust responsibility that accompanied state ownership remains."[77]

Many states have gone further, though, and simply ignore the part of the *Hughes* decision that holds that states do not own their wildlife. State courts and legislatures have argued that the *Hughes* opinion applies only when a state statute conflicts with federal law; absent such a conflict, the state still owns the wildlife in its borders.[78] In fact, more than 30 states explicitly claim in their wildlife statutes to have state ownership of wildlife.

Requirements of the Public Trust

Whether states legally own the wildlife in their borders or not, the public trust duty still requires the individual states to manage their wildlife for the benefit of all the citizens of the state. Under the public trust duty, the states must balance the use of natural resources with protection of those resources for present and future generations. A majority of states have passed statutes that implicitly or explicitly describe wildlife as being part of a public trust managed by the state. For example, a Georgia statute asserts that "wildlife is held in trust by the state for the benefit of its citizens and shall not be reduced to private ownership except as specifically provided for in this title."[79] At least 44 states have passed the Interstate Wildlife Violator Compact, which is an agreement among participating states that a person who loses hunting or fishing rights in one state loses those rights in all member states. The statutes passed by the individual states to implement the compact declare that "wildlife resources are managed in trust by the respective states for the benefit of all their residents and visitors."[80] Consequently, almost all the states have declared that they intend to manage wildlife in trust for the benefit of the public.

This raises a difficult question, though: what constitutes the public interest, and how must states manage wildlife to benefit the public? It is useful to begin by explaining the workings of a trust. In general, a trust is a legal entity that owns property. The trust must be managed by a trustee for the benefit of individual beneficiaries. Most trusts have a document stating the terms of the trust—the parameters within which the trustee must manage the trust. A beneficiary may sue the trustee if the trustee does not act in the best interest of the beneficiaries.

Under the public trust duty, the state is the trustee and the citizens of the state are the beneficiaries. However, there is no written document laying out the terms of the trust. This means that in most states, what constitutes both the boundaries of the public trust and how a state must manage wildlife for that trust are not entirely clear. Does the public trust require that each state produce as much economic value from the wildlife in the state as possible?[81] If so, should species that produce the most revenue, such as fish species valuable for the fishing industry or species valued by hunters, be managed to the detriment of less economically valuable species? For instance, a Kentucky statute announces that the state must act "to protect and conserve the wildlife of the Commonwealth to insure a permanent and continued supply of the wildlife resources of this state for the purpose of furnishing sport and recreation for the present and future residents of this state."[82] How are ecosystem services to be valued? Should those services be managed for the benefit of the citizens of the state even if that reduces biodiversity within the state or the ecosystem services of nearby states?

The answers to these questions may be slowly answered as state legislatures pass statutes more clearly defining the requirements of the public trust. Court decisions will also likely play an important role in determining the requirements of the trust. The public trust duty began as a common law doctrine. Courts use common law to "fill gaps" left by statutes. This means that where existing state laws do not regulate how wildlife should be managed, courts can create common law to flesh out the duties of the public trust.

The courts in many states have recognized and applied the traditional public trust doctrine requiring the state to protect navigable waterways for the benefit of the public. Very few state courts, though, have recognized a public trust duty for wildlife. When they have done so, state courts have typically applied the public trust duty retroactively to allow recovery of monetary damages, such as when industrial pollution leads to fish die-offs. State courts have rarely applied the public trust duty to wildlife conservation proactively.

The view that the public trust duty cannot be applied proactively may slowly change, however. A possible template for this comes from a public trust doctrine case in California. In 1983 the California Supreme Court laid out the requirements of the traditional public trust doctrine in that state. The case concerned the state of California's authorizing the diversion of water from streams in the Mono Lake watershed. The diversion had caused the water level in Mono Lake to fall, making Negit Island into a peninsula. Negit Island was a major breeding ground for California gulls, and when the island became a peninsula, coyotes could suddenly hunt the gulls. The California Supreme Court ruled that the state must reconsider the authorized water diversion. The court held that under the public trust doctrine, the state must (1) consider public trust values before approving an action that will affect a natural resource, (2) preserve trust values when feasible, and (3) continually supervise actions that affect natural resources.[83] The Mono Lake decision concerns the traditional public trust doctrine and therefore applies only to waterways. However, the reasoning of the court in the case may eventually come to influence how courts in other states begin to apply the public trust duty. As more and more states statutorily recognize the public trust duty for managing wildlife, state courts may become bolder in ruling that the public trust duty of the states applies proactively to protect all wild species.[84]

Hunting

With the individual states claiming to own the wildlife within their borders, or at least the authority to manage them, the states have the power to regulate where and when hunting of wildlife may occur. All the states in the United States have statutes regulating the hunting of wildlife. Courts have held that a state needs only a reasonable basis for deciding where hunting is allowed. The means that states have the authority to prohibit a landowner from hunting the wildlife on his or her own land.

State statutes usually put wildlife into one of three categories: game animals, protected animals, and unprotected animals. Game animals must be killed according to the laws of the state, and a hunting license is required before killing such animals. Protected animals may not be killed at any time. Unprotected animals are those species the state considers to be pests and may therefore be killed at any time. A license is not necessary to kill pest species.

Hunting can have numerous ecological and evolutionary effects on game species. One of the most obvious effects is that hunting decreases the size of game populations. A game animal that is killed is one less member of the population (see box 2.3). Removing individuals is not the only way hunting effects population size, though. Hunters often prefer to kill reproductive-aged adult animals that have reached a large size. These individuals are usually the ones with the greatest reproductive value. This often differs from

---

**BOX 2.3. DEFINITIONS OF COMMON
ECOLOGICAL TERMS**

CARRYING CAPACITY: The largest population of a particular species that the resources in an area can support.

COMMUNITY: All the living organisms in an area.

ECOLOGY: The study of the number and distribution of organisms in the environment.

ECOSYSTEM: The organisms in an area plus their interactions with the abiotic surroundings.

ENVIRONMENT: All the surroundings of an organism.

HABITAT: The area where an organism lives, including the physical and biotic factors of that area.

LANDSCAPE: Typically a large geographical area encompassing several interacting ecosystems.

POPULATION: The individuals of one species living in a particular area.

SPECIES: Most often defined as a group of organisms that can interbreed and produce fertile offspring. No single definition of *species* works well in all situations.

TROPHIC LEVEL: A level within a food chain, such as primary producer (e.g., plants), herbivore, or carnivore.

the individuals chosen by natural predators. For example, a study of elk (*Cervus elaphus*) in northern Yellowstone found that hunters killed female elk with a mean age of 6.5 years, while wolves (*Canis lupus*) killed females with a mean age of 13.9 years.[85] The prime reproductive ages for elk are 2–9 years, meaning that hunters tended to kill female elk with the greatest reproductive value. Conversely, wolves tended to prey on more vulnerable individuals, preferring to kill calves and older females with less reproductive value. The study concluded that hunting had a greater reproductive impact on the elk herd than did wolf predation.

Besides influencing the population ecology of game species, hunting can also create strong evolutionary pressures. As already mentioned, hunters prefer to kill reproductive-aged adults that have reached a large body size. This preference can create strong evolutionary pressure on a population.[86] Morphological traits, such as body and horn size, may both decrease within a population. At the same time, life-history traits, such as age at first reproduction, may shift to earlier in life. Rates of evolutionary change caused by harvesting, of which hunting is a prime example, are often greater than rates caused by natural phenomena.

The evolutionary pressure on game species may not be solely on physical morphology—hunting can also select for behavioral traits. A study of elk in Alberta, Canada, found that hunters were more likely to kill elk that exhibited bolder behavior, such as moving more and using open spaces, than shyer elk.[87] Hunting by humans thereby selected for shyer elk. Selection for shy game animals could potentially influence other ecological interactions, such as making game populations less competitive with nongame species, or more difficult to catch by predators. More studies are needed that explore how hunting pressure that selects for behavioral traits influences ecological interactions among the competitors and predators of game species.

A final effect of hunting is that it may act as an anthropogenic Allee effect (these effects are discussed in general in box 2.4).[88] Some humans have a greater desire to hunt a species as that species becomes rarer. The thrill of killing a very rare animal brings joy to some people. As a result, the rarer a species becomes, the more hunting pressure may be placed on the species, thereby making the species even rarer. If hunting pressure is allowed to intensify with rarity, the species may be hunted to extinction.

Regulation by the states should prevent hunting from becoming an anthropogenic Allee effect. Each state usually attempts to track the population sizes of the game animals within that state. The number of hunting licenses issued for a game species in the state may then be increased or decreased as

BOX 2.4. ALLEE EFFECTS

An Allee effect occurs when the growth rate of a population declines as the size of the population declines. The decline in growth rate can result in a threshold population size below which the population growth rate becomes negative, and the population inevitably goes to zero. An example is when a population gets very small and the individuals in the population have difficulty finding each other to mate. The inability to find mates reduces the growth rate of the population, which reduces the size of the population, further reducing the growth rate, and so on, until the population goes extinct.

population numbers rise or fall. State regulation of hunting may also be used to mitigate some of the other ecological and evolutionary consequences of hunting. For example, hunters may be prohibited from killing game animals whose horns are below a minimum size.

Some of the consequences of hunting may be more difficult to mitigate, though. Requiring that hunters kill equal numbers of shy and bold elk seems unreasonable. States may need to use hunting regulations in more subtle ways to influence the behavioral traits that face selection pressure from hunting. States already specify different times of the year for hunting by archery equipment, muzzleloaders, and modern firearms. There may also be restrictions on where and what type of hunting blinds hunters can use. If the ways in which hunters are allowed to hunt influence which behavioral traits are more likely to result in a game animal being killed, then altering hunting regulations may also alter selection pressures on those traits. Additional research on how hunting regulations influence selection pressure may help in eventually reducing those pressures.

While state regulation of hunting can mitigate the ecological and evolutionary consequences of hunting, such mitigation is rarely of central concern for the states. State management of game species is often based on what is most beneficial for hunters or the economic interest of the state. For example, the licenses sold or auctioned to hunters to kill mountain sheep can bring in considerable revenue to a state. Consequently, mountain sheep populations are often managed primarily to produce large-horned trophy rams.[89] As another example, a state may use sport hunting to keep deer populations low as a means of reducing grazing by deer on agricultural crops.[90] States expend little effort determining the population sizes and densities of game species that are best for the functioning of ecosystems in the state. Independent ecologists are usually only peripherally involved in influencing the hunting regulations in a state.[91] As a consequence, the ecological and evolutionary effects of hunting are usually of secondary importance to a state, trumped by the economic value to the state of hunting.

CHAPTER THREE

Threatened and Endangered Species

For many members of the public, and likely many scientists as well, the most familiar species facing extinction are charismatic megafauna such as the Florida panther (*Puma concolor coryi*) or leatherback sea turtle (*Dermochelys coriacea*). For those who deal with environmental law, however, one of the most familiar species facing extinction is the distinctly noncharismatic snail darter (*Percina tanasi*, figure 3.1).

The snail darter is a small fish native to the Little Tennessee River, where it feeds primarily on aquatic snails.[1] In 1967 the Tennessee Valley Authority began construction of the Tellico Dam on the Little Tennessee River to slightly increase the hydropower capacity of a nearby dam already in operation. It was in 1973, after the Tellico Dam was virtually complete, that the snail darter was first discovered. In that same year, President Richard Nixon signed into law the Endangered Species Act (ESA). Under the authority of the ESA, the US Fish and Wildlife Service (FWS) listed the snail darter as "endangered" in 1975.

One provision of the ESA requires that federal agencies not undertake any action that will jeopardize the existence of an endangered species.[2] Allowing the Tellico Dam to begin operation would have destroyed the entire habitat of the snail darter—certainly an action that would jeopardize the existence of the species. This was an enormous test for the new law. On the one hand was a dam that had cost millions of dollars to build, had begun construction before Congress had even passed the ESA, and was ready to begin operation. On the other was a small fish that had only recently been discovered and had no known economic value, but would be driven to extinction if the dam began operation.

FIGURE 3.1. Snail darter. Photograph from FWS.

The Supreme Court weighed in on the dilemma in the case *Tennessee Valley Authority v. Hill*.[3] In its decision, the court stated that the intent of the ESA is to "halt and reverse the trend toward species extinction, whatever the cost." The court went on to rule that the Tellico Dam could not begin operation because it would violate the ESA by jeopardizing the existence of the snail darter. As discussed in box 2.2, when the Supreme Court interprets a statute, that interpretation also becomes part of the statute (unless Congress later changes the statute, or the court decides on a different interpretation). By interpreting the ESA to require species extinction be stopped "whatever the cost," the court had interpreted the ESA in such a way as to make it a very powerful law.[4]

Following the decision by the Supreme Court, however, Congress quickly amended the ESA. The amendments allowed for the creation of a committee that could grant exemptions to the requirement that federal actions not jeopardize the existence of endangered species.[5] The committee (which will be discussed in more detail later in the chapter) met in 1979 to consider the fate of the Tellico Dam and the snail darter. Amazingly, the committee unanimously refused to grant an exemption for the Tellico Dam, stating that there were reasonable alternatives to the dam.

Clearly frustrated by this result, Congress passed a law in late 1979 expressly authorizing completion of the Tellico Dam, notwithstanding the

ESA. The gates of the Tellico Dam were closed on November 28, 1979, flooding the habitat of the snail darter. While this destroyed the native habitat of the snail darter, the species was introduced into other streams in the Tennessee River valley, and reproducing populations currently exist in the Hiwassee and French Broad Rivers.[6]

As this story illustrates, the ESA is one of the toughest environmental laws on the books, with the goal of preventing species extinction no matter the cost. However, despite the idealism of the ESA, species and their habitats do not always receive the level of protection the ESA seems to afford them.

This chapter first explains how ecologists determine whether a species is near extinction or not, and compares that to the process used by the federal government. Then it explains the protections the species receives from the ESA once a species has been declared "endangered" or "threatened" with extinction under the act. The chapter continues by describing how landowners and others may obtain exemptions from some of the requirements of the ESA, and what that means ecologically for species facing extinction. The chapter concludes with a brief assessment of the success of the ESA in protecting species facing extinction.

DETERMINING EXTINCTION RISK

Over the last century, the extinction rate in well-documented groups of species has been 100–1,000 times larger than average rates in the past.[7] This rate is similar to the rates during the biggest episodes of mass extinctions in the fossil record. Astonishingly, some have estimated that the extinction rate will continue to increase in the near future by a factor of ten or more.[8] Such a dramatic loss of species has already led to degradation of many ecosystem services, such as air and water purification, genetic resources for biochemical and pharmaceutical research, and aesthetic quality of natural lands, among many others.

The main causes of the extremely high extinction rate are habitat loss, overexploitation, the introduction of invasive species, pollution, and climate change.[9] All these causes are almost exclusively the result of human activities. The way in which these human activities begin the extinction process will be discussed later in the chapter.

As all of the above suggests, in the coming years and decades, many additional species will face an increased risk of extinction. Deciding which species to protect is not always an easy task. Every species that is alive today will eventually go extinct (up to 99% of all species that have ever existed have gone extinct). This is not meant to be a statement of existential depression, but merely to suggest that the designation of a species as endangered and facing the prospect of extinction is ultimately a value judgment. Should a species be classified as endangered if it has an elevated risk of going extinct in 10 years, 100 years, 1,000 years?

Scientists attempt to make the classification process as objective as possible, often by predicting the probability of extinction in the near term using population viability analyses (see box 3.1). Population viability analyses require information about the life history and demography of the species of interest, as well as the threats facing the species. A population viability analysis is only as good as the data used in estimating its parameters.[10] Species that may be facing extinction, though, usually have just a few small populations, making it very difficult to obtain good-quality long-term data. Additionally, each population viability analysis must be tailored to the specific threats facing that species. As these threats are often caused by humans, it may be difficult to predict how the threats will change over time.

BOX 3.1. POPULATION VIABILITY ANALYSIS

A population viability analysis is a model that estimates the probability that a species will go extinct over a given period of time. For example, a population viability analysis may estimate that in the next 50 years the probability that a species will go extinct is 10%. As it is difficult to predict exactly when the last member of a species will die, a researcher may conclude that a species is facing extinction if a population viability analysis suggests that the species will fall below a minimum population size over a certain period of time.

The International Union for Conservation of Nature Red List of Threatened Species (IUCN Red List) is widely considered the most comprehensive database of plant and animal species facing the risk of extinction around the world.[11] The IUCN Red List places species in the categories of vulnerable, endangered, or critically endangered, depending on the level of extinction risk the species face. Species are often placed in a category based on population viability analyses. For a species to be placed in the endangered category, a population viability analysis must show that the probability of extinction in the wild is "at least 20% within 20 years or five generations, whichever is longer."[12] A population viability analysis may not always be possible, however. The IUCN Red List describes several other criteria for listing species as endangered; examples include a recent reduction of more than 50% in population size of the species, or a total population size of fewer than 250 mature individuals. The probability of extinction or reduction in population size that qualifies a species as endangered may ultimately be a value judgment, but the criteria are quantitative and clear.

The requirements for listing species as endangered in the IUCN Red List are very different from the requirements under the ESA. The quantitative criteria of the IUCN Red List are replaced by much less well-defined criteria in the ESA. This presents opportunities to list species that do not meet strict quantitative criteria, but also to avoid listing species that would be listed under stricter criteria.

Listing Species as Endangered or Threatened

Under the ESA, a species may be listed as endangered or threatened. The ESA defines an endangered species as "any species which is in danger of extinction throughout all or a significant portion of its range."[13] A threatened species is "any species which is likely to become an endangered species within the foreseeable future throughout all or a significant portion of its range."[14] When is a species "in danger of extinction"?

The ESA states that a species is in danger of extinction when one of five factors is met: "(A) the present or threatened destruction, modification, or curtailment of its habitat or range; (B) overutilization for commercial, recreational, scientific, or educational purposes; (C) disease or predation; (D) the inadequacy of existing regulatory mechanisms; or (E) other natural or manmade factors affecting its continued existence."[15] As can be gleaned from these factors, the ESA emphasizes protecting species that are at risk of extinction due to human-caused threats. As can also be gleaned, the factors are so broad and vague as to provide very little guidance in deciding whether a species is in danger of extinction.

The federal agencies tasked with interpreting this lack of guidance are FWS and the National Oceanic and Atmospheric Administration Fisheries office (NOAA Fisheries). NOAA Fisheries lists marine and anadromous species (such as salmon), while FWS lists all other species.[16] (The rest of this chapter refers solely to FWS for the sake of brevity, but FWS and NOAA Fisheries follow nearly identical rules in listing and protecting endangered and threatened species.)

The process of deciding whether a species is in danger of extinction, and should therefore be listed under the ESA, begins when FWS initiates a review of a species or when a citizen petitions FWS to list a species.[17] A petition needs the support of biological data, such as trends in population size and distribution of the species. According to the ESA, once a petition has been received, FWS has 90 days to determine whether the petition presents "substantial scientific or commercial information" indicating the species should be listed.[18] An FWS regulation defines "substantial information" as the amount of information "that would lead a reasonable person to believe" that listing is warranted.[19]

If the petition does present substantial information, the ESA then gives FWS a year to decide whether listing is warranted or not. The listing decision must be made "solely on the basis of the best scientific and commercial data available."[20] The term "commercial data" refers to information about commercial harvesting of species. Consequently, the ESA requires that FWS not weigh economic considerations when deciding whether to list a species—FWS must decide whether to list a species based entirely on scientific data.

To fulfill this requirement, it is FWS policy to require agency biologists to evaluate all scientific information and to document their evaluation.[21] FWS also prefers listing decisions be based on primary sources. Finally, it is FWS policy to seek expert opinions from three independent species specialists.[22] After a year of consideration, FWS will ideally make the final decision on whether to list a species as threatened or endangered.

FWS has another option if it determines that listing is warranted, but does not actually want to list the species. FWS may decide that other species are a higher priority for listing, and therefore not list the species at the current time.[23] This is called a warranted-but-precluded finding. After such a finding, the species moves into a pool of other "candidate" species that are awaiting a final decision.

FWS argues that warranted-but-precluded findings are necessary because of the limited resources of the agency. FWS has only so many staff, and a limited budget, so species that appear closer to extinction should receive higher priority in being listed than species facing less of a threat. Perhaps not surprisingly, political considerations frequently play an even larger role in warranted-but-precluded findings. Candidate species receive no protection from the ESA. That means that FWS does not have to spend time and money protecting those species, and does not have to face the pushback of interest groups that are negatively affected by a species' being listed.[24]

The ability to make warranted-but-precluded findings, along with the vagueness of the listing factors in the ESA, means that listing decisions can verge on being arbitrary. FWS is most likely to list charismatic species and those with vocal public support, regardless of whether other species are at greater risk of extinction or whether protecting other species would be of greater benefit to the environment.[25] A study of listing decisions made through 1996 found that mammals were more likely to receive a positive listing decision than any other taxon.[26] The study also found that there was no consistent standard for when a listed species was placed in the threatened versus endangered category.

The arbitrary nature of listing decisions is best illustrated by the number of species that are actually listed. There are currently more than 1,500 species in the United States listed under the ESA as threatened or endangered. The problem is that the number of species at risk of extinction in the United States is likely more than ten times that number.[27] The species listed under the ESA can be compared to the species listed in the IUCN Red List. The ESA lists only 60% of the bird species listed by the IUCN Red List, 50% of the mammal species, 20% of the amphibian species, and approximately 10% of the invertebrate species.[28] The ability to make warranted-but-precluded findings, plus the vagueness of the listing criteria, has given FWS the cover to be extremely slow in listing species. As an example, the Dakota skipper butterfly (*Hesperia dacotae*) has been waiting for a listing decision since 1984.[29]

There are signs of improvement, though. In 2011 FWS entered into a settlement agreement with conservation groups that had sued FWS to force the agency to list hundreds of species. In the settlement, FWS agreed that by the end of 2017 it will decide whether to list or not list over 200 candidate species. FWS also agreed to move forward with listing decisions on over 550 other species.[30] FWS finally proposed listing the Dakota skipper butterfly as threatened in 2013.

The arbitrary nature of listing decisions would be greatly reduced by using quantitative criteria to make listing decisions. At least one scholar argues there would be no need to change the language of the ESA to allow a move to quantitative criteria.[31] FWS could rewrite its regulations to incorporate criteria such as those used by the IUCN Red List. Of course, writing new regulations with quantitative criteria would not remove all value judgments from the listing process—setting the cutoffs at which species are considered threatened or endangered would still require deciding the probability of extinction that brings protection from the law.

What May Be Listed

The ESA commands FWS to list only "species," but then goes on to define species very broadly. Species include "any subspecies of fish or wildlife or plants, and any distinct population segment of any species of vertebrate fish or wildlife which interbreeds when mature."[32] As a result, the "species" FWS lists as threatened or endangered need not be ones that fit a formal taxonomic definition of species.

The most remarkable inclusion in the definition is "distinct population segment." A distinct population segment is a single population that may be listed as threatened or endangered while the entire species is itself not listed. Listing these populations allows FWS to protect populations it thinks are important to the survival of a species, even if the entire species does not appear to be at risk of extinction. FWS may decide to list a distinct population segment based on three criteria: (1) the discreteness of the population; (2) whether the population, when looked at by itself, is threatened or endangered; and (3) the significance of the population to the species.[33] Significance to the species might mean that the population is in a unique ecological environment for that species or differs markedly in its genetic characteristics, or that potential loss of the population would result in a significant gap in the range of the species. In the ecological literature, a distinct population segment is often called an evolutionarily significant unit.

An example of a distinct population segment is the northern sea otter (*Enhydra lutris kenyoni*).[34] Since the 1980s, the size of the southwest Alaska population of the sea otter occurring among the Aleutian Islands has fallen by an order of magnitude. Although the reasons are still controversial, FWS stated the main reason for the decrease in population size appears to be predation from killer whales (*Orcinus orca*). Killer whales who previously fed on large whales switched to feeding on sea otters after whale populations were decimated by commercial whaling following World War II. FWS determined that the sea otter population in southwest Alaska is physically separated from other populations of the sea otters, that loss of the population would result in a large gap in the range of the sea otter, and that loss of the population would also result in a significant loss of genetic diversity for the species. Due to the continued population decline, in 2005 FWS listed the southwest Alaska sea otter population as a threatened distinct population segment.

Congress indicated that FWS should list distinct population segments sparingly.[35] Additionally, the ESA explicitly states that only vertebrate populations may be listed as distinct population segments.[36] There seems to be no ecologically justifiable reason for distinct population segments to include only vertebrate populations. However, FWS is required to follow the exact language of the law, so as a result only vertebrate populations qualify.

Finally, the ESA has a special provision for listing species that are not facing extinction, but closely resemble a listed species. In situations where differentiating between two such species would be a substantial difficulty, the ESA states that species not facing extinction may be listed as threatened or endangered.[37] For example, the pallid sturgeon (*Scaphirhynchus albus*) is an endangered fish occurring in the Missouri and Mississippi river basins.[38] Overlapping its range is the shovelnose sturgeon (*Scaphirhynchus platorynchus*), a fish that looks very similar to the pallid sturgeon, but not facing extinction. The shovelnose sturgeon was being harvested commercially, and because of the difficulty in differentiating between the two species of fish, the endangered pallid sturgeon was also being occasionally killed. To stop this, FWS declared the shovelnose sturgeon threatened, to prevent it from being harvested, thereby protecting the endangered pallid sturgeon.

PROTECTIONS OF THE ENDANGERED SPECIES ACT

Once FWS lists a species as threatened or endangered, the protections of the ESA kick in. The ESA is meant to "provide a means whereby the ecosystems upon which endangered species and threatened species depend may be conserved [and] to provide a program for the conservation of such endangered species and threatened species."[39] The act defines "conservation" as all those methods and procedures needed to bring a listed species to the point at which it no longer needs protection.[40] As an appellate court has written, the purpose of the ESA "is to enable listed species not merely to survive, but to recover from their endangered or threatened status."[41]

For all the good intentions of the ESA, though, species near extinction confront a multitude of difficulties. In the United States, habitat loss contributes to the risk of extinction for roughly 85% of imperiled species.[42] Nonnative species add to the possibility of extinction for 49% of imperiled species. Pollution affects 24%, and overexploitation affects 17%. Just because a species is listed as endangered or threatened under the ESA does not mean that these influences suddenly disappear.

The above influences can also result in greatly reduced population sizes, which may in turn cause a species to be sucked into what is called an extinction vortex.[43] In an extinction vortex, small population sizes create ecological or evolutionary dynamics that form positive feedback loops that cause populations to get smaller and smaller until they go extinct.

Extinction vortices can take several forms. One form is caused by inbreeding depression and genetic drift. Inbreeding depression occurs when related individuals breed with each other, resulting in reduced fitness for the population. Genetic drift is random change in the frequencies of alleles in a population. Drift may lead to disadvantageous alleles becoming fixed in the population. Both inbreeding depression and genetic drift are more likely to occur in small populations, and can result in populations becoming even smaller.

Another extinction vortex is due to demographic stochasticity. Demographic stochasticity is the variability in population growth rates that is due to random differences in individual survival and reproduction.[44] For example, imagine a population where reproductive females have a 50% chance of giving birth to a female offspring, and a 50% chance of giving birth to a male. In a large population, the number of female and male offspring born in a given year will be nearly equal. In a small population, however, simply by chance all the females in the population may give birth to males one year. Such a skew in the sex ratio could be difficult for a small population to recover from.

Extinction vortices also include a reduced ability to respond evolutionarily to changing conditions, or populations that become patchy as population size decreases, leading to increased risk of extinction of each patch. The Allee effects discussed in chapter 2 are another type of extinction vortex.[45]

To prevent species from going extinct, the ESA must overcome influences such as habitat loss and nonnative species, and prevent the formation of extinction vortices. The ESA attempts to do this through several provisions. First, it requires FWS to designate "critical habitat" necessary for the recovery of each listed species.[46] It also requires FWS to devise a recovery plan for each species that maps out the steps that must be taken for the listed species to no longer face extinction.[47] The ESA requires all federal agencies to consult with FWS to ensure that each agency's actions do not jeopardize the existence of any endangered or threatened species.[48] In legal discussions and court decisions, this part of the ESA is often simply referred to as Section 7. Section 7 also prevents any agency from adversely modifying habitat designated as critical.[49] The ESA prohibits the "take" of species listed as endangered.[50] This part of the ESA is often referred to as Section 9. There are exceptions to the take prohibition, but to take advantage of the exceptions, a person must often first obtain a permit. Finally, if a person wants to develop land where a listed species exists, the ESA usually requires habitat compensation for the developed land, such as through conservation banking.[51] The following sections will discuss critical habitat designation, recovery plans, sections 7 and 9 of the ESA, exceptions, and conservation banking.

Critical Habitat Designation

For every species FWS lists, the ESA requires the agency to designate habitat critical for its recovery.[52] The ESA defines "critical habitat" as geographical areas within the current range of the species "on which are found those physical or biological features (I) essential to the conservation of the species and (II) which may require special management considerations or protection."[53] Both of these factors must be met for FWS to designate habitat as critical. According to the ESA, FWS may also designate critical habitat that is outside the current geographical range of the species, if such habitat is "essential for the conservation of the species."[54]

Once an area has been designated as critical habitat, the ESA states that federal agencies may not take any action that will result in the "destruction or adverse modification" of such habitat.[55] Critical habitat designations affect only federal agency actions; however, what constitutes federal agency action is very broad. If an action is carried out by a nonfederal agency but has a federal nexus—it is funded or permitted by a federal agency—then that is considered a federal activity, and the activity legally may not destroy or adversely modify critical habitat.

Critical habitat designation must be based on the "best scientific data available," but must also take into account the "economic impact" of the designation.[56] An area may be excluded from designation if the economic benefits of exclusion outweigh the ecological benefits of designation. An area may not be excluded from designation, though, if failure to designate will result in extinction of the species.[57] Finally, except under special circumstances, the ESA requires that the designated critical habitat not include the entire geographical area that the species can occupy.[58]

A regulation describes the criteria that FWS uses for designating critical habitat: FWS focuses on the "primary constituent elements" that are essential to the species.[59] The regulation indicates that these primary constituent elements include "roost sites, nesting grounds, spawning sites, feeding sites, seasonal wetland or dryland, water quality or quantity, host species or plant pollinator, geological formation, vegetation type, tide, and specific soil types." At least one of these elements must be found on the land for FWS to designate it as critical habitat.[60]

Notice that all these primary constituent elements describe physical aspects of the habitat—none of the elements describe ecological or evolutionary processes that may be crucial to a species' recovery. For example, habitat where adaptive evolution is most likely to occur, or where natural disturbances occur at a rate most conducive to the population's survival, are not mentioned as primary constituent elements. On the other hand, at least one of the primary constituent elements as currently listed will likely occur on habitat that hosts an important ecological or evolutionary process for a species. Additionally, physical elements are much easier to find and describe when deciding what habitat to designate as critical.

The courts have had several opportunities to interpret the critical habitat requirement of the ESA. One federal appellate court has stated that the purpose of designating critical habitat is to carve out territory necessary not only for survival of an endangered or threatened species, but also for the recovery of that species.[61] The court further stated that it "is logical and inevitable that a species requires more critical habitat for recovery than is necessary for species survival."

Courts have also held that making a critical habitat designation after a species has been listed is a mandatory requirement of the ESA.[62] FWS, however, typically delays making such designations until forced to do so by court order.[63] The most famous example of this is the northern spotted owl (*Strix occidentalis caurina*).[64] At first, FWS refused to even list the spotted owl as threatened or endangered. FWS was well aware that listing the spotted owl could necessitate designating millions of acres of old-growth forest in the Pacific Northwest as critical habitat, and they knew that such a listing would create significant backlash from the timber industry. A court held that the decision not to list the owl was arbitrary and capricious, and FWS eventually listed the spotted owl as threatened.[65] FWS then refused to designate critical habitat, claiming it had inadequate information to do so. The court held that a critical habitat designation may be deferred only in "extraordinary circumstances."[66] FWS finally designated nearly seven million acres as critical habitat for the spotted owl.

Despite the history of the spotted owl, as of 2013 FWS has not made critical habitat designations for 55% of all listed species. FWS has argued in the past that its resources are better used in determining which species should be listed as threatened or endangered. FWS further argued that designating critical habitat often harms listed species because of the backlash in public sentiment against designating habitat, and because if an area is outside the designated critical habitat area, other federal agencies tend to assume it is unimportant to the species. FWS also claimed that designating critical habitat is not even necessary, because federal agencies are already required to consult with FWS to ensure that their actions do not "jeopardize" listed species.

Despite the arguments of FWS, designating critical habitat may be beneficial for listed species. If FWS does not officially designate an area as critical habitat, then the area does not receive the formal protections of the ESA discussed below in the section on Section 7. This may be particularly important if an area outside of the current range of the species should be designated because it is essential to species conservation. If such an area were not designated, then federal agencies would likely not consult with FWS before undertaking activity on such land, simply because the listed species does not occur there yet. Without official designation, federal activity could have the potential to destroy an area essential to the conservation of the species.

Finally, the second requirement for designating critical habitat is that the area must "require special management considerations or protection."[67] FWS defines special management as any procedure useful in protecting the physical and biological features of the environment that help with the conservation of the listed species.[68] A court has held that the special management need not be immediately necessary; it could occur in the future.[69]

Recovery Plans

Along with designating critical habitat, the other action the ESA requires of FWS is the creation of recovery plans for every listed species. A recovery plan is a set of objectives meant to help a species recover to such an extent that it is no longer listed as threatened or endangered. The ESA states that priority in writing recovery plans should be given to those species that are facing extinction because of development projects or other types of economic activity.[70] Conversely, the act gives FWS leeway to not write a recovery plan for a species if the agency determines that such a plan will not benefit the species.[71] When FWS does decide to write a recovery plan, it takes on average six years to write it.[72]

The main objectives found in most recovery plans are (1) mitigating the threats causing the species to decline, (2) taking steps to increase the size of the species' populations, and (3) increasing the range of the species.[73] These objectives are meant to help overcome the ecological and evolutionary difficulties faced by species nearing extinction, such as extinction vortices.

If there are two or more threatened or endangered species in the same ecosystem, FWS may create a multispecies recovery plan. The main ecological benefit of a multispecies plan is that it forces FWS to consider ways to help the entire ecosystem that both species occur in, so that those benefits accrue to each species. Multispecies plans allow FWS to think at the ecosystem level, and not simply at the level of a single species.[74] There has been an increasing realization within FWS that protecting the functioning of ecosystems is the best way to help listed species recover. The primary drawback of multispecies plans is that they tend to have less concrete objectives and recommendations for each species, often resulting in reduced recovery for each species compared to single-species plans. Separate from a multispecies plan, FWS may simply create an ecosystem recovery plan, describing steps to help an entire ecosystem recover. In practice, though, such plans end up being very similar to multispecies plans.[75]

The ESA requires that recovery plans describe measurable criteria that, when met, would allow a species to be removed from the list of threatened or endangered species.[76] In recovery plans, these criteria tend to focus on the conservation biology concepts of representation, resiliency, and redundancy. Representation means conserving the genetic diversity of the species to allow continued adaptation. Resiliency means ensuring that the populations of the species are large enough to withstand stochastic events. And redundancy means ensuring that there are enough populations of the species so that if one disappears, the species will not go extinct. The level FWS sets for each of these criteria depends on the individual species. A population viability analysis indicating a sufficiently high long-term survival of a listed species may be required by a recovery plan as evidence that a criterion has been reached. However, FWS states that a population viability analysis alone cannot be used to decide that a species has recovered enough to be taken off the list. FWS reasons that population viability analyses must make assumptions about the future threats facing the listed species. Consequently, FWS argues that the criteria for recovery should clearly state what the acceptable levels of those threats are, not make assumptions about what they may be in the future.[77]

The ESA also explicitly states that FWS may create experimental populations of listed species to help aid in their recovery.[78] An experimental population of a listed species is formed by humans outside the current range of the species. For example, FWS has introduced populations of black-footed ferrets (*Mustela nigripes*) onto both public and private land outside of the previous range of the ferrets.[79]

Although recovery plans list detailed steps that should be taken to help a listed species recover, recovery plans are merely documents meant to guide the actions of FWS. Unlike regulations, recovery plans do not have the force of law. As a result, they require the cooperation of state and local governments, and other groups such as landowners and conservation organizations, in order to succeed. Consequently, one of the most important aspects of species recovery is getting everyone with an interest in the species to agree with and support the steps of the recovery plan.

Section 7 and Jeopardy

Section 7 of the ESA requires all federal agencies to consult with FWS to ensure that their actions do not "jeopardize" the existence of any threatened or endangered species.[80] Section 7 also requires federal agencies to consult with FWS to ensure that their actions do not result in the "destruction or adverse modification" of critical habitat.

FWS defines "jeopardize" in a regulation as any federal agency action that is expected, directly or indirectly, to "reduce appreciably the likelihood of both the survival and recovery of a listed species in the wild by reducing the reproduction, numbers, or distribution of that species."[81] The regulation then defines "destruction or adverse modification" as an action that "diminishes the value of critical habitat for both survival and recovery of a listed species." The regulation also states that modifying any of the primary constituent elements used in designating critical habitat constitutes adverse modification. Several courts, however, have stated that the regulatory definition of adverse modification does not provide enough protection to critical habitat.[82] The courts have held that the ESA requires consultation not only if a proposed agency action will diminish critical habitat to such an extent that it threatens survival of a listed species, but also if the action will adversely affect recovery of the species.

If a proposed action by a federal agency may jeopardize a listed species or adversely modify its critical habitat, the agency and FWS enter into a formal consultation process. In fiscal year 2010 FWS had over 30,000 consultations with other federal agencies.[83] After consultation, FWS issues a biological opinion describing the impact of the proposed agency action. If FWS concludes that the action will result in jeopardy to the listed species or adverse modification of critical habitat, FWS must list any reasonable and prudent alternatives to the action.[84]

If a federal agency receives an FWS opinion that its proposed action will result in jeopardy or adverse modification, the agency may proceed with its original plan only if it receives an exemption to Section 7 of the ESA. Exemptions come from the Endangered Species Committee composed of high-ranking department heads and cabinet members.[85] This is the committee mentioned at the opening of the chapter that refused to give an exemption for the Tellico Dam to begin operation. The committee is almost always referred to as the God Squad. It has the power to grant an exemption to Section 7 for an action that would drive a species to extinction, giving them final say over the very existence of species. The God Squad has rarely been convened, and has never granted a complete exemption from Section 7.[86] As a consequence, the ESA is a powerful brake on federal agencies that would undertake actions negatively affecting threatened or endangered species.

Section 9 and Take

Section 7 of the ESA applies only to federal agencies. State and local governments, as well as other nongovernmental entities and individuals, do not have to consult with FWS to avoid jeopardizing listed species. Section 9 of the ESA, however, does provide protections for listed species that all persons in the United States must follow.

Section 9 prohibits the sale, import, export, or transport of any listed species.[87] This part of the ESA is meant to stop the trade in live animals, skins, and other portions of listed species.[88] Section 9 also prohibits the take of species FWS has listed as endangered.[89] Similarly, an FWS regulation written under the authority of the ESA prohibits the take of species listed by FWS as threatened.[90] As frequently happens in the law, how to interpret the meaning of a single word, in this case "take," has turned out to be very complicated.

According to the ESA, "take" means to "harass, harm, pursue, hunt, shoot, wound, kill, trap, capture, or collect, or to attempt to engage in any such conduct."[91] In a regulation, FWS further defined "harm" to "include significant habitat modification or degradation where it actually kills or injures wildlife by significantly impairing essential behavioral patterns, including breeding, feeding or sheltering."[92] This regulation has been upheld by the Supreme Court.[93] However, the court noted that "actual death or injury of a protected animal is necessary for a [harm] violation."[94] In other words, Section 9 prohibits many different actions that in some way negatively affect a listed species, but there must be proof that the action directly or indirectly results in the death or injury of at least one individual of that species.

As mentioned earlier, Section 9 applies to any "person." A person is defined as including individuals, corporations, and state or local government agencies.[95] This means that Section 9 prohibits a landowner from doing anything on his or her private property that will result in take of a listed species, such as directly killing or degrading its habitat. This is important because the majority of endangered species have at least part of their ranges on private land, and some endangered species occur entirely on private land.[96] It should be noted, though, that Section 9 does not force private landowners to undertake any activity on their properties that will benefit a listed species—landowners are only prohibited from harming listed species.

Exceptions

The take prohibition in Section 9 seems to be fairly comprehensive; however, there are exceptions to the prohibition that allow for the take of many threatened and endangered species. The first major exception is the prohibition as applied to plants. Under the ESA persons may not import or sell threatened or endangered plants. There is no prohibition in the act, though, on taking plants.[97] Threatened and endangered plants species may not be killed on federally owned land, but unlike animals, an individual may kill listed plant species if they occur on his or her private property. Because so many endangered species are found on private land, the fact that the take provision does not apply to plants on private land is a huge hole in the ESA.

The second major exception is for take defined as "incidental."[98] The ESA defines "incidental take" as an activity where taking of a listed species occurs, but the taking is not the purpose of what is an otherwise lawful activity. Thus, a threatened or endangered species may be taken as long as the take is incidental. Before incidental take is legal, though, FWS must first issue an "incidental take permit" (see box 3.2 for a discussion of research permits).

The person hoping to undertake a project that will result in inciden-

---

BOX 3.2. PERMITS FOR RESEARCH ON
THREATENED OR ENDANGERED SPECIES

Scientists are persons as defined by the ESA, so they are also prohibited from taking threatened or endangered species. The ESA, however, states that FWS may allow acts prohibited by the ESA, if done for scientific purposes.* If a research project will involve the take of a threatened or endangered species, the researcher must request a permit from the regional office of FWS where the listed species is located. FWS requires a detailed description of the proposed research. If the research will cause the death or removal of organisms from the wild, FWS expects the researcher to conclusively demonstrate that existing specimens are unavailable from museums, from nurseries, or in captivity, or that organisms must be taken from the wild for the research to succeed. FWS also requires a fee ($100 in 2015) for most permit applications. FWS recommends contacting it for guidance on applying for a permit at least three months before the proposed research project is set to begin.

*16 U.S.C. § 1539(a)(1)(A).

tal take must submit a habitat conservation plan (HCP) to FWS.[99] An HCP specifies the impact the proposed taking will have on the listed species, and what steps the person will undertake to minimize or mitigate those impacts. Additionally, the ESA requires assurances that the taking will not significantly reduce the survival and recovery of the listed species.[100] Chapter 6 discusses the regulation of private lands in detail. It is worth noting here, though, that the requirement that landowners submit HCPs before being issued incidental take permits is a significant way in which the federal government regulates the use of private lands in the United States.

Incidental take permits issued by FWS are bound by a "no surprises" rule.[101] Under the no surprises rule, if FWS has issued an incidental take permit and unforeseen circumstances make the person's HCP inadequate to mitigate harm to the listed species, the person is not required to take any additional steps to help conserve the species. The federal government would have to undertake any additional conservation activities. The no surprises rule means that persons know the cost of conservation efforts, without worrying about changing circumstances leading to additional costs. But the rule also means that if circumstances do change, FWS loses much of its ability to protect listed species, because persons do not have to do anything more than what was agreed upon in their HCPs.

The no surprises rule can lead to certain perverse incentives for landowners. When a landowner applies for an incidental take permit, his or her HCP may be in the form of a multispecies HCP. A multispecies HCP contains conservation actions for two or more species. Multispecies HCPs are similar to multispecies recovery plans discussed earlier in that both contain conservation actions for multiple species. At least one of the species in a multispecies HCP must be the species for which the incidental take permit will be issued, but the other species in the plan may be unlisted species. FWS encourages multispecies HCPs because the plans ideally allow for greater ecosystem planning than would occur in single-species HCPs. Landowners like multispecies HCPs because of the no surprises rule. If a landowner lists several species in a HCP, and one of those species eventually becomes a listed species under the ESA, the landowner would not have to undertake additional conservation measures because the newly listed species is already covered by the HCP. As a consequence, the no surprises rule creates an incentive for landowners to list as many species in an HCP as possible. Landowners take this so far as to include species in their HCPs even when there is little evidence that those species actually occur on their land.[102] This is problematic because it is difficult to assess the effectiveness of a conservation plan for species that may or may not be present. Additionally, multispecies HCPs often lack specific conservation actions for many of the species covered by the HCP.[103] This is especially true for those species that may or may not actually be present on the land covered by the HCP. Similar to multispecies recovery plans, multispecies HCPs often appear to do a poorer job of protecting species than single-species HCPs.[104]

The no surprises rule is meant to increase certainty for landowners about the conservation actions they will have to undertake. The rule does so by protecting landowners from unexpectedly being forced to undertake extra activity. FWS has implemented two other programs that are also meant to increase certainty for private landowners, but only in return for their volunteering to help protect species habitat: safe harbor agreements and candidate conservation agreements with assurances.

A safe harbor agreement is an arrangement between FWS and a private landowner who volunteers to manage or improve her land to help in the recovery of a listed species.[105] In return, FWS gives formal assurances that the agency will not place land-use restrictions on the property or require any additional management activities by the landowner without her consent. Further, once the agreement period is over, the landowner may return her property to the baseline condition that existed before the agreement began, even if that results in incidental take of listed species. A study of safe harbor agreements in North Carolina found that the agreements created stepping-stone corridors that lead to increased population connectivity for endangered red-cockaded woodpeckers (*Picoides borealis*).[106]

A candidate conservation agreement with assurances (CCAA) is very similar to a safe harbor agreement, but applies to candidate species that FWS has not yet listed as threatened or endangered.[107] Under a CCAA, a private landowner volunteers to engage in conservation activities that are beneficial to a candidate species. In return, FWS provides assurances that the landowner will not be required to perform additional conservation measures. FWS also agrees that it will not impose further restrictions on the landowner's property if the candidate species does become listed. On top of that, FWS agrees to authorize a specific level of incidental take of the newly listed species. The hope behind the CCAA program is that enough landowners will agree to undertake conservation activities on their land that the listing of candidate species is no longer necessary. Other than the research on the red-cockaded woodpecker, though, there has been very little peer-reviewed ecological study of the effectiveness of safe harbor agreements or CCAAs in benefiting species.

Returning to incidental take permits and HCPs: Regardless of whether an HCP is for one species or for several, FWS often requires that HCPs include an arrangement for habitat compensation.[108] Habitat compensation occurs when a person compensates for the incidental take of a listed species by conserving land where the listed species currently exists, or by restoring previously degraded land to a state where the listed species will again occur on that land. FWS frequently insists on habitat compensation when a landowner wants to develop a piece of land where a listed species occurs. Habitat compensation may take the form of conservation or restoration of a small piece of land in the vicinity of the proposed development, or conservation or restoration of a different piece of land separate from the proposed development.

There are often problems with developers' attempting to conserve habitat for a listed species. Conserving a piece of land, or restoring a previously degraded piece of land, requires considerable biological expertise. Most developers do not have that expertise. Additionally, different development projects often occur across a wide geographical area, with the actions of one developer unconnected to the actions of other developers. Consequently, the habitat set aside by one developer for habitat compensation is often small and not connected to other such conserved habitats.[109] As a result of these problems, FWS created another option for developers to fulfill the requirement for habitat compensation: buying credits from a conservation bank.

Conservation Banking

Conservation banking is a way for developers to compensate for the incidental take of a listed species by buying credit for land that has already been conserved for that species.[110] A conservation bank is created when a landowner conserves the habitat of a listed species and then markets that habitat as credit to an individual who needs to compensate for developing land where incidental take of the species will occur. For example, the Ohlone Preserve Conservation Bank in California contains land where the threatened California red-legged frog (*Rana draytonii*), California tiger salamander (*Ambystoma californiense*), and Alameda whipsnake (*Masticophis lateralis euryxanthus*) species all occur.[111] If a person develops land where one of those species exists, he or she can buy credits from the Ohlone Preserve Conservation Bank to compensate for the developed land. Once all the credits in a conservation bank have been sold, the bank land simply becomes a species reserve.

The ESA does not specifically mention conservation banking, but FWS has interpreted the ESA to allow the creation of conservation banks.[112] FWS requires that the conservation bank owner have a conservation easement, so that the bank land will be conserved in perpetuity. (Conservation easements are discussed in more detail in chapter 6.) The conservation bank must also have a management plan for the conservation of the listed species on the bank land, and enough capital in an endowment that the management plan can be implemented in perpetuity.[113]

Establishing a conservation bank is the first task. The next task is determining the degree to which the habitat on the bank land can compensate for habitat being developed someplace else. To do this, FWS works with the conservation bank owner to set a mitigation ratio.[114] This is the ratio of the number of acres conserved in the conservation bank to the number of acres being developed. For example, if an individual develops one acre of land and buys credit for the conservation of one acre in the conservation bank, then the mitigation ratio is 1:1. Most (65%) conservation banks have mitigation ratios of 1:1.[115] If a conservation bank protects high-quality habitat for a particular listed species, however, and most other habitat where the species exists is of lower quality, the conservation ratio may be smaller than 1:1 (e.g., 1:2, a ratio of one bank acre to two developed acres). To return to the Ohlone Preserve Conservation Bank, the mitigation ratio for the California tiger salamander and the Alameda whipsnake is 1:1, while the ratio for the California red-legged frog is 1:1.667.

FWS hopes that conservation banking will overcome many of the problems associated with developers' trying to conserve isolated plots of land as a way of providing habitat compensation. Conservation banks are typically much larger than the individual plots of land conserved by developers. The larger areas of conservation banks means they are more likely to support functioning ecosystems. Additionally, individual plots conserved by developers are often isolated, while conservation banks are likely less fragmented and more connected to other existing habitat. Finally, conservation banks have management plans and endowments that should provide better management of listed species than would occur on isolated plots. By 2011 FWS had approved 105 conservation banks in the United States.[116]

There is a different problem with habitat compensation, however, that conservation banks do little to overcome. The entire concept of habitat compensation, and conservation banks in particular, is based on an assumption of fungibility.[117] Fungibility in this context is the notion that for any piece of land with a listed species on it that is to be developed, a corresponding piece of land can be found or created that is of equivalent ecological value for the listed species. Habitats and the species occupying those habitats are in essence tradable. Ecologically, the assumption of fungibility may rarely be met. For example, the population of a listed species that occurs on land that is to be developed may be very different genetically from the population that occurs in a conservation bank. As another example, the fragmentation that results from developing several pieces of land throughout an ecosystem may have a much larger negative effect on that ecosystem than can be made up for by preserving the land in a conservation bank. The ecology of habitat compensation is discussed in greater detail in chapter 8, which describes stream and wetland compensatory mitigation schemes.

Despite the dubious assumption of fungibility, markets for the trading of habitats or other ecosystem services are becoming increasingly popular with federal agencies. There are now markets for air quality, carbon emissions, water quality, wetlands, and the habitats discussed in this section. Federal agencies like these trading schemes because they replace top-down regulations from the agencies with market mechanisms. In fact, in 2008 the US Department of Agriculture announced the establishment of the Office of Environmental Markets to help develop markets for ecosystem services. The creation of these markets, however, is a vast experiment with the ecosystems of the United States. Federal agencies are hoping that the markets will be win-win: allowing continued anthropogenic activity and development, but also protecting and restoring high-quality ecosystems. Unfortunately, there is relatively little ecological research on the cumulative effects of these markets.

ASSESSMENTS OF THE ESA

Scholars have leveled criticisms at the very structure of the ESA. One criticism is that the incentives in the ESA are perverse.[118] As discussed earlier, most threatened and endangered species occur on private land. By not allowing landowners to take listed species, but also not giving them anything in return for their compliance, some landowners may decide to illegally destroy the habitat on their land that supports a listed species, so that they are not constrained by the ESA. This is sometimes referred to as "shoot, shovel, and shut up." Critics have argued that the federal government should pay landowners to protect the listed species on their land. FWS has attempted to reduce the temptation to shoot, shovel, and shut up by offering safe harbor agreements and CCAAs. Despite the existence of these agreements, some landowners may still decide it is easier to simply destroy the habitat on their land and not worry about the ESA.

A second criticism of the ESA is that it takes a species-by-species approach to conservation.[119] Some have argued that a more holistic approach is needed to confront the massive extinction rate we currently face. There have been calls for a new law to replace the ESA that focuses more on conserving ecosystems, with conserving individual species as a side benefit.

In answer to this second criticism, others have argued that a focus on ecosystems may be ecologically ideal, but would not work in reality. While the definition of "species" is still an open question in the scientific literature, it is much easier for scientists to agree on what constitutes a species or subspecies than it is for them to agree on where the borders of an ecosystem are.[120] The law thrives on certainty. Remember, laws are mostly written and interpreted by nonscientists. A fuzzy definition for what constitutes an ecosystem boundary would be much more difficult to decipher than the current definition in the ESA of what constitutes a species. Additionally, politicians or interest groups may be able to use the fuzziness of ecosystem boundaries to find ways around a law meant to protect those ecosystems. On the other hand, politicians and interest groups have already found ways to prevent the listing and protection of species.

Perhaps the best way of assessing the ESA is to examine its success rate.[121] Since 1973, when the ESA was signed into law, less than 1% of listed species have gone extinct while 10% of candidate species that were waiting to be listed went extinct. Listing has also helped increase the population size, or helped stabilize the population size, of at least 35% of listed species. Complete recovery, though, has occurred for just 2% of listed species. These numbers make it difficult to argue that the ESA has been a complete success, or a complete failure.

CHAPTER FOUR

International Environmental Laws Protecting Biodiversity

Atlantic bluefin tuna (*Thunnus thynnus*) is a remarkable species (figure 4.1). Bluefin tuna are the largest of the tuna species, averaging 250 kilograms (550 pounds) in weight and 2 meters (6.5 feet) in length. Despite their large size, the powerful crescent-shaped tail of the bluefin tuna allows them to reach speeds of 70 kilometers per hour (43 miles per hour). They also have pineal windows in their heads that enable them to navigate over thousands of miles of ocean. Additionally, bluefin tuna can thermoregulate, meaning that they can keep their body temperature above the temperature of the surrounding water.[1]

Atlantic bluefin tuna are also very good in sushi and sashimi. Since the 1970s, total Atlantic bluefin tuna numbers have fallen by at least 51%.[2]

In 2010 representatives of countries from around the world met in Doha, Qatar, as part of the Convention on International Trade in Endangered Species of Wild Fauna and Flora (CITES). CITES is a treaty that attempts to regulate international trade in wild animal and plant species. A vote of two-thirds of the countries present at a CITES meeting (called a conference of the parties) is necessary to place a species in the strictest category of the convention, essentially ending international commercial trade of that species. In the 2010 meeting, Atlantic bluefin tuna was put up for a vote.

The potential fate of the bluefin tuna should not have come down to a vote in Doha. The International Commission for the Conservation of Atlantic Tunas was meant to be the body managing Atlantic bluefin tuna harvesting. Unfortunately, the commission had mismanaged bluefin stocks, regularly setting the total allowable catch above the level the commission's scientists

FIGURE 4.1. Atlantic bluefin tuna in Italy. Photograph from NOAA/Antonio Pais.

considered sustainable. Additionally, illegal and underreported fishing of bluefin tuna resulted in a catch that was nearly double the total allowable catch set by the commission in 2007.

The precarious nature of bluefin tuna populations was well accepted by most scientists, and experts at the United Nations Food and Agriculture Organization, the scientific advisers to the CITES secretariat, recommended that bluefin tuna be protected by CITES. Several countries, including the United States, supported listing the tuna. The European Union, whose ships catch much of the bluefin tuna harvest, supported listing the species, but wanted the listing delayed until May 2011.[3]

Other countries vociferously opposed listing the bluefin tuna under CITES. Japan, whose citizens consume nearly 80% of the total bluefin tuna harvest, argued that listing bluefin tuna would burden coastal countries that rely on bluefin tuna fishing for income.[4] Japan also argued that the International Commission for the Conservation of Atlantic Tunas was the proper body to continue regulating the bluefin tuna catch. Before the conference of the parties began, Japan had begun building a coalition to vote against listing the bluefin tuna. In fact, the Japanese embassy had served bluefin tuna sushi hours before the vote.[5] Libya also opposed listing the bluefin tuna. Libya harvested a large percentage of bluefin tuna, and was suspected of harvesting more than its legal quota. During the debate on listing, the delegate from Libya heatedly exclaimed that the desire to list bluefin tuna was politics trumping science.

After the close of debate, the parties voted on whether to list the Atlantic bluefin tuna under CITES. A secret ballot commenced, at the end of which 68 countries voted against listing the bluefin tuna, while 20 voted for listing.[6] Thirty countries abstained from voting, most of them European Union nations that fish for bluefin tuna. As a result, bluefin tuna is not listed under CITES. Despite the trouble Atlantic bluefin tuna populations face, bluefin tuna can still be legally traded internationally, with no protection from CITES.

Attempts to list the Atlantic bluefin tuna under CITES illustrates some of the difficulties international law faces in tackling ecological issues. First, because every country is sovereign, countries must be persuaded to join environmental treaties, and then choose to abide by those treaties. There is no world organization that can pass laws and then require other countries to follow those laws. Second, trade between countries plays a very important role in how countries respond to ecological issues. Countries are frequently unwilling to undertake any environmental protections that may disrupt their international trade. Finally, there is often disagreement between countries over whether threatened species are best protected by being banned from all international trade, or whether sustainable harvesting of the species creates an incentive for countries to manage the species, thereby protecting them better.

This chapter briefly introduces international law concepts before moving on to an examination of treaties that are primarily concerned with protecting threatened species and biodiversity. The chapter continues with a look at treaties that are primarily concerned with promoting trade between the United States and other countries, but that contain provisions for protecting the environment. (Chapter 8, on laws relating to wetlands, discusses the Ramsar Convention; chapter 9 examines the United Nations Convention on the Law of the Sea; and chapter 11 discusses treaties related to global climate change.) The chapter concludes with a discussion of whether international law is currently capable of protecting biodiversity hot spots around the world.

INTERNATIONAL LAW

When thinking about international affairs, we often tend to assume that countries do whatever is in their best interest, and short of tough economic sanctions or military intervention, there is little that one country can do to influence the actions of another country. We also tend to assume that while a country may sign a treaty, it will disregard that treaty or other international law whenever it suits its interests. After all, there is no international police force making sure that countries abide by international laws. While it is certainly true that countries act in their best interest, and may occasionally ignore the requirements of treaties they have signed, they are much more likely to comply with international law than to ignore it. Despite the story of the Atlantic bluefin tuna in the beginning of this chapter, international law often constrains and influences how countries interact with the species and ecosystems within their borders.

To enforce international law, some treaties create bodies that have authority to judge whether a country has violated the obligations of that particular treaty, and may allow sanctions against the country if there has been a violation. Even if a treaty does not create such a body, there are other reasons countries abide by international law. Countries trust that "every treaty in force is binding upon the parties to it and must be performed by them in good faith."[7] This is known in international law as *pacta sunt servanda*. As a consequence of *pacta sunt servanda*, breaching international law comes with a reputational cost for a country. For example, if a country routinely violates its treaty obligations, other countries may be less likely to enter into new treaties with that country. Countries may also be less likely to give aid or trade with a country that breaches international law. While there may be no police enforcing international laws, there are costs to violating those laws.

The two most important sources of international law are treaties and customary international law. Treaties most often take the form of a written agreement that creates a set of continuing obligations between two or more countries. Some treaties are called agreements, while a multilateral treaty between several countries is often called a convention. Many multilateral conventions do not go into effect until a predetermined number of countries have ratified the convention. It should be noted that when discussing international law, countries are often referred to as states. (See box 4.1 for the way treaties affect US law.)

Negotiating treaties among several countries can be very difficult, with each country having a slightly different goal. As a result, there has been the relatively recent emergence of "framework" conventions. A framework convention has vague or very generalized obligations, but with a commitment to negotiate more elaborate regulatory provisions in the future. The hope is that specific obligations will be more easily agreed upon once countries have committed to being bound by the general framework of the convention. The Convention on Biological Diversity, discussed below, is a framework convention.

Customary international law results from general and consistent practice by countries out of a sense of legal obligation. If many countries are engaged in a practice for a significantly long period of time, and they are doing so because they believe they are legally bound to do so, then the practice is considered to be a part of international law without the need for a written treaty. Obviously, with no written document, there are frequent arguments between countries as to what constitutes customary international law.

The above forms of international law are often called hard law because they are legally binding on nations. There is also soft law—measures such as declarations or recommendations issued by a country or international organization that are not legally binding but that may influence how nations act. A common form of soft law is the memorandum of understanding. As its name implies, a memorandum of understanding is a document describing an agreement between parties, but an agreement that is not necessarily meant to be legally binding. A memorandum of understanding does not have to be between two countries; it can be between government agencies in two or more countries, or between an international body and another party, or even between a nongovernmental organization and a country. As only countries can sign treaties, memoranda of understanding increase the range of actors that can be brought together to cooperate on a topic. Also, because soft law documents are not legally binding, they are often easier to negotiate.

There is often no sharp dividing line between soft law and hard law. The wording of a document often makes it clear whether it is meant to be legally

BOX 4.1. TREATIES AND THE US CONSTITUTION

The United States is currently bound by over 10,000 treaties and international agreements. The treaty clause of the US Constitution appears to lay out exactly how a treaty must come to bind the United States. The treaty clause, in Article II, Section 2 of the Constitution, states that the president "shall have Power, by and with Advice and Consent of the Senate, to make Treaties, provided two thirds of the Senators present concur." A treaty that receives a two-thirds vote in the Senate is known as an *Article II treaty*.

While this is the path many treaties take, there are two other possible paths, leading to *congressional-executive agreements* or *presidential executive agreements*. These agreements have the same force of law as Article II treaties.

To bind the United States, congressional-executive agreements do not require a two-thirds vote by the Senate; instead they require only a majority vote in both houses of Congress. Presidential executive agreements, however, require only the signature of the president to bind the United States. Some scholars have argued that both of these types of agreements are unconstitutional, but such agreements have a long history of use in the United States, and have become accepted as a valid means of binding the United States.

The content of the treaty often determines which path it takes. For example, treaties dealing with human rights, arms control, and military alliances are often submitted to the Senate as Article II treaties. Agreements dealing with trade, finance, and energy are often concluded as congressional-executive agreements. Presidential executive agreements are most often used for technical matters.

Article VI of the Constitution states that treaties, federal statutes, and the Constitution are the supreme law of the land. Consequently, treaties and federal statutes have essentially equal status under the Constitution, and both preempt inconsistent state laws. There is a complication to this seemingly straightforward constitutional statement, however. Treaties may be either self-executing or non-self-executing. A self-executing treaty immediately becomes judicially enforceable federal law upon ratification. A non-self-executing treaty, even though it has been ratified, is not enforceable federal law until Congress passes additional implementing legislation. Frequently it is clear from the language of a treaty whether it is meant to be self-executing or non-self-executing. Occasionally, however, US courts must determine whether a treaty has the force of federal law without implementing legislation from Congress.

binding or not, but such wording may be open to interpretation. Additionally, there may be instances when a country is more diligent in carrying out the agreements in a soft law document than in a treaty it has signed.

TREATIES FOR PROTECTING BIODIVERSITY

Wild species do not respect national boundaries—they cross them through both natural migration and international trade. Conflicts between countries over species that cross borders may require resolution through international law.[8] There are at least five types of processes involving animal or plant species where a treaty may be useful: (1) species that move across national borders; (2) species that exist in areas not owned by any country, such as the high seas; (3) species that are part of international trade; (4) species that are associated with other international resources, such as a fish population in one country that depends on stream flow from a river originating in a second country; and (5) species that are considered to have intrinsic value as part of the global commons.[9]

The first treaty this chapter examines is the Convention on International Trade in Endangered Species of Wild Fauna and Flora (CITES). CITES focuses on species that are a part of international trade, but the overarching motivation of the treaty, as for many wildlife treaties, is that species have intrinsic value and international law should be made to help prevent their extinction.

Convention on International Trade in Endangered Species of Wild Fauna and Flora (CITES)

International trade in wild species is estimated to be worth $240 billion annually, and involves the trade of hundreds of millions of organisms. CITES went into force in 1975 as an attempt to protect endangered plant and animal species through restrictions on international trade of those species. CITES currently regulates the trade of nearly 34,000 species. The convention has 175 parties, making it one of the largest environmental treaties.[10]

CITES puts plant and animal species into one of three categories: Appendix I, II, or III. The appendix that a species is placed in determines the level of trade restrictions on that species, with Appendix I species receiving the greatest protection.

Appendix I species are all those that are "threatened with extinction which are or may be threatened by trade."[11] Trade of an Appendix I species is allowed only in "exceptional circumstances," and is not allowed for commercial purposes.[12] To trade an Appendix I species, the trader must obtain both an export and an import permit.[13] CITES requires that each party to the treaty establish a management authority and a scientific authority to administer the treaty. Before issuing an export permit, the scientific authority of the exporting country must find that exporting the species "will not be detrimental to the survival of that species." This is often called a "nondetriment" finding. Additionally, the management authority of the exporting country must find that there has been no violation of its domestic law in acquiring the organism. Before issuing an import permit, the scientific authority of the importing country must also find that the import will not be for a purpose detrimental to the survival of the species. The management authority of the importing country must confirm that the organism will not be used primarily for commercial purposes. Being forced to obtain both an export and an import permit makes trade in Appendix I species difficult and time consuming, thereby greatly restricting such trades. In practice, trade in Appendix I species is usually limited to specimens for scientific or educational purposes, or for hunting trophies. CITES allows exchanges of specimens of listed species between scientists or scientific institutions without the need for permits as long as the scientists or institutions have been registered with the management authorities of the countries in which they are located.[14] Over 900 species are listed in Appendix I.

Appendix II species are not necessarily threatened with extinction now, but may become threatened in the future without regulation of trade.[15] Unlike Appendix I species, Appendix II species may be traded commercially under CITES. Trade in Appendix II species requires only an export permit, although the scientific authority and management authority of the exporting country must go through the same steps as for an Appendix I species, including a determination that trade is not detrimental to the survival of the species.[16] Additionally, quotas on the amount of trade in an Appendix II species may be approved by a conference of the parties. Appendix II contains over 34,000 species.

Appendix III species are those that a party identifies as being subject to regulation within that country to prevent or restrict exploitation, and as needing the cooperation of other countries to control trade.[17] Each party to CITES can decide which species it wants to list in Appendix III. If a trader is exporting a species from a country that has listed the species in Appendix III, then the trader must receive an export permit. The permit does not require a finding by the country's scientific authority that export will not be detrimental to the survival of the species. The permit does require, though, that the management authority find that no domestic laws were violated in obtaining the organism from that country.[18] If a country has not listed a species in Appendix III that another country has listed in the appendix, then no export permit is required from the country not listing the species. CITES, however, requires that all trade between parties to CITES in Appendix III species be accompanied by a "certificate of origin," even if the trade is between countries that have not listed the species in Appendix III. Appendix III contains only 160 species.

Like the Endangered Species Act, CITES takes a broad view of what may be listed under the convention. CITES applies to "specimens of species" listed in one of the appendices.[19] The convention defines a "specimen" as any living or dead animal or plant, or "any readily recognizable part or derivative" of such an animal or plant.[20] CITES then defines species as "any species, subspecies, or geographically separate population thereof."[21] Different populations of a species may be listed in different appendices if well-managed populations in one country are at less risk of becoming endangered relative to populations in other countries. For example, the African elephant (*Loxodonta africana*) is in Appendix I, except for the populations in Botswana, Namibia, South Africa, and Zimbabwe, which are in Appendix II. This is generally avoided, though, as it increases the difficulty of enforcing trade restrictions on that species.

Criteria for Adding Species to an Appendix

The parties to CITES meet every two to three years in a conference of the parties, at which time they may vote to add species to Appendix I or II. After being criticized for not using sufficiently scientific criteria for deciding when species should be listed, a conference of the parties in 1994 adopted specific guidelines for listing species.[22] Under the guidelines, a species should be listed in Appendix I if it may be affected by trade and meets at least one of three biological criteria: (1) the wild population of the species is small, (2) the wild population has a restricted area of distribution, or (3) there has been a marked decline in the wild population size, or a decline may be inferred from decreased habitat area.

The biological criteria are then further defined in the guidelines. The definition of a "small wild population" is a population of less than 5,000 individuals for a low-productivity species. The number may be higher for higher-productivity species. The definition of restricted "area of distribution" is taxon-specific, but should take into account habitat specificity, population density, and endemism. The definition of a marked "decline" can be based on the long-term extent of decline, or a recent rate of decline. A long-term decline is a reduction in population size to between 5 and 30% of a species' historical baseline size, depending on the productivity of the species. A recent decline is a reduction of 50% or more in the last ten years or three generations, whichever is longer.

According to the guidelines, a species should be listed in Appendix II if it is in danger of meeting the criteria for Appendix I in the "near future."[23] "Near future" is defined as greater than five years and less than ten years. Additionally, a species may be listed in Appendix II if it looks sufficiently like a species listed in Appendix I such that an enforcement official would likely not be able to distinguish the two species. If the parties to CITES vote to do so, they may also downlist a species from Appendix I to Appendix II.

The guidelines note repeatedly that numerical values are only meant as examples because they will likely not apply equally to all taxa. Despite the attempt at quantitative objectivity in determining when species should be listed under CITES, the example of the Atlantic bluefin tuna suggests that politics still play a large role in listing decisions.

Listing a species in Appendix I or II requires a two-thirds majority of the parties present and voting at a conference of the parties. A species may also be listed in Appendix I or II in-between conferences of the parties if half the parties to CITES vote and, of those, two-thirds vote to list the species. For a species to be placed in Appendix III, a party must simply indicate that it wishes to add the species to that appendix.

Loopholes and Compliance

There are several loopholes in CITES that undercut some of the protections of the convention. The most important is that any party to the convention may file a reservation against the listing of a species in Appendix I, II, or III.[24] If a country files a reservation, it is as though the country is not a party to CITES for that particular species. The species will still be added to the appendix, but the country filing the reservation completely avoids the permit system for that species. This means that the country may continue freely trading in a CITES-listed species with other countries that are not parties to CITES, or that have made a reservation to the listing. As a consequence, parties to CITES are bound by the convention only to the degree they wish to be bound.

Another loophole is that a party to CITES may continue to trade listed species with nonparties. The country that is a party to CITES is bound by the convention's requirements, but the nonparty simply has to issue documents "comparable" to those required for parties to CITES.[25] The protection of CITES thus depends on how diligent the nonparty country is in producing scientifically informed import or export documents. Fraud is obviously a possibility in such a practice.

A third loophole is that an individual's "personal or household effects" may be imported or exported without a CITES permit.[26] For example, a person with crocodile-skin boots would not need an export or import permit to visit another country. Fraud is again a possibility, with travelers claiming that items made from CITES-listed species are personal effects when they really plan to sell those items in a different country. There are two ways in which this loophole has been partially plugged, however. First, a specimen of an Appendix I species acquired by an individual outside of his or her usual country of residence may not be brought back to the country of residence without a CITES permit.[27] Second, a specimen of an Appendix II species taken from the wild and acquired by an individual outside of his or her usual country of residence may not be brought back to the country of residence without a CITES permit if the country in which it was acquired requires an export permit for that species.[28]

The final loophole concerns species that are bred in captivity or artificially propagated. An Appendix I animal specimen that is bred in captivity, or an Appendix I plant specimen that is artificially propagated, are treated as Appendix II specimens. As a result, these Appendix I animal or plant specimens may be commercially traded.[29] Species listed in Appendix II or III that have been bred in captivity or artificially propagated require only a certificate from the management authority of the country of export indicating such, and the specimens may be traded without permits.[30] Fraud is again a possibility for a specimen being declared as captive-bred or artificially propagated when it has in fact been taken from the wild.

Despite these loopholes, CITES does provide mechanisms to induce parties to comply with its regulations. A country may be asked to produce a special report on a species, or may be asked for a compliance action plan to bring the country back into compliance with CITES. If there is persistent noncompliance, the standing committee of CITES may recommend that other countries suspend all trade in CITES species with the offending country. For example, in 2005 the standing committee for CITES recommended that all parties should stop trading in listed species with Nigeria because it was not implementing the convention.

Nondetriment Findings

The CITES requirement that scientific authorities in each country make a nondetriment finding before allowing trade in Appendix I or II species is an important tool in preventing overharvesting of endangered species. Unfortunately, there is often very little ecological information on the species listed in CITES. While there is often information on the distribution of listed species, there is frequently no peer-reviewed literature on the population demography, response to harvesting, or levels of sustainable use for listed species.[31] A study in South Africa found that for more than 90% of the species listed under CITES in that country, the only published information on those species related to taxonomy and distribution.

More research on how harvesting affects populations of CITES-listed species would likely help in making CITES a more powerful tool for protecting species. Ideally, controlled experiments would be performed to determine how populations of listed species respond to harvesting. For instance, an experiment on the Appendix II medicinal plant *Nardostachys grandiflora* found that populations of the plant recovered from harvesting at different rates in different habitats, suggesting different sustainable harvesting rates in those habitats.[32] Some scientists have advocated that ecologists use CITES-listed species as model taxa for their research, so that even if research is not done directly on sustainable harvesting rates, at least there will be more basic biological information for countries to use in making nondetriment findings.[33]

Experiments on CITES-listed species may often be too costly or difficult, though, making mathematical and simulation models more feasible. Such models could also be used to incorporate other factors that interact with harvesting to influence populations, such as habitat loss and climate change. Additional research on when generalizations about how species respond to harvesting may be made across taxa would also help countries determine when trade in a listed species may threaten the survival of that species. Beyond effects on individual populations, harvesting species also affects the ecosystems in which those species occur. Much more research is needed on how the harvesting of species listed under CITES affects entire ecosystems.

In the United States, the courts and Congress have weighed in on what constitutes sufficient scientific information to support a nondetriment finding for an Appendix II species. In *Defenders of Wildlife, Inc. v. Endangered Species Scientific Authority*, Defenders of Wildlife argued that the US scientific authority made a nondetriment finding for the proposed level of bobcat (*Felis rufus*) pelt exports in the 1979–80 season without adequate scientific information.[34] A federal appellate court ruled that the export quota set by the scientific authority was indeed invalid because the authority did not have a reliable estimate of the population size of bobcats, or information on the number of animals to be killed in that particular season.

Despite the commonsense suggestion that the scientific authority should know the population size of a species before it makes a nondetriment finding, the US Congress took a different view. In a 1982 amendment to the Endangered Species Act, Congress overturned the effect of the appellate court ruling. The amendment first gives the secretary of the interior the duties of the management and scientific authorities, to be carried out by the US Fish and Wildlife Service.[35] The amendment then states that the secretary shall use the best biological information available in making a nondetriment finding under CITES, but is not required to estimate the population size of a species in making such a determination.[36] While the amendment does not require the secretary to know the size of populations before issuing a nondetriment finding, the requirement that the secretary use the best biological information available suggests that information on population size must be used if it is available. As a result, the need for more published scientific information on the species listed in CITES will still influence how those species are traded in the United States.

International Convention for the Regulation of Whaling

The International Convention for the Regulation of Whaling (Whaling Convention) is the primary source of international law controlling whaling around the world. The convention came into force in 1948 and has 89 parties. When it was originally negotiated, the Whaling Convention was intended to manage the exploitation of whales, but over time it has evolved to become the primary treaty conserving whale species.

The heart of the Whaling Convention is a "Schedule" of regulations setting the exploitation level of each whale species.[37] The schedule states, among other things, which whale species are protected and which are unprotected, when and where whales may be hunted, and approved methods of whaling. The supervising body of the Whaling Convention is the International Whaling Commission, made up of a representative from each party to the convention. The commission meets every year, at which time the schedule may be amended. Each party to the convention has a vote on the commission, and each amendment to the schedule must obtain a three-fourths majority to pass.

The Whaling Convention was clearly meant to regulate large cetacean species such as blue whales (*Balaenoptera musculus*), but it is less clear whether it was meant to protect small cetacean species such as dolphins and porpoises. The Whaling Convention does not actually provide a definition of "whale." Some member countries, among them Denmark and Norway, argue that the Whaling Convention protects only large cetaceans, while other countries argue that the convention also protects small cetaceans. The commission does sponsor continuing scientific research in small cetacean conservation.

Initially, most parties to the Whaling Convention were pro-whaling, and the International Whaling Commission set relatively high catch limits for many whale species. This resulted in the populations of several commercially hunted whale species being decimated by the 1970s. During the 1970s and early 1980s public outrage over whaling increased in many countries. By 1982 a large majority of parties in the commission were nonwhaling countries. As a result, in that year the commission approved a moratorium on all commercial whaling to begin in 1985-86. The moratorium continues to this day.

While for ethical or moral reasons many people are opposed to harvesting whales, there are also ecological and economic reasons to oppose commercial harvesting of whales. There is growing scientific evidence for the central role of whale populations in many ocean ecosystems. For example, research suggests that whales play an important part in recycling nutrients. They tend to feed in deep waters, but they then bring nutrients to the surface of the ocean in their feces, which float. This "upward biological pump" is especially important in bringing nitrogen to the surface of the ocean, where it is often a limiting nutrient.[38] In the Gulf of Maine, whales input more nitrogen than the combined input of all the rivers feeding the gulf. The nutrient input from whales tends to increase primary productivity, which in turn increases secondary productivity, ultimately resulting in bigger fisheries. Whale species may therefore be very important in ecosystem productivity, and in increasing the populations of fish species that humans harvest.

The whaling moratorium contains two significant loopholes. First, a party to the Whaling Convention may object to any amendment to the schedule. Consequently, if a party objected to the moratorium when the International Whaling Commission first passed it, then it does not apply to that country. Indeed, Norway and the Soviet Union (now the Russian Federation) objected to the amendment and are not bound by the moratorium. As a result, commercial whaling is ongoing today. Since 1985 over 22,000 whales have been harvested through commercial whaling.

Second, the Whaling Convention allows whaling for "scientific research."[39] Iceland, Japan, and Norway have all taken large numbers of whales for the stated purpose of scientific research. The commission has no authority to ban the killing of whales by a country for scientific research. Since the moratorium has been in effect, over 13,000 whales have been killed under special permits issued for scientific research. Japan had ongoing research programs in the Southern Ocean (Antarctic Ocean) and western North Pacific with plans to take 850 Antarctic minke whales (*Balaenoptera bonaerensis*), 340 common minke whales (*Balaenoptera acutorostrata*), 100 sei whales (*Balaenoptera borealis*), 50 humpback whales (*Megaptera novaeangliae*), 50 fin whales (*Balaenoptera physalus*), 50 Bryde's whales (*Balaenoptera edeni*), and 10 sperm whales (*Physeter macrocephalus*) every year. The stated objectives of Japan's research program were to determine the demographic structure of the different whale populations, examine the role of whales in their ecosystems, and assess the effects of climate change on cetaceans. Many scientists argued that these objectives could be met through nonlethal research. The real purpose of killing whales for research appeared to be an attempt to keep Japanese whaling industries intact.

The International Court of Justice (ICJ) agreed. The ICJ is the judicial organ of the United Nations, and one of its purposes is to settle legal disputes between countries. Importantly, countries must agree to accept the jurisdiction of the ICJ—the ICJ does not automatically have jurisdiction over legal disputes between countries. In 2010 Australia sued Japan in the ICJ to stop the Japanese scientific whaling program in the Southern Ocean.[40] (Australia and Japan had already accepted the jurisdiction of the ICJ over such disputes.) In its ruling in 2014, the ICJ noted that Japan had killed 3,600 minke whales since 2005, but had produced only two peer-reviewed papers related to the lethal sampling. The ICJ observed that lethal sampling of whales was not an unreasonable means of accomplishing Japan's research objectives. However, the ICJ stated that the number of whales that Japan had killed under the auspices of its Southern Ocean research program was not reasonable in relation to the scientific objectives Japan hoped to achieve. As a result, the ICJ concluded that Japan's killing of whales was "not for purposes of scientific research."[41] The ICJ therefore held that Japan should immediately stop the killing of whales in the Southern Ocean under the cover of its scientific whaling program.

Japan agreed to be bound by the ruling of the ICJ, and stopped killing whales in the Southern Ocean in 2014. However, Japan continued to kill whales in the Pacific as part of its research program in that ocean.[42] Further, Japan announced that it would create a new scientific program for the

Southern Ocean, and would likely resume killing whales in the Southern Ocean in 2015.[43]

Despite the loopholes, the moratorium on commercial whaling has proved successful in helping several whale populations increase in size. It remains controversial, though, whether any of the populations have increased enough to support a resumption of commercial whaling. The Scientific Committee of the International Whaling Commission has created a revised management procedure to manage whale stocks if commercial whaling is allowed to recommence. The management plan aims for all whale stocks to reach 72% of their unexploited stock level. If a species is below 54% of its estimated carrying capacity, the management procedure would not allow the species to be harvested.

Although the International Whaling Commission has yet to relax the whaling moratorium, the moratorium may not last indefinitely. Some countries that do not have an interest in whaling but have signed the Whaling Convention may be willing to vote against the continuation of the moratorium if it allows them to curry favor with countries that are interested in commercial whaling. Currently, a majority of countries on the commission are pro-whaling and against the moratorium. If the number of countries that are pro-whaling reaches three-fourths, the moratorium will likely be reversed, and commercial whaling will resume.

Biodiversity Treaties to Which the United States Is Not a Party

There are several important treaties meant to conserve biodiversity that have been signed and ratified by many countries, but that the United States is not a party to. Two of the most important are the Convention on Biological Diversity and the Convention on the Conservation of Migratory Species of Wild Animals (frequently referred to as the Bonn Convention, for the name of the German city in which it was signed). The next sections briefly touch on these two conventions.

Convention on Biological Diversity

The Convention on Biological Diversity was originally envisioned as an overarching environmental treaty that would consolidate existing biodiversity treaties to provide greater coherence. This proved to be politically unattainable. Entering into force in 1993, the Convention on Biological Diversity acts as a framework treaty containing several lofty objectives, but requiring meetings of the conference of the parties to fill in the details. There are 193 parties to the convention, which is nearly universal participation by the world community. The United States has signed but has not ratified the convention, and is thus the only major nation that is not a party to the convention.

The Convention on Biological Diversity has three primary objectives: (1) conserving biological diversity, (2) the sustainable use of that diversity, and (3) fair and equitable sharing of the benefits arising from human utilization of genetic resources.[44] To realize these objects, the convention takes an "ecosystem approach."[45] This means the convention recognizes that integrated management of land, water, and living species is the best means for achieving the objectives of the convention.

Unfortunately, the Convention on Biological Diversity includes few concrete obligations to realize those objectives. Most of the convention consists of goal setting. Countries that are party to the convention are encouraged to manage their wild resources at the national level. Parties to the convention are also required to prepare a national biodiversity strategy.[46] There is no provision in the convention, however, for independent inspectors to monitor how the actions of a country are affecting its biological diversity. Additionally, the convention contains no method to mandate actions at the international level. On the other hand, the Convention on Biological Diversity does bring the parties to the convention together every two years to discuss biodiversity. As more details are slowly filled in at these conferences of the parties, the Convention on Biological Diversity may become a more powerful tool for protecting biodiversity.

The convention does make explicit what countries often considered to be a part of customary law: that nations have "the sovereign right to exploit their own resources pursuant to their own environmental policies."[47] The convention goes on to say, though, that countries have the responsibility to make sure that the activities within their country "do not cause damage to the environment of other States or of areas beyond the limits of national jurisdiction." In other words, a country has a sovereign right to do what it wants with its own biological resources, as long as it does not damage the resources of any other country.[48] The convention contains no provision, however, for assigning liability if the actions of one country negatively affect the biodiversity of another country.[49]

Bonn Convention

Earlier in the chapter it was mentioned that treaties may be necessary to deal with species that move across national borders. One country may have an excellent management plan in place for a species, but if that species migrates to another country where it is overharvested or killed for other reasons, then the management of the first country will be pointless. The Bonn Convention is the primary treaty for protecting migratory species. Entering into force in 1983, the Bonn Convention attempts to protect most migratory animal species, including birds, fish, reptiles, and mammals. The convention currently has 117 parties, not including the United States.

Similar to CITES, the Bonn Convention splits species into appendices, but the Bonn Convention uses only two. Appendix I consists of species that are migratory and endangered. An endangered species is defined as one that "is in danger of extinction throughout all or a significant portion of its range."[50] In a subsequent conference of the parties, "endangered" was interpreted to mean "facing a very high risk of extinction in the wild in the near future."[51] As with CITES, individual populations of a species may be listed in Appendix I, allowing other well-managed populations to be exploited.[52] A migratory species is then defined as one with a population where a significant proportion of the members of the population "cyclically and predictably" cross one or more national boundaries.[53]

The meaning of "cyclically and predictably" appears to be rather open to interpretation by the parties to the convention. For example, the mountain gorilla (*Gorilla gorilla* and *Gorilla beringei*) is listed under Appendix I because it crosses the borders between Rwanda, Uganda, and Democratic Republic of the Congo, even though its movements are not necessarily cyclical and predictable. Conversely, the endangered Asian elephant (*Elephas maximus*) is not listed under the Bonn Convention because its movements between national boundaries were not considered to be cyclical and predictable enough.

Appendix II species are migratory species with an "unfavorable conservation status" that require an international agreement for their conservation, or species that would significantly benefit from international cooperation for their conservation.[54] Unlike CITES, the Bonn Convention may list species concurrently under both Appendix I and II.

Species may be added to the appendices with a two-thirds majority vote during a conference of the parties.[55] Similar to CITES, a party may make a reservation to any species added to an appendix, thereby making it a non-party for that species.

For Appendix I species, parties must prohibit the "taking" of those species.[56] Parties are also required to "endeavor" to conserve and restore habitats for the species, minimize obstacles that impede migration, and reduce further endangerment to the species, such as by introduced species.[57] The requirements that parties restore habitat, minimize obstacles to migration, and reduce endangerment are less binding than the prohibition on taking, as they are only required when "appropriate."[58]

The Bonn Convention does not provide direct protection for Appendix II species. Instead, Article IV(3) of the convention encourages parties to enter into cooperative "agreements" to help conserve these species. Unfortunately, only four such agreements have been completed. The agreements protect European bats, African-Eurasian waterbirds, albatrosses and petrels, and gorillas. The agreements arising from Article IV(3) are legally binding treaties under international law. The fact that they are binding law makes their negotiation difficult, likely accounting for why there are so few of them.

There is an additional means for parties to make agreements under the Bonn Convention. Article IV(4) of the convention calls for agreements between countries for any species that periodically crosses one or more national boundaries, even if the species is not listed in Appendix I or II. As with agreements under Article IV(3), agreements under Article IV(4) may be legally binding international law. There are three such agreements under Article IV(4). However, Article IV(4) also allows for less formal agreements that are not legally binding. An example of a less formal agreement would be a memorandum of understanding between countries. There have been 19 such memoranda between countries, likely because the memoranda are easier to negotiate as they are not legally binding. The memoranda are also useful because they can include entities that would not be able to enter into formal agreements, such as nongovernmental organizations.

TREATIES FOR PROMOTING TRADE

As several of the conventions above suggest, trade between countries plays a huge role in shaping international environmental law. Trade is both an inducement for countries to enter into environmental treaties, and an important part of why many species and ecosystems are threatened and need the protection of treaties. As this section shows, treaties that focus on trade unsurprisingly emphasize promoting trade far more than protecting the environment.

In 1948 the General Agreement on Tariffs and Trade (GATT) came into force, focused primarily on tariff reductions (a tariff is a tax on imported goods). It was hoped that increased trade ties between countries would help prevent another world war. The GATT was meant to be a component of a later agreement establishing an international trade organization, but the negotiations stalled. Eventually, in 1995 the World Trade Organization (WTO) came into being after lengthy negotiations, providing a forum for resolving trade disputes. The WTO has 155 members, including the United States.

The GATT's primary objectives are to reduce trade barriers among members, and to prevent countries from discriminating against goods and services on the basis of their national origin. The GATT contains five articles of particular importance. Article I prevents members from discriminating between similar products from two or more other GATT members. For example, the United States cannot favor a product from China over a "like product" from Germany. Article III prevents members from discriminating against imports from other member countries in favor of domestic products. Article XI forbids quotas that limit the amount of imports from other member countries. Article XIII prohibits trade restrictions on products from a member country unless like products from all other countries face similar restrictions.

Finally, Article XX allows members to make exceptions to Articles I, III, and XI for environmental reasons, as long the exceptions are not arbitrary or discriminatory against other member countries. Specifically, Article XX(b) allows for exceptions that are "necessary to protect human, animal or plant life or health." Article XX(g) then allows for exceptions "relating to the conservation of exhaustible natural resources."

Article XX would seem to be a strong tool for one country to influence the ecological practices of other countries—if we do not like how you harvest or manufacture a product, we will invoke Article XX and refuse to import that product. The ability of countries to use Article XX to influence the ecological practices of other countries is severely limited, however, by the dispute settlement process under the GATT and the WTO.

If a member country is unhappy with the laws or practices of another member, it can bring the country before the WTO Dispute Settlement Body. The Dispute Settlement Body will rule on whether the law or practice violates the trade rules. The losing country may appeal the ruling of the Dispute Settlement Body to an appellate body for a review of issues of law. If the appeal is unsuccessful, the country has two options: it can change the offending law or practice, or it can keep the law or practice in place. If the country keeps the law or practice in place, the member country that brought the dispute before the Dispute Settlement Body may raise countervailing tariffs against imports from the losing country equal in value to the harm it is suffering. Trade sanctions are therefore the penalty for a country not following the trading rules.

The ability of the WTO to permit one country to raise tariffs on another makes the WTO very powerful in resolving trade disputes. This power is potentially troubling. The Dispute Settlement Body is a three-member panel generally drawn from the trade community. It does not necessarily have expertise in environmental issues. As a consequence, a member country that discriminates against another member for an environmental reason under Article XX may lose before a Dispute Settlement Body that is more interested in reducing trade barriers than in protecting the environment.

This concern is especially relevant when a member country is concerned with how a product is being produced in another country. In 1988 Congress amended the Marine Mammal Protection Act (discussed in chapter 9) to require the US secretary of commerce to ban imports of tuna products from countries that did not use dolphin-friendly nets. The following year the secretary banned tuna imports from several countries, including Mexico. Mexico challenged the law before the GATT dispute panel (which preceded the WTO Dispute Settlement Body), claiming that it violated Articles III, XI, and XIII. Mexico argued that the GATT had been violated by the US law because once in the can, tuna from Mexico was identical to tuna from the United States and was thus a "like product." The United States argued that the two countries' tuna were not like products because their method of production was different. The United States also argued that the ban was valid under Article XX as an exception to protect animal life and conserve a natural resource.

The GATT panel ruled in favor of Mexico, finding the United States had violated Articles III and XI.[59] The panel stated that the process and production method of a product may not be the basis for a trade restriction. The panel also stated that Article XX did not apply because a country may not suspend trade rules for environmental harms occurring outside its borders. The GATT panel clearly felt that allowing trade discrimination based on environmental harms outside of the member nation would invite protectionism hidden behind environmental concerns.

Mexico ultimately dropped the case for political reasons involving the negotiation of the North American Free Trade Agreement. Additionally, subsequent rulings by the WTO have softened the requirement that the way in which products are produced or harvested is not a justification for invoking Article XX, and that all the environmental harm must happen within the borders of the country to form the basis of a trade exception.[60] However, the GATT and the WTO seem inclined on a philosophical level against allowing exceptions under Article XX.

It should be remembered, though, that if the WTO Dispute Settlement Body rules against a member country, the country does not have to stop the offending law or practice: it merely has to suffer the resulting tariffs. If a country believes strongly in not allowing the importation of products produced in an environmentally unfriendly way, then it has the option to maintain that belief and pay the economic consequences.

BIODIVERSITY HOT SPOTS

How well has international environmental law done at protecting species and ecosystems? One way to answer that question is to look at how well international law has protected those ecosystems around the world that contain high diversity of species and are facing a great threat to their survival. These ecosystems are often termed biodiversity hot spots. Protecting hot spots can potentially save a huge number of species from extinction, as well as safeguarding some of the most fascinating ecosystems in the world.

The ecologist Norman Myers first described the concept of the biodiversity hot spot in 1988. To be a hot spot, an ecosystem must meet two criteria: (1) it must contain at least 1,500 endemic vascular plant species; and (2) it must have lost at least 70% of its original vegetation. There are 34 biodiversity hot spots in the world, covering just 2.3% of the earth's land surface. However, these 34 hot spots contain over 50% of plant species and 42% of terrestrial vertebrate species that are regional endemics. The United States is home to one biodiversity hot spot, the California Floristic Province.

There are certainly critics of the hot spot concept. Criticisms include that the definition of hot spots focuses too much on vascular plants, and does not take into consideration whether previous land loss is continuing at the hot spot or whether other areas with high endemic species richness are facing even faster species loss but have not yet reached the 70% cutoff. Other criticisms include that hot spots do not take into account ecosystem services or phylogenetic diversity, and do not consider the cost of protecting one hot spot instead of another. Despite these criticisms, this section focuses on how well international law has done in protecting those areas categorized as hot spots.

Roughly 38% of the area of global hot spots is protected by parks and reserves created by the nations in which they are located. This is reflected in the California Floristic Province, where 37% of the land area is protected by state and federal parks and refuges, along with other forms of protection such as landownership by nongovernmental organizations. This protection, however, accounts for only about 5% of the original extent of global hot spots. On top of that, the parks and reserves in many countries offer very little protection to the species they contain. There is thus ample room for international environmental law to help protect hot spots. Unfortunately, international law does little to protect them.

The Convention on Biological Diversity urges signatory countries to protect ecosystems "containing high diversity, large numbers of endemic or threatened species, or wilderness."[61] This would surely include biodiversity hot spots. However, the convention also states that a country should protect ecosystems "in accordance with its particular conditions and capabilities"[62] and "as far as possible and as appropriate."[63] The Convention on Biological Diversity places no real requirements on parties to protect their hot spots because it is up to each country to decide when protecting its ecosystems is "appropriate."

CITES also does not directly help conserve biodiversity hot spots. CITES takes a species-by-species approach to conservation, regulating the trade of only those species that have been listed in one of its appendices. The convention does little to protect the ecosystems where those individual species occur. In fact, CITES does little to prevent a country from completely eliminating species listed in the convention if the extirpation is done domestically and not through trade. Also, as discussed above, a country may take a reservation to the restriction in trade of any species, allowing the country to be treated as a nonparty with regard to that species. CITES was not meant to protect ecosystems, and thus does a poor job of directly protecting hot spots.

The Bonn Convention faces similar problems as CITES in protecting hot spots. The Bonn Convention takes a species-by-species approach to conservation, and only migratory species at that. The Bonn Convention also does not provide direct protection to biodiversity hot spots.

One convention that may help protect biodiversity hot spots is the Convention concerning the Protection of the World Cultural and Natural Heritage (World Heritage Convention). The World Heritage Convention is meant to protect natural and cultural areas of "outstanding universal value."[64]

To be named a natural heritage site, an area must meet three requirements. First, the potential site must consist of physical and biological formations that are of outstanding universal value because of their aesthetic beauty; because of their value to science or conservation; or because they contain significant biological diversity, including habitats for threatened animal or plant species important to science or conservation.[65] Second, the site must meet the condition of integrity, meaning that the biophysical processes and landform features of the site should be relatively intact.[66] Third, the site must have adequate protection and a management system in place.[67] The World Heritage Convention states that it is the duty of the entire international community to protect heritage sites.

Countries that have signed the World Heritage Convention must conduct scientific research and create technical reports describing any dangers threatening their heritage sites. The World Heritage Fund provides limited financial support to countries that need help in protecting their heritage sites. Countries must also not take any "deliberate measures" that might directly or indirectly damage the heritage sites of another country.[68]

Coming into force in 1975, the World Heritage Convention has 189 parties, including the United States. There are currently 962 heritage sites in 157 countries. Of the heritage sites, however, only 188 are natural heritage sites, with an additional 29 mixed natural and cultural heritage sites. There are 21 heritage sites in the United States, including such natural sites as Yellowstone National Park, Grand Canyon National Park, and Everglades National Park. Two heritage sites, Yosemite National Park and Redwood National and State Parks, are within the California Floristic Province hot spot.

Although there are certainly heritage sites around the world that are biodiversity hot spots, many hot spots do not qualify as heritage sites under the World Heritage Convention. Biodiversity hot spots almost certainly meet the first requirement of being of outstanding universal value. The other requirements, for integrity and for adequate protection and management, may be problematic because by definition a hot spot is facing danger to its existence. Additionally, an application for a site to be named a heritage site can come only from the country where the site is located—no country can nominate a site located within another country.

Even if a hot spot is named a heritage site, the World Heritage Convention does relatively little to protect the site from harm. Signatory nations must submit reports detailing the steps they are taking to comply with the World Heritage Convention. If the information is not provided, there is no penalty or sanction that can be applied to the country. Further, if a heritage site is facing dangers to its existence, the World Heritage Committee may list the site as in "danger," or may even remove the site from the list of heritage sites. Indeed, Everglades National Park in the United States is listed as a site in danger. However, neither of these actions forces a signatory country to do anything to protect the heritage site within its borders. At best, listing a site as in danger may shame a nation into providing better protection for the site. The World Heritage Convention therefore has little ability to protect biodiversity hot spots if the country in which that hot spot occurs does not already wish to protect it.

International environmental law seems to have little ability to protect biodiversity hot spots. At best, the international community can help convince countries to protect their hot spots by creating incentives for protecting those ecosystems. Laws passed by individual countries to provide funding and technical help for other nations to protect their biodiversity hot spots would potentially assist in protecting many such ecosystems.[69] A model for such a statute is the Tropical Forest Conservation Act passed by the United States in 1998. The act helps protect tropical forests by forgiving debt owed to the United States in countries that undertake activities aimed at conserving their tropical forests.

Although international law appears to be inadequate in protecting biodiversity hot spots, very little ecological research directly addresses this topic. The protection of certain species by existing treaties may be having indirect effects that benefit biodiversity hot spots to a greater degree than it would seem on the surface. As mentioned above, very little scientific information is available for many of the species listed under the CITES convention, making it difficult to say whether CITES is indirectly helping protect ecosystems. If current international environmental laws are providing indirect benefits to hot spots and other ecosystems, elucidating the pathways of those benefits may allow them to be strengthened, thereby helping protect ecosystems even more.

PART II

Land

CHAPTER FIVE

Federal Public Lands

Desert bighorn sheep (*Ovis canadensis mexicana*) in the Kofa Wilderness Area in Arizona seemed to be in trouble (figure 5.1). There had been 813 bighorn sheep in 2000, but by 2006 that number had dropped to 390. Drought conditions in Arizona seemed the most likely culprit for the steep decline in numbers. The US Fish and Wildlife Service (FWS), which manages the wilderness area, considers bighorn sheep to be the symbol of the Kofa Wilderness Area; as a result, FWS decided to take quick action. The agency created a plan to construct two 49,200-liter (13,000-gallon) water tanks to provide water to the bighorn sheep; the tanks were then installed in 2007 by FWS and the Yuma Valley Rod and Gun Club.[1]

Soon after, environmental groups began to question the installation of the water tanks. In the 1960s and 1970s the bighorn sheep population in Kofa was between 200 and 375 individuals, suggesting that a size of 390 may be within the natural range of variation for that population.[2] The groups also questioned FWS's motivation for attempting to increase bighorn sheep numbers. Besides installing water tanks, FWS had also proposed killing any mountain lion (*Puma concolor*) that preyed on more than one bighorn sheep in a six-month period.[3] This was proposed even though the agency had no knowledge of how mountain lion predation was affecting the bighorn sheep population. Hunting of bighorn sheep by humans is allowed in Kofa, however, and FWS specifically stated that it would allow continued hunting of bighorn sheep. This created the appearance that FWS was constructing water tanks and killing mountain lions in an attempt to keep bighorn sheep hunters happy.

FIGURE 5.1. Desert bighorn ram. Photograph from FWS/Lynn B. Starnes.

Finally, the environmental groups argued that the Wilderness Act, which controls the management of wilderness areas, seemed to have been violated. The Wilderness Act tries to ensure that wilderness areas, of all the federal public lands, receive the greatest protection from human interference. As a guiding philosophy, the act states that federal agencies that manage wilderness areas are responsible for "preserving the wilderness character of the area."[4] To put this philosophy in action, the Wilderness Act declares that there are to be no structures or installations in wilderness areas, except as necessary to meet the "minimum requirements" of managing such an area.[5] Water tanks are certainly structures, and it did not seem that keeping one species at a high population level was the minimum requirement of managing the Kofa Wilderness Area. Several environmental groups sued to force removal of the water tanks.

In 2010 the Ninth Circuit Court of Appeals held that FWS did not provide adequate reasoning to demonstrate why the construction of the water tanks was necessary for the recovery of bighorn sheep in Kofa.[6] The court stated that the agency did not explain why taking actions other than building structures, such as prohibiting hunting or occasionally hauling in water, would have been insufficient to rebuild the bighorn sheep population. The court ordered a lower court to determine whether FWS should dismantle the water tanks. The court of appeals did find, however, that the desire of FWS to increase the size of the bighorn sheep population in Kofa was perfectly fine. If FWS had done a better job of explaining why the water tanks were the best way to increase the bighorn sheep population in Kofa, the court likely would have decided the case in the agency's favor.

Wilderness areas ideally receive the greatest protection from human interference, but even in wilderness areas there is still considerable pressure to manage them in ways that favor human interests. This pressure is even greater in the other types of federal public land that will be discussed in this chapter. In these other lands, the controlling statutes make clear that species and ecosystem protection is only one use of the lands, with recreation and commercial activities receiving equal consideration. Balancing the competing pressures of protecting land while allowing it to be used for human activities is the challenge that defines the management of every parcel of federal public land.

This chapter briefly describes the main categories of federal public lands, and then moves into a more detailed description of the statutes and ecological processes affecting each category. The chapter concludes with a description of the National Environmental Policy Act (NEPA). NEPA is a federal statute that requires federal agencies to write an environmental impact statement (EIS) for all major federal actions. All the federal agencies discussed in this chapter are required to write EISs for almost all the major actions they take in managing federal lands.

CATEGORIES OF FEDERAL PUBLIC LANDS

The federal government owns approximately 263 million hectares (650 million acres) of land, covering roughly 28% of the surface of the United States. In general, this land may not be used for residences or cropland, and must be managed at least in part to protect species and ecosystems. The federal public lands act as the most important reserve for native species and habitats in the United States. For this reason, understanding the requirements and constraints on federal land management is critical for understanding environmental protection in the United States.

The history of the federal lands often makes them less than ideal as reserves. First, most of the federally owned land is in the western states and Alaska, meaning that ecosystems in the eastern United States are less likely to be protected as federal public lands. Second, the federal government used to own considerably more land, but sold or gave away 607 million hectares (1.5 billion acres) to the states, railroads, homesteaders, and others.[7] In choosing land to buy or homestead on, individuals logically chose the land that was most valuable. As a result, much of the land currently owned by the federal government is arid, contains relatively less biological diversity, and tends to be at higher elevations.[8] Additionally, the land grants to private railroad companies were often made in a checkerboard pattern.[9] Railroads were given every other section of land surrounding the path where the train tracks were to be laid, often for ten miles in either direction perpendicular to the railway line. The federal government kept the other sections, with each section of land being 259 hectares (640 acres). The government believed that the construction of the railway would increase the value of the land that the government owned, bringing in more revenue when sold. However, much of the checkerboard land kept by the federal government was never sold. As a result, there is still considerable checkerboarding of federal land in the western United States. Third, many federal public lands contain inholdings of private land. These inholdings are completely surrounded by public land, but may contain residences or other land uses that are not allowed on the public lands. These three factors make the federal public lands imperfect as potential reserves for species and ecosystems.

The land that the federal government owns can be managed in almost any way that the government decides is best. The property clause of the Constitution declares that Congress has the power to "make all needful Rules and Regulations respecting the Territory or other Property belonging to the United States."[10] The Supreme Court has long held that the federal government can regulate federal lands however it sees fit, and that federal law preempts state law.[11] State law does apply on federal lands, but only to the extent that it is not inconsistent with federal law.[12] One of the most important activities that state law continues to regulate on federal public lands is hunting and fishing. The federal government can and does designate areas where hunting and fishing are not allowed, but when it is allowed, state law usually sets such things as hunting season and catch limits.

The property clause also gives the federal government the power to protect federal public lands from threats that exist outside those lands. For instance, the Supreme Court held in 1927 that the federal government may prohibit fires on private lands that are near national forests.[13] A federal appeals court has stated that Congress is acting constitutionally when it restricts activities to protect the "fundamental purpose" for which federal land has been reserved, and the restrictions "reasonably relate to that end."[14] One of the most important examples of Congress's restricting such an activity is found in the Unlawful Inclosures Act of 1885. The act prohibits persons from enclosing public land with a fence or other barrier, including building a fence on private property if it denies access to public land.[15] A federal appeals court has held the Unlawful Inclosures Act also prohibits fencing that keeps migrating wildlife from reaching federal lands.[16] In that ruling, the court held that a fence that prevented antelope from migrating to their winter foraging ground on federal land had to be removed.

Nearly all federal public lands fall into one of five categories:

- National Park System, managed by the National Park Service within the Department of the Interior. The system got its start in 1872 when Congress reserved Yellowstone National Park from settlement or sale.
- National Forest System, managed by the US Forest Service within the Department of Agriculture. This system was formed in 1891 when Congress authorized the president to reserve federal forested land.
- BLM lands, managed by the Bureau of Land Management within the Department of the Interior. In 1934 the Taylor Grazing Act took federal public lands that had not been reserved for conservation and assigned them to grazing districts. These lands eventually became BLM lands.
- National Wildlife Refuge System, managed by FWS within the Department of the Interior. Beginning in 1903, several presidents and Congresses reserved different federal public lands as wildlife refuges. These refuges were consolidated into one system by Congress in 1966.
- National Wilderness Preservation System, existing within the above four categories, and managed by the above four agencies. Congress created this system in 1964 to protect the wilderness character of exceptional federal public lands.

Although all the lands in these five categories are owned by the federal government, their management can be very different depending on the category they are in. The different management policies reflect the tension between the conservation and preservation philosophies of land use. The conservation philosophy contends that the natural resources on public lands should be sustainably used to produce the greatest benefit to the greatest number of people, both today and in the future. Conservationists such as Gifford Pinchot argued that public lands have considerable instrumental value, meaning that if used wisely, the lands have the ability to benefit a large number of humans. The preservation philosophy, conversely, contends that the natural resources on public lands should never be consumed, and that nature should be left in its natural state. Preservationists such as John Muir argued that public lands have value beyond how they benefit humans—the lands also have intrinsic value.

Most federal public land is managed in a way that is closer to the conservation philosophy than the preservation philosophy. This can be seen in the "multiple-use" approach statutorily required for most federal public lands. For example, the Forest Service must manage its lands for uses including recreation, logging, and grazing. The Forest Service, however, must also protect the diversity of plants and animals on its lands. This requirement is a recognition that there is intrinsic value in nature. The National Wilderness Preservation System was created with the intent of prohibiting the consumption of natural resources on at least some federal public lands and thereby fulfilling the ideals of the preservation philosophy. Commercial enterprise and permanent roads are not allowed in the system. As the introduction to this chapter suggests, however, even lands in the National Wilderness Preservation System may be managed in part for the benefit of humans, such as allowing sport hunting. The next five sections examine each category of federal public land in detail, and illustrate how it fits within the conservation and preservation philosophies.

National Park System

The National Park System consists of the public lands most familiar to many Americans. This familiarity is also the greatest threat to the species and ecosystems in the park units. The US National Park Service must balance protecting the plants and animals in the parks with the desire of millions of us to use the parks for recreation. In the past, recreation almost always won; today, the Park Service strives for a greater balance. As will be discussed below, the greatest emerging threat to the national parks may be the desire of people to live near the parks.

The units in the National Park System are not all labeled national parks; they also include national monuments, memorials, preserves, battlefields, historic sites, and recreation areas. There are 401 units in the National Park System, encompassing 34 million hectares (84 million acres). To put this in perspective, the state of Arizona covers roughly 30 million hectares (73 million acres).

Most national parks were created by specific legislation for each park. For example, in 1980 Congress passed the Alaska National Interest Lands

Conservation Act (ANILCA). The act doubled the size of the National Park System, adding 17.6 million hectares (43.6 million acres) and ten units to the National Park System.[17] Most national parks were carved out of existing federal lands, but some were created by buying private lands. The Great Smoky Mountains National Park was created in large part by cobbling together private lands.[18] Through the Antiquities Act of 1906, the president has the power to unilaterally create national monuments. The Antiquities Act will be discussed in greater detail in chapter 9.

The National Park Service Organic Act of 1916 (Organic Act) created the Park Service in order to manage the units in the National Park System. The act states that the Park Service must regulate the National Park System to "conserve the scenery and the natural and historic objects and the wild life therein."[19] The Park Service must also provide for recreational enjoyment of the National Park System, but in such a way that the system will remain "unimpaired for the enjoyment of future generations." The Organic Act clearly sets two tasks for the Park Service: to protect the wildlife in the parks and to provide for recreation.

If the specific legislation creating a park conflicts with the general legislation of the Organic Act, the specific legislation wins. For instance, hunting is not allowed in national park units unless specifically authorized by Congress.[20] ANILCA allows for subsistence uses of resources by rural Alaska residents.[21]

Resource extraction, such as mining, is not allowed in the National Park System unless the specific law creating a park allows such activity. The same is true of timber harvesting. Fishing is allowed in most parks, but must be done in accordance with specific regulations for that park and the laws of the state where the national park is located.[22] Scientific research is allowed, and even encouraged, but not surprisingly there are some hoops to jump through. The Park Service must issue a scientific research and collecting permit before research can begin. Before giving a permit, the superintendent of the park must approve a written research proposal. Taking plants or wildlife for research may be allowed under a permit, but will be prohibited if removing a specimen will adversely affect the environment, or if the specimen is readily available outside the park.[23] Additionally, if the legislation establishing a particular park area prohibits the killing of wildlife, no permit for taking plants or wildlife will be issued.

The Organic Act requires the Park Service to write general management plans for each unit of the National Park System.[24] The act does not include many specifics for those plans, however. The act requires only that general management plans include four parts: (1) "measures for the preservation of the area's resources," (2) type and intensity of development for public use of the park, (3) "visitor carrying capacities for all areas" of the park, and (4) reasons for potential modifications to the external boundaries of the park.

In the early years of the National Park System, the Park Service did a poor job of integrating ecological principles into its management of park units. For instance, from the creation of Yellowstone National Park in 1872 until the 1930s, the Park Service focused on increasing the population sizes of "good" species (such as elk and deer) in Yellowstone, while decreasing the population sizes of "bad" species (such as wolves and coyotes).[25] As a result, gray wolves were extirpated from the park by the 1920s, while the elk population exploded. As the elk population increased, the Park Service began to fear the elk would overgraze Yellowstone vegetation.[26] To combat the perceived problem, the Park Service began killing elk in large numbers. This culminated in the winter of 1961, when the Park Service shot over 4,000 elk. The killing of so many elk produced a public outcry that led to Senate hearings and the Park Service's commissioning of a study by a scientific advisory board.[27] The advisory board released a report in 1963 entitled *Wildlife Management in the National Parks* (usually called the Leopold Report, after the chair of the advisory board, A. Starker Leopold). The Leopold Report had a revolutionary impact on how the Park Service managed its lands. The report called for managing national parks according to ecological principles, recommending that "the biotic associations within each park be maintained, or where necessary recreated, as nearly as possible in the condition that prevailed when the area was first visited by the white man." The report stated that the goal of Park Service management should be national parks that "represent a vignette of primitive America." The recommendations in the Leopold Report convinced the Park Service to reduce its intervention in ecological processes. For example, the Park Service began letting wildfires burn in park units. The Park Service also implemented a strategy of "natural regulation" of the elk population in Yellowstone. Natural regulation meant the service would no longer shoot elk, but would instead allow elk populations to fluctuate naturally, assuming that density-dependent factors would prevent the elk population from getting too large. Natural regulation was bolstered when the Park Service reintroduced wolves into Yellowstone in 1995.

The ecological issues the Park Service must consider when determining how best to manage its lands continue to evolve and are not specifically addressed by the Organic Act. To be effective, general management plans for park units should confront the three greatest threats currently facing the

National Park System: visitors to the parks and the roads necessary for those visitors, land use around the parks, and global climate change. As the following sections suggest, the Park Service has not always done a stellar job of planning for those threats.

Visitors and Roads

The greatest threat to the national parks is almost certainly human visitors. In 2011 there were nearly 279 million visitors to the National Park System. In the early days of the national parks, the Park Service viewed its role as encouraging tourists to visit park units.[28] To do this, the Park Service built roads, hotels, and other facilities throughout many of the national parks. The Park Service now does a better job of balancing the requirement in the Organic Act that the national parks be used both for recreation and for protecting wildlife. For instance, it is now the policy of the Park Service to construct facilities outside the boundaries of national parks. Whenever possible this includes letting the private sector meet visitor needs, such as lodging, in the towns surrounding the parks.[29] While this policy certainly helps reduce disturbance inside of the national parks, the next section illustrates how human land use outside of parks can also negatively affect ecosystems within the parks.

The people who venture inside a national park have the potential to negatively affect many aspects of its ecology. One of the most obvious disturbances is trampling vegetative groundcover. Trampling may also lead to erosion and compaction of soil. It is especially problematic when visitors create unofficial trails (called social trails) through sensitive vegetation where park managers would not normally put an official trail. Less obviously, visitors often pollute water bodies and disturb wildlife.[30] All these impacts may result in degraded and fragmented habitats.

There is considerable evidence that the effect of humans on parks is curvilinear and asymptotic.[31] This means that when human use of an area is low, any additional use tends to greatly increase the impact on the ecology of the area. However, once the use of an area is high, any additional use will have only a small added impact.

This is why it is so important that the Organic Act requires general management plans to include visitor carrying capacities for all areas in a park. Managers may use carrying capacities to limit the number of visitors in ecologically sensitive areas of the park. Remembering the curvilinear nature of human impacts, park managers may attempt to concentrate human use in popular areas of a park and on official trails, while limiting human use in other areas.

The official guidance the Park Service gives to individual park managers in setting carrying capacities comes from two documents, *Management Policies* and the Visitor Experience and Resource Protection (VERP) handbook. The Park Service has written *Management Policies* to help guide park managers. The document defines visitor carrying capacity as the "level of visitor use that can be accommodated while sustaining the desired resource and visitor experience in the park."[32] *Management Policies* makes clear that visitor carrying capacity is based not only on the ecological impacts of visitors, but also on the desire to protect visitors from an experience of overcrowding. Managers are to use natural and social science research in setting visitor carrying capacities. The VERP handbook suggests that in setting visitor carrying capacities, managers must determine the "limits of acceptable change" in park resources.[33] This entails allowing environmental conditions in the park and the quality of the visitor experience to degrade, but only to a previously established acceptable level. To implement visitor carrying capacities, *Management Policies* requires that each park's general management plan delineate management zones that correspond to "desired resource and visitor experience conditions."[34] For instance, sensitive ecosystems in the park may be zoned for very little or no visitor use, while other less sensitive areas may be zoned for much heavier use. While *Management Policies* and the VERP handbook give guidance on setting visitor carrying capacities and management zones, individual park managers have a great deal of discretion in determining carrying capacities and zones.

This discretion is potentially problematic because the ecological research being done on how humans affect national parks tends to be very limited. Most studies focus only on effects at the local scale.[35] The studies often examine the effect of human use on vegetation cover and soil characteristics at the campsite or trail level at one point in time, but rarely examine effects on the population dynamics of species or functions of ecosystems. Indeed, very few studies attempt to study effects at multiple scales, such as the local and landscape scale. Additionally, there is little research on spatial patterns in impacts. If looking only at the local scale, trampling of vegetation or disturbing wildlife may seem to be unacceptably large, whereas at a larger scale the impacts may not be particularly important. Conversely, recreation activities may seem to have little effect at a small scale, but at a larger scale the cumulative impact may be significant.[36] For example, light human use of an area may result in very little trampling of vegetation, but the occasional presence of humans may stress several animal species, resulting in reduced reproduction for those species, and thereby affecting the population dynamics of many species in the area. Research on the effects of humans on the ecology of national parks is needed at multiple scales and through time in order for the visitor carrying capacities and management zones set by park managers to be more than nearly blind decision making.

The primary way visitors get around in national parks is by driving on roads. There are roughly 8,855 kilometers (5,500 miles) of paved roads, and 7,245 kilometers (4,500 miles) of unpaved roads in the National Park System.[37] In a survey of park managers, a study found that 36% of parks with endangered species have roads that bisect the critical habitats for those species.[38] Roads can have several effects on species and ecosystems. The roads themselves take up space, reducing habitat for organisms. Vehicles traveling on roads may strike and kill many individuals, which may affect population sizes. This seems to be especially true for small and medium-sized mammals. Roads also fragment ecosystems and create edge effects. Additionally, individuals from a species may be unable or unwilling to cross roads; this unwillingness may hamper gene flow between populations within a park. Roads can also have other effects, such as facilitating illegal hunting and the movement of invasive species, and increasing the stress levels of resident species.

Park managers can mitigate some of the effects of roads by taking actions such as fencing along the roads where collisions with wildlife are most common, or the removal of particularly damaging roads. Very few park managers, though, actively study how roads influence the species in their park unit, or create plans to mitigate the effects of roads.[39] The ways in which different types of roads fragment habitats in national parks is still an open question.[40] Perhaps surprisingly, there are also few studies on how roads that go through the critical habitat of threatened and endangered species influence those species. For managers to make better decisions regarding how to mitigate the effects of roads on ecosystems, additional research on roads in the national parks is needed.

Land Use outside of Parks

Human disturbances inside national parks are not the only factor that determines the state of natural resources inside those parks. Human disturbances outside of a park can also affect the environment inside the park. Many national parks were created for their scenic beauty or recreational opportunities, not to protect ecosystems. As a consequence, park boundaries rarely follow ecosystem boundaries.[41] This means that human land uses outside of park boundaries can influence the species and ecosystems inside of parks.

In the past the national parks were often surrounded by agricultural land or undeveloped land, but much of this land is now being converted to residential developments. The increase in human population density and housing may lead to the loss of top predators that require large ranges that often go beyond park boundaries. Increasing housing density may also decrease connectivity with other protected areas. Former connections between parks and protected areas may be lost as land is developed for housing. Additionally, rising human population density increases both air and water pollution outside of parks, and this pollution may frequently move into the parks. Finally, increasing population density outside of parks may swell the introduction of nonnative species into parks.

The land-use issue is especially important because the land around national park units is changing at a more rapid pace than other areas in the United States.[42] The increase in human population density from 1940 to 2000 was twice as high around national parks as in the rest of the United States. Housing density is also increasing faster around national park units than in the rest of the country. This pattern is understandable, and likely to continue, as many people desire to live near the scenic beauty and recreational opportunities provided by national parks.

It should be noted that the increase in human population density adjacent to national parks is more of a problem for parks in the eastern half of the United States. Many western national parks are surrounded at least in part by public land managed by agencies other than the Park Service. The land surrounding these national parks may be subject to uses such as forestry or grazing, but increased population density directly adjacent to national parks is less of an issue.

Even if other public lands surround a national park, though, the geography of the lands and increases in human density elsewhere may still conspire against species in national parks. For example, the federal government owns 44% of Montana and northern Idaho, constituting 30.2 million hectares (74.6 million acres). There are national park units, as well as land managed by BLM, Forest Service, and FWS. Such extensive land holdings would seem to provide excellent protection to the species living on those public lands. However, most of the publicly owned land is located at high elevations in the mountains, while the land in the valleys is privately owned. The human population in those valleys is growing rapidly, and there are already 16,000 kilometers (9,942 miles) of roads in the area, running primarily through the valleys.

A study modeled 105 hypothetical species dispersing through this combination of public and private lands.[43] As might be expected, species that primarily use higher-elevation habitat are better protected than species that use lower elevations where more private lands occur. However, the study showed that movement between federal public lands was severely constrained for many species because individuals had to move through valleys to reach other public lands, and that movement was made more difficult by human development and roads. As a result, while the government owns considerable land in the northern Rocky Mountains, the pattern of ownership results in considerable habitat fragmentation. The study also suggested that no species was well protected by national park units alone. All relied on public lands managed by other agencies to at least some extent.

Species living in national parks and relying on other federal public land for persistence can run into problems because, as discussed below, the Forest Service and BLM place less emphasis on the protection of species than does the Park Service. There is an obvious solution: annex parts of the other federal public lands and add them to the National Park System.[44] Congress could move federal public lands from other agencies and add them to the existing national park units. Of course, commercial interests that rely on Forest Service or BLM lands for mineral or other resources would oppose such an annexation, making annexation politically difficult.

For national parks surrounded predominantly by private lands, or for species that move through private lands, a different strategy is required. Potentially the best way to counter increasing population and housing densities around national parks is for conservation groups to purchase private land before it is converted into residential developments.[45] The conservation groups may then create conservation easements on the private land. This would keep the land undeveloped in perpetuity. This strategy may be especially important if an easement can be used to connect two or more protected areas, thereby creating corridors for the movement of species between protected areas. Conservation easements are discussed in the next chapter.

Climate Change

The gravest threat facing the National Park System is global climate change. A 2007 report to Congress by the Government Accountability Office (GAO), however, found that Park Service general management plans and park managers have done little to address the threats posed by climate change.[46] If this lack of planning were confined to the Park Service, it would be regrettable but perhaps easy to fix. However, the GAO report found a lack of planning for climate change in all the agencies that manage federal public lands. After interviewing the managers of several public lands, the GAO found several reasons for the lack of planning. Managers and management plans tend to focus on near-term problems such as wildfires, or administrative tasks such as approving permits. Focusing on near-term problems leaves little time to consider long-term problems such as climate change. The managers also stated that there are often few baseline data for the species on federal lands. This means that it is difficult to determine whether the population sizes and ranges of species are changing over time. Finally, the managers reported that they did not have local- or regional-scale models predicting how climate change will affect the lands they manage. There appears to be a desperate need for high-resolution computer simulations showing the impact of climate change on the ecosystems in the public lands. As a consequence of all these factors, the GAO report concluded that managers felt that they could not begin to plan for climate change because they did not know what to plan for.

National Forest System

If management of the National Park System is made difficult by having a dual mandate for natural resource protection and recreation, then pity the Forest Service and BLM for both having multiple-use mandates. They must manage their lands for uses including recreation, commercial interests, and natural resource protection. The statutes controlling the management of Forest Service lands are considered first.

The National Forest System began to take shape in 1891 when Congress gave the president authority to reserve forested areas from settlement.[47] Created in 1905, the Forest Service had as its primary concern the management of the national forests to create a steady supply of timber for harvesting. The Weeks Act of 1911 appropriated federal funds to purchase forested lands in the watersheds of major rivers as a way to protect those rivers. Consequently, the lands the Forest Service manages are also meant to help protect waterways in the United States. Today, the National Forest System comprises 155 national forests and 20 national grasslands, totaling 78 million hectares (193 million acres). For comparison, California and Montana together cover approximately 80 million hectares (198 million acres).

The guiding principle of the National Forest System is multiple-use, which is codified in the Multiple-Use, Sustained-Yield Act of 1960. The act requires that the Forest Service manage forests to protect five uses: (1) outdoor recreation, (2) range, (3) timber, (4) watersheds, and (5) wildlife and fish.[48] The act defines "multiple use" as management of all the renewable surface resources of the national forests so that they are "utilized in a combination that best meets the needs of the American people."[49] "Sustained yield" is defined as "maintenance in perpetuity" of a high-level output of the renewable resources in the national forests without damage to the productivity of the land.[50] The act does not require that every acre of national forestlands be managed for all five uses.

Critics have argued that the principle of multiple use is little more than a vague ideal that does little to guide management of the national forests. The multiple-use principle could conceivably allow the Forest Service to manage the forests however it wants, and claim the resulting decisions are a product of multiple-use management.

To help remedy the vagueness of the Multiple-Use, Sustained-Yield Act, in 1976 Congress passed the National Forest Management Act (NFMA). Limitations and guidance on how the Forest Service must manage the national forests are clearer in the NFMA. The NFMA requires the Forest Service to create land and resource management plans (forest plans) for each unit of the National Forest System.[51] The forest plans must provide for multiple use and sustained yield by coordinating recreation, range, timber, watershed, wildlife, fish, and wilderness.[52] Forest plans must be revised when conditions in a unit significantly change, or at least every 15 years.[53]

Importantly, the NFMA requires that forest plans protect the "diversity of plant and animal communities" on forestlands.[54] Additionally, each forest plan must preserve, "where appropriate" and "to the degree practicable," the diversity of tree species in a national forest similar to that existing in the region.[55]

In practice, the protection-of-diversity requirement often means that the Forest Service manages forests to create a diversity of habitats, assuming that they will produce a diversity of species. Recall from chapter 1 the discussion of the *Marita* case.[56] In *Marita*, the Forest Service wrote forest plans for the Nicolet and Chequamegon National Forests that proposed to protect diversity by creating many different habitat types. The problem was that the habitats that were to be created were small and extremely fragmented because of plans to build roads and allow logging. The Sierra Club sued the Forest Service, claiming that the agency had ignored the principles of conservation biology in drafting its forest plans, and therefore was not adequately protecting diversity as required by the NFMA. A federal appellate court disagreed with the Sierra Club and ruled that the Nicolet and Chequamegon forest plans were sufficient to protect diversity. The court also wrote that the Forest Service did not have to apply the principles of conservation biology when drafting its forest plans, and that the service was "entitled to use its own methodology, unless it is irrational."[57]

Part of the reason the Forest Service did not include any conservation biology principles in its plans for the Nicolet and Chequamegon forests was that the regulations that controlled those plans were promulgated in 1982. In fact, all the forest plans that are currently in place were written according to the requirements of the 1982 regulations. The Forest Service promulgated new regulations controlling the creation of forest plans in 2005, but the regulations were invalidated by a federal district court.[58] The same thing happened to another set of regulations the agency promulgated in 2008.[59] The 2005 and 2008 regulations attempted to water down the environmental protections required in forest plans written under the 1982 regulations.[60] As a result of the action by the district court, no forest plans were written under the 2005 or 2008 regulations. The Forest Service finally promulgated regulations in 2012 that (so far) appear to have passed judicial review. The 2012 regulations are set to meaningfully change how the Forest Service writes forest plans, and thereby how it manages diversity in the National Forest System.

The 2012 regulations maintain the environmental protections of the 1982 regulations, but differ in substantial ways from the older regulations. The 1982 regulations relied on management indicator species (MIS) to fulfill the requirement of the NFMA to protect diversity. The MIS were a set of species that the Forest Service believed would indicate the viability of all other vertebrate populations in a planning area. As long as the MIS in a forest were doing well, the Forest Service assumed that all the other species in the forest were also doing well. The Forest Service took this a step further, though, and argued that it could monitor the condition of the habitat that the MIS used as an indicator of the condition of the MIS. As a consequence, habitat was an indicator of an MIS, and an MIS was an indicator of diversity (often called a "proxy-on-proxy" approach).

Not surprisingly, there were significant problems in using a proxy-on-proxy approach to managing diversity. For example, a forest plan for the Beaverhead-Deerlodge National Forest named sage grouse as an MIS for the rangeland part of the national forest. The Forest Service then monitored sagebrush habitat as an indicator of the sage grouse. The only problem was that although sagebrush habitat was abundant, there were no actual sage grouse in the Beaverhead-Deerlodge National Forest. In a rather obvious 2010 decision, a federal appeals court held that sagebrush habitat was an inadequate proxy for sage grouse and thus for vertebrate diversity in Beaverhead-Deerlodge National Forest.[61]

In promulgating the new 2012 regulations, the Forest Service admitted that the MIS approach was not sufficient for managing species diversity, and was not based on good science.[62] The new regulations state that the Forest Service will now monitor "focal species." The regulations define focal species as a small subset of species in an area whose status allows an inference on the condition of the larger ecosystem.[63] The species to be chosen as focal species are those that play a functional role in the ecosystem. A species may also be chosen as a focal species because it is thought to be particularly sensitive to potential threats to the ecosystem. The Forest Service assumes that if an ecosystem is functioning well, the diversity requirement of the NFMA will be met.

Although moving from an MIS and proxy-on-proxy approach to a focal species approach appears to be a step in the right direction, some ecologists argue that relying on focal species is also a flawed management method.[64] There will rarely be enough long-term studies to determine which species are the best focal species, so choosing focal species may often be little more than an educated guess. As a result, the focal species that are chosen may not do a good job of representing the entire ecosystem. Additionally, management actions to improve the condition of the focal species may harm other species that the focal species do not adequately represent. The focal species approach will be most beneficial if focal species are only one factor the Forest Service considers in making management decisions, and are part of a broader understanding of how levels of connectivity, natural and anthropogenic disturbance, and landscape heterogeneity interact to affect national forest ecosystems.[65]

Despite the ruling in the *Marita* case, the 2012 regulations actually incorporate several conservation biology principles into the requirements for forest planning. The new regulations require that forest plans include guidelines for maintaining or restoring the "ecological integrity of terrestrial and aquatic ecosystems and watersheds in the plan area."[66] Forest plans must also maintain or restore the "structure, function, composition, and connectivity" of ecosystems in the plan area. Finally, forest plans must maintain or restore the "diversity of ecosystems and habitat types throughout the plan area." Reflecting the scientific understanding gained between 1982 and 2012, the new regulations also require the Forest Service to consider the processes of succession, the impact of invasive species, and global climate change when writing forest plans. Although the *Marita* case allowed the Forest Service to ignore the principles of conservation biology when writing forest plans, the Forest Service has now voluntarily integrated those principles into its new planning regulations.

One of the most important issues each forest plan must confront is the extent and type of timber harvesting the Forest Service will allow in the national forest units. The chapter now turns to commercial logging on national forestlands.

Commercial Logging

There are numerous methods for harvesting timber, each with its own goal for what the forest should look like after the harvesting is done. The two classes that are most important to the Forest Service are regeneration methods and variable retention methods.

Regeneration methods are mainly concerned with how new trees will grow in the forest stand after logging. For these methods, the age classes of trees that are produced after logging are of considerable importance. Regeneration methods that result in a forest stand where all the trees are the same age after logging are called even-aged methods. Clear-cutting is the main even-aged method. In clear-cutting, all the trees in a stand are cut, so all the trees that recolonize that stand will be the same age. Regeneration methods that result in stands with two or more different age classes are called uneven-aged methods.

Clear-cutting is a favored form of harvesting by commercial loggers because it is economically efficient. Removing all the tress in one stand is more efficient than taking only a percentage of the trees from several different stands. Also, the trees that grow after the clear-cut are all the same age, making harvesting of the stand more efficient the next time.

As one can imagine, though, clear-cutting has dramatic effects on a forest stand.[67] Most obviously, clear-cutting changes the physical structure of the forest, destroying habitat for the species that rely on that structure. The loss of structure also allows increases of 15 to 50% in precipitation hitting the forest floor, and increases in temperature on the forest floor of up to 30°C. Such changes can often lead to considerable plant mortality. Other effects of clear-cutting include increasing fragmentation and increasing the percentage of standing trees that are on the edge of the forest. Harvesting trees promotes soil erosion and removes much of the nutrients contained within the harvested trees, disrupting nutrient cycling. Also, commercial logging often requires the construction of roads to reach and transport timber, the ecological effects of which are discussed in the section below on the Roadless Rule. Finally, clear-cutting tends to promote the growth of hardwood species, while reducing the growth of conifer species. Hardwood species often sprout faster after disturbance than relatively slow-growing conifer seedlings. This can convert a conifer forest that has been clear-cut into a hardwood forest.

The other class of harvesting methods is variable retention methods. These methods are primarily concerned with retaining structural elements of the stand (i.e., trees) after logging. By retaining trees after logging, these methods aim to achieve ecological goals such as minimizing losses of species from the stand. Green-tree retention cutting is the main variable retention method. In green-tree retention, living trees are left in a stand after logging to promote the retention of species and to increase the rate of recovery in the stand.

Green-tree retention cutting began being used in forests around 30 years ago. It lessens the drastic ecological effects of clear-cutting through three main mechanisms.[68] First, the trees remaining after green-tree retention act as a lifeboat for species during regeneration of the forest. The living trees provide habitat for species that will colonize the rest of the stand. Second, the living trees help retain some of the physical structure and function of the forest ecosystem. For example, there is less loss of nitrogen and better nitrogen cycling after green-tree retention cutting, compared to clear-cutting.[69] Third, the remaining trees promote landscape connectivity. Species that refuse to move through an area that has been clear-cut may move through an area still containing some living trees.

There is some overlap between clear-cutting and green-tree retention. Clear-cutting often leaves some mature trees to produce seeds to restore the cut areas. The trees left standing in clear-cutting, however, are meant to regenerate the forest. In green-tree retention, the trees left standing are meant to retain some of the structure of the logged forest.

The NFMA, and through it the Forest Service, determine where commercial logging, and what method of logging, may occur in the National Forest System. After World War II when demand for timber increased, the Forest Service increasingly allowed clear-cutting on national forestlands.[70] Stopping clear-cutting soon became a primary goal for several environmental organizations. In 1975 a federal appeals court affirmed a lower-court decision prohibiting clear-cutting in the Monongahela National Forest in West Virginia.[71] This seemed to be the break the environmental organizations had been hoping for, with a ban on clear-cutting in national forests a real possibility. In response to the court decision, however, Congress quickly passed the NFMA, which specifically allows clear-cutting to continue in the National Forest System. The NFMA, however, does place several limits on commercial logging.

The NFMA requires the Forest Service to identify in its forest plans those lands that are not suitable for timber production, considering "physical, economic, and other pertinent factors."[72] Lands not suitable for timber production must be reviewed every ten years, with land returned to timber production whenever conditions have changed enough to make it suitable for harvesting. The NFMA states that forest plans may allow harvesting of timber only where soil, slope, or other watershed conditions will not be irreversibly damaged.[73] Harvesting may also occur only if the land can be adequately restocked within five years after harvest. Additionally, where timber harvests are likely to seriously affect water conditions or fish habitat, there must be protection from detrimental changes in water temperature, blockages of watercourses, and deposits of sediment. Finally, the NFMA requires that the Forest Service limit the sale of timber from each national forest to the amount equal to or less than the quantity that can be removed from the forest annually in perpetuity.[74] This is another way of saying that there must be sustained yield of timber.

The NFMA places some restrictions on the harvesting methods that may be used. The act states that the harvesting system prescribed in a forest plan must not be selected primarily because it will produce the greatest profit or number of harvested trees.[75] The act also announces that clear-cutting may be used only when the Forest Service determines it to be the "optimum method."[76] Clear-cutting and the other even-aged cutting systems may be used only if the "potential environmental, biological, esthetic, engineering, and economic" effects have been considered. Finally, the cut areas must be blended to the extent practicable with the natural terrain.

One of the most important, and most fought over, forest plans that specifies where and what type of commercial logging may occur is the Northwest Forest Plan. It describes how the Forest Service and BLM will manage the forests on their lands in the Pacific Northwest. The Northwest Forest Plan is significant because it was the compromise plan meant to help protect the habitat of the northern spotted owl.[77] The Northwest Forest Plan is also significant because it requires the green-tree retention method on large areas of national forestlands.

To implement green-tree retention cutting, someone must first decide what percentage of trees to retain, and whether to group the remaining trees into islands or disperse them throughout the harvested area. Grouping trees promotes the lifeboat effect discussed above, but dispersing trees promotes forest connectivity. Another consideration is that retaining trees in a dispersed pattern makes it more difficult to harvest the other trees in the stand, and therefore leads to more mechanical damage to remaining trees than leaving trees in groups.[78] The Northwest Forest Plan requires that at least 15% of green trees be retained on timber harvesting units in national forest land. Of the retained trees, 70% must be in groups of 0.2–1.0 hectares (0.5–2.5 acres), with the rest dispersed throughout the harvested area.

Subsequent research suggests that the 15% requirement for green trees may not be sufficient to maintain forest structure and function. In a long-term study of green-tree retention cutting in Pacific Northwest forests, researchers created plots where 15% of trees were retained in harvested areas, and other plots where 40% were retained. Half the plots had trees left in groups, with the other half containing trees dispersed throughout the harvested area.[79] Ten years after the start of the experiment, there appeared to be little or no benefit to retaining only 15% of trees. The few remaining trees were unable to maintain the light, temperature, and humidity microclimate that are necessary for many forest species. Additionally, the isolated trees were more likely to be knocked over by wind, undoing any structural benefit that comes from retaining living trees. Conversely, retaining 40% of trees helped mitigate many of the effects of logging. As a consequence, there seems to be little reason to retain 15% of trees in a harvested area instead of simply clear-cutting the area. Grouping versus dispersing retained trees also affected the resulting forest ecosystem. When trees were aggregated into groups of one hectare, they retained most forest-dependent plant species, although with considerable edge effects. However, natural regeneration was greater when retained trees were dispersed.[80] Dispersed trees were better at getting seeds to the harvested areas than grouped trees.

The results of this long-term research suggest that for green-tree retention cutting to work, the Northwest Forest Plan should be amended to require retention of more than 15% of trees in a stand. Additionally, the research suggests that the plan should allow greater flexibility in grouping or dispersing retained trees, depending on whether the objective is maintaining forest species or ensuring fast regeneration.

To end this section, it is worth thinking about the larger considerations of logging in the national forests. The NFMA and the Multiple-Use, Sustained-Yield Act have built into their structure the assumption that commercial logging in the National Forest System is beneficial for the American people. Many ecologists hold the opposite view. In 2002 a letter to President Bush signed by 220 scientists called for an end to logging in the national forests. The scientists argued that commercial logging greatly damages forest ecosystems, including the other ecosystem services that the forests provide, such as fish and wildlife, clean water, and recreation opportunities. Commercial logging accounts for only 2.7% of the value of ecosystem services derived from national forests, while fish, wildlife, and recreation account for 84.6%. They also pointed out that the national forests provide only 4% of the nation's timber. The scientists concluded that it does not make sense to endanger all the other ecosystem services provided by national forests by allowing commercial logging to continue.

Roadless Rule

The Forest Service has set a large swath of the National Forest System off-limits to commercial logging. In 2001 the Forest Service issued a rule prohibiting timber harvesting and the construction of roads in national forest roadless areas (known as the Roadless Rule).[81] There are 23.7 million hectares (58.5 million acres) in the National Forest System classified as roadless areas, which is approximately one-third of all national forest lands. Although roads and logging are not allowed in roadless areas, motorized vehicles may still be permitted in such areas. Additionally, grazing and oil and gas development may also be permitted in roadless areas.

The Forest Service issued the Roadless Rule on January 12, 2001, eight days before the end of President Clinton's second term.[82] After taking office, the Bush administration decided not to implement the new rule, instead issuing its own rule for managing roadless areas. The new rule allowed each state to petition the federal government for permission to create a management plan for the roadless areas in national forests in that state. Management plans for individual states were assumed to be a way to allow states to open up roadless areas for commercial development.[83] A federal appeals court in 2009, however, held that the Bush administration rule had violated NEPA by not considering how the new rule would impact the environment.[84] The court reinstated the Roadless Rule, and the rule is still in effect.

Although roadless areas may face some anthropogenic pressures such as motorized vehicle use or grazing, the ecosystems in those areas should remain fairly intact. As a result, roadless areas may act as species reserves within national forest units. However, there do not appear to be any long-term ecological studies examining in what ways roadless areas in national forests influence the other areas in national forests, or in surrounding ecosystems.

Healthy Forests Restoration Act

While deciding what type of logging to allow in the national forests and where are important concerns of the Forest Service, perhaps its greatest concern is wildfires. In fiscal year 1991 the Forest Service spent 13% of its budget on wildfire-related activities; by 2012 that number was 40%.[85] The Forest Service spends a huge amount of time and money trying to prevent, and often fighting, wildfires on the national forest lands.

For most of the last 100 years, the official policy for dealing with wildfires was to suppress them. Suppression, though, often led to the buildup of fuels in forests, such as dense tree stands, understory vegetation, and dead trees. The buildup of fuels creates the potential for catastrophic wildfires. At the same time, as for national parks, people increasingly chose to live near national forests.[86]

The combination of a potential for catastrophic wildfires on national forest lands and an increase in people living near those forests prompted Congress to pass the Healthy Forests Restoration Act in 2003. The purpose of the act is to reduce wildfire risk to communities, address the impact of insect infestations on forests, and protect and restore forest ecosystems.[87] The act attempts to reduce wildfires by reducing fuel on federal lands; reducing fuel is also intended to restore forest ecosystems to historic conditions. Additionally, communities that are at the "wildland-urban interface" with federal lands at high risk of wildfires are instructed to create a community wildfire protection plan. The plans are intended to identify areas for hazardous-fuel reduction treatments on federal public lands.[88] The act applies to both the National Forest System and BLM lands.

The two main goals of the Healthy Forest Restoration Act, reducing wildfire risk and restoring forest ecosystems to their historic conditions, are often in direct conflict. Removal of hazardous fuel in forests is usually achieved through two activities: harvesting trees with trunks below a certain diameter, and prescribed burns. Forest managers often claim that reducing fuels will also help restore forest ecosystems. The argument goes that forests have evolved in concert with frequent wildfires, and that logging and prescribed burns can be used instead of wildfire to replicate the ecological effects of such fires.

This argument rests primarily on studies of ponderosa pine forests in the southwest United States.[89] These ponderosa pine forests historically had low-intensity surface fires every two to twenty years. Surface fires kill low-lying vegetation and small trees, but rarely kill large trees. Conversely, crown fires are more intense fires that may kill most or all large trees in a stand. The research on the ponderosa pine forests in the Southwest showed that fire suppression increased the stand density of trees and enhanced the probability of severe crown fires. Thinning trees in such a forest reduced the likelihood of intense crown fires, while helping restore the forest to its historic structure.

Other forests are different, though. In the ponderosa pine ecosystem of the Colorado Front Range, for example, severe crown fires occurred naturally before fire suppression began, and dense stands of ponderosa pine (*Pinus ponderosa*) and Douglas fir (*Pseudotsuga menziesii*) were common.[90] Unlike in the dry ponderosa pine forests of the Southwest, dense stands of trees in the Front Range do not necessarily indicate a forest that differs from its historic condition. A study of ponderosa pine and Douglas fir forests in Boulder County, Colorado, examined the percentage of forest that would need to be thinned to reduce wildfire risk, and the percentage of forest that would need to be thinned to restore the forest to historic conditions.[91] The study found that nearly 50% of national forest land in Boulder County would need to be thinned to reduce the risk of intense wildfires, but only 6% would need to be thinned to return the forest to its historic condition. Different research suggests that subalpine forests composed of Engelmann spruce (*Picea engelmannii*), subalpine fir (*Abies lasiocarpa*), and lodgepole pine (*Pinus contorta*) historically burned at intervals greater than a century, and that those fires were high-intensity crown fires that killed most trees.[92] The time between fires in these forests is longer than current fire-suppression practices.

As the above research suggests, the occasional high-intensity crown fire may be exactly what is necessary to maintain the historic condition of many forests. This is precisely what the Healthy Forest Restoration Act is attempting to prevent to protect the people and houses that are near the national forests.

A logical suggestion would be that instead of merely thinning such forests, clear-cutting or green-tree retention cutting could replicate the intense crown fires that kill all or most of the trees in a stand. Unfortunately, clear-cutting and green-tree retention do not do a good job of replicating an intense wildfire.[93] Much more of the structure of the forest remains after a wildfire than after logging, and this structure is important in helping regenerate forest habitats. Additionally, many tree species have evolved to take advantage of wildfires. For example, lodgepole pinecones open and release their seeds only after being exposed to high temperatures. By waiting for a wildfire before releasing its seeds, lodgepole pine can take advantage of the reduced competition from adult trees that were killed by the fire. If there is no fire, only clear-cutting, then lodgepole pinecones would not open.

Perhaps surprisingly, clear-cutting and then prescribed burns appear even worse than clear-cutting alone at producing a forest ecosystem similar to the one that develops after a wildfire.[94] Having two disturbances in succession seems to alter the forest ecosystem to a greater extent than clear-cutting or wildfire alone. The research on the inability of clear-cutting and prescribed burns to replicate crown fires is still in the early stages, though, and may prove to be a valid technique for some forests. Long-term studies are needed that examine how well green-tree retention cutting followed by prescribed burning does at replicating naturally occurring wildfires.

At present there does not seem to be a good alternative to intense wildfires in maintaining the structure of some forest ecosystems. Of course, intense wildfires may imperil humans living near such forests, and so the Forest Service does not want to increase the possibility of such wildfires. Forest Service managers and local communities creating fire management plans have no easy task in balancing a reduction in the risk of severe wildfires with allowing forests to have a natural fire regime.

BLM Lands

Although the above discussion of the National Forest System would seem to suggest that the Forest Service manages nothing but forests, that is not true the National Forest System includes considerable rangelands. In fact, in 2004 the Forest Service issued permits to allow grazing on almost half of national forest lands.[95] As remarkable as this is, grazing on BLM lands is even more impressive, occurring on almost two-thirds of its lands. This section delves into grazing on BLM lands, but first it discusses the statutes controlling BLM.

There are roughly 99 million hectares (245 million acres) of federal public lands managed by BLM, making BLM the largest manager of federal public lands. For comparison, the state of Texas covers approximately 69 million hectares (172 million acres). The primary statute controlling the BLM is the Federal Land Policy and Management Act (FLPMA) of 1976. The

FLPMA is similar to the NFMA and the Multiple-Use, Sustained-Yield Act in that it requires BLM to manage federal public lands by applying the principles of multiple use and sustained yield.[96] Under the FLPMA, "multiple use" means managing the lands so that they are "utilized in the combination that will best meet the present and future needs of the American people"; "sustained yield" means "maintenance in perpetuity of a high-level annual or regular periodic output of the various renewable resources of the public lands consistent with multiple use."[97] The FLPMA defines the multiple uses of the BLM lands as including "recreation, range, timber, minerals, watershed, wildlife and fish, and natural scenic, scientific and historical values."[98]

The FLPMA also requires land-use planning for BLM lands, similar to the requirement for forest plans in the National Forest System.[99] The plans must achieve "integrated consideration" of physical, biological, economic, and other sciences.[100] The plans must also give priority to designation and protection of areas of critical environmental concern. The FLPMA defines "areas of critical environmental concern" as places where special management attention is necessary to prevent irreparable damage to important historic or scenic values, fish and wildlife resources, other natural systems, or human life.[101] Despite the requirements just listed, the FLPMA is not nearly as specific as the NFMA in detailing what each management plan must contain. Also, there is no specific requirement for BLM to maintain the diversity of species on BLM lands. As a consequence, BLM does much less formal land-use planning than the Forest Service.[102]

Complicating management of many tracts of BLM land is the presence of wild horses and burros. In the past, the animals were simply killed so that they would not compete with domestic livestock grazing on BLM lands. In response to public pressure to stop the killing, Congress passed the Wild, Free-Roaming Horses and Burros Act in 1971. The act directs BLM and the Forest Service to manage wild horses and burros on BLM lands and the National Forest System, respectively. The act requires that the horses and burros be managed to maintain a "thriving natural ecological balance" on the public lands.[103] To meet this goal, the agencies may designate specific ranges on the public lands as sanctuaries for wild horse and burro protection.[104] The act does allow BLM and the Forest Service to remove excess horses or burros and put them up for adoption or sale, or to kill them.[105] The act defines "excess animals" as horses or burros that must be removed from an area to preserve and maintain a "natural ecological balance" and multiple-use relationship in the area.[106] If the horses or burros stray onto private land, the act prohibits the landowner from killing the animals.[107]

Any person who removes a wild horse or burro from public lands, or kills or harasses such an animal, may face fines or imprisonment.[108]

Grazing

The presence of wild horses and burros complicates management of federal public lands because livestock grazing is the single greatest commercial use of such lands.[109] A considerable amount of BLM lands consists of rangeland, and much of it is available for livestock grazing. BLM allows livestock grazing on 62.7 million hectares (155 million acres), which is more than 63% of BLM lands.

Livestock grazing on federal public lands was virtually unregulated before 1934. This led to intensive overgrazing on large areas of BLM lands. In 1934 Congress passed the Taylor Grazing Act to regulate grazing on federal lands. The act allowed the federal government to issue permits for grazing in return for a fee.

The FLPMA currently controls grazing on both BLM lands and Forest Service lands. It authorizes BLM to grant permits to individuals to graze domestic livestock in return for a grazing fee.[110] Most of the permits BLM issues are for cattle and sheep grazing. Permits usually last for ten years, but in practice they last much longer because the holder of an expiring permit is given first priority for a new grazing permit.[111] The FLPMA requires BLM to specify the number of animals that can graze an area, and the seasons when grazing is allowed.[112] The act goes on to state that if BLM finds that the condition of a rangeland has deteriorated, BLM may decrease the allowable number of animals grazing on that rangeland.

As might be expected, grazing by domestic animals can significantly deteriorate rangeland. Cattle prefer to graze in riparian areas, especially during the hot summer months.[113] This grazing can result in the loss of riparian vegetation, which allows erosion of soil into the stream or river, and increases the temperature of the water. The result may be the loss of habitat and aquatic species.

Additionally, livestock grazing promotes the spread of woody species through three different ecological processes. First, cattle prefer to graze on grass species, while woody species are often unpalatable. Intensive grazing on grass species allows the establishment of woody species, with the grass species no longer able to competitively exclude the woody species. Second, livestock grazing displaces native herbivores, such as prairie dogs, that help limit the growth of woody species. Third, in many undisturbed grassland ecosystems, periodic wildfires kill woody species, allowing grass species to remain dominant. Just as for forests, however, rangeland wildfires were suppressed by federal agencies beginning in the early part of the 1900s. In addition, intensive grazing reduces the fuel necessary for wildfires to spread over large areas. As a consequence of the synergistic effects of these three processes, intensive grazing on rangeland has allowed the expansion of woody species into what were formerly grasslands.[114] Now that woodlands have been established where grasslands formerly stood, prohibiting continued grazing will likely not reverse the process. The ecosystems have crossed a threshold, making it extremely difficult to return the woodlands to their former state as grasslands.

BLM regulations create guidelines for grazing that are meant to limit the potential damages described above. The regulations require that grazing standards consider watershed function, nutrient cycling and energy flow, water quality, habitat for endangered and threatened species, and habitat quality for native plant and animal populations.[115] The regulations further require that grazing guidelines must maintain "adequate amounts of vegetative ground cover" to promote soil moisture storage and to stabilize soils. The guidelines must maintain or restore riparian-wetland functions, and maintain the appropriate species to maintain the hydrologic cycle, nutrient cycle, and energy flow. They must promote the opportunity for seedling establishment of appropriate plant species, and maintain the physical and biological conditions to sustain native populations and communities, while emphasizing native species "in the support of ecological function." The regulations also state that season-long livestock grazing is allowed only when it is consistent with achieving "healthy, properly functioning ecosystems." To meet these guidelines, activities such as vegetation manipulation, water development, and reductions in grazing allotments are allowed.[116]

Although current BLM regulations do help protect rangeland ecosystems, the damage from overgrazing in the past has already extensively degraded many rangelands. Unfortunately, reducing livestock grazing now would likely do little to repair the damage from the previous overgrazing. This can be best understood by examining the most crushing legacy of overgrazing on BLM lands: cheatgrass.

Cheatgrass (*Bromus tectorum*) is an invasive annual grass that has become widespread on BLM lands in the western United States (figure 5.2). Cheatgrass invaded after intensive grazing by domestic livestock. Cattle prefer to graze on native perennial species, reducing those species and allowing cheatgrass to spread. Once established, cheatgrass can outcompete many native perennial species by producing more seeds and germinating earlier

FIGURE 5.2. Cheatgrass invading a rangeland. Photograph from USDA/Jaepil Cho.

in the year. Cheatgrass is most problematic in the arid region of the Great Basin, stretching from the Sierra Nevada in California to the Rocky Mountains in Utah.

In a twist on the ecological processes allowing the spread of woody species, cheatgrass is *more* likely to support wildfires than native species are. More frequent wildfires actually work to spread cheatgrass. For example, sagebrush ecosystems often burn every 32 to 70 years. Ecosystems dominated by cheatgrass may burn every 5 years if not suppressed.[117] The wildfires fueled by cheatgrass also tend to be more intense and larger. Such frequent fire return times and intense fires place native species at a disadvantage to cheatgrass, further increasing its dominance.

Invasion by cheatgrass reduces rangeland productivity and the diversity of rangeland species. Cheatgrass has a solid claim to being the most disruptive invasive plant in the United States.[118] Many hectares of rangeland have become virtual monocultures, growing almost nothing but cheatgrass. While considerable research has been done to stop the spread of cheatgrass and restore native vegetation, so far it has not been successful.

Simply stopping livestock grazing on rangelands where cheatgrass has spread would likely do little good. A study examined 16 exclosure sites on public lands in the Great Basin that were established from 1936 to 1939 in response to the Taylor Grazing Act.[119] The study sites had been exposed to intensive livestock grazing, but after being established, the exclosures kept out all livestock. The land around the exclosures continued to be moderately grazed by livestock. A study of the exclosures and surrounding land in 2001, 62 years after the exclosures were created, found that total vegetation cover was slightly greater inside the exclosures while species richness was slightly greater outside the exclosures. More importantly, there was no difference inside or outside the exclosures in cheatgrass cover. While exclosures are not a return to the original dynamics of the rangeland, as they exclude larger herbivores that would naturally graze there, they suggest that eliminating livestock grazing on BLM lands in the Great Basin will not by itself reverse the spread of cheatgrass. A separate simulation model of cattle grazing on BLM lands bolstered this conclusion by showing that prohibiting grazing would do little to help convert rangelands infested with cheatgrass back to natural grasslands.[120] The model did suggest, however, that prohibiting grazing would slow the conversion of grasslands to woodlands. The message in these studies is that the intensive grazing of the past may have created new stable rangeland communities dominated by cheatgrass that will be extremely difficult to restore.

Wildlife Services

Another important effect of livestock grazing is the impact it has on the management of mammalian predators. To protect livestock grazing on rangelands, the federal government engages in large-scale killing of predators. Wildlife Services is a division in the Department of Agriculture's Animal and Plant Health Inspection Service.[121] In 2011 Wildlife Services killed 365 gray wolves, 398 mountain lions, 565 black bears, 1,237 bobcats, 2,404 gray foxes, and 83,195 coyotes.[122] Methods for killing predators include traps, snares, poison, and aerial gunning. The year 2011 was typical, with similar numbers of predators being killed most years. Most of these predators were killed in the belief that doing so would prevent predation on livestock.

Wildlife Services also kills many other mammals and birds that it considers nuisances, totaling 3.7 million organisms in 2011. It should be noted that only a portion of the animals were killed on BLM lands, with many of the killings done on other public or private lands.

Killing so many large predators can have several ecological effects, a significant one being mesopredator release. Mesopredator release occurs when apex predators that are harassing, killing, and eating smaller predators are removed, resulting in an increase in the populations of smaller predator species. When wolves are removed, for example, coyote populations increase; and in areas where coyotes are the top predators and are removed, bobcat, gray fox, badger, skunk, and domestic cat populations increase.[123] The mesopredator species are more likely to prey on birds and even smaller predators, depressing those populations.

Ranchers tend to believe that killing predators is necessary to protect the profits of the livestock industry; however, the evidence for this is equivocal. Studies by Wildlife Services researchers suggest that killing predators is very helpful in protecting the livestock industry. Studies by independent scientists, though, tend to disagree. For example, a study by Berger found that hay prices, wage prices, and lamb prices contributed much more to sheep numbers than did predator control efforts.[124] Either predator control efforts did not work to lower predation on sheep, or all those other factors were much more important in determining sheep numbers. A different study found that gray wolf (*Canis lupus*) predation on livestock in Idaho, Montana, and Wyoming accounted for less than 0.01% of the annual gross income from livestock operations in those states.[125] These studies suggest that killing predators may help reduce predation on livestock, but that the effect is so weak that it does little to influence the productivity of the livestock industry. Unfortunately, the ecological consequences of killing so many predators every year have received very little independent scientific study. Long-term studies on the effectiveness of Wildlife Services' program of killing predators to protect the livestock industry, along with the ecological consequences of such killings, are needed.

Just as for logging, it is worth considering the larger consequences of livestock grazing on federal public lands. A US Government Accountability Office report to Congress found that in 2004 the federal government spent $144 million managing federal public lands for livestock grazing.[126] The fees charged for grazing permits for that year amounted to $21 million. Importantly, the $144 million price tag did not include the damage done by livestock to the other ecosystem services provided by the public rangelands. American taxpayers spend a great deal of money every year subsidizing livestock grazing on the public lands.

National Wildlife Refuge System

While the lands managed by BLM and the Forest Service are multiple-use lands, the lands in the National Wildlife Refuge System are best thought of as dominant-use lands. Recreation, grazing, oil and gas exploration, and other commercial uses may occur on wildlife refuge lands, but they must be compatible with the dominant use of the refuges: protecting plants and animals.[127]

President Theodore Roosevelt created the first national wildlife refuge, Pelican Island in Florida, in 1903 by executive proclamation. Presidents and Congress have created many national wildlife refuges since then. In creating a refuge, the president or Congress often stated that the refuge was designated to protect some specific purpose, such as maintaining a particular population for hunting. Most refuges were created to protect wildlife, but some were created to allow grazing for domestic livestock.

In 1966 Congress passed the National Wildlife Refuge System Administration Act (NWRSAA) to consolidate all the national wildlife refuges under one statute. The act gives FWS responsibility for the National Wildlife Refuge System. Congress later extensively amended the NWRSAA by passing the National Wildlife Refuge System Improvement Act of 1997. In 1980, passage of the Alaska National Interest Lands Conservation Act added 21.7 million hectares (53.7 million acres) to the National Wildlife Refuge System. The refuge System currently consists of 60 million hectares (150 million acres).

The purpose of the National Wildlife Refuge System, as declared in the NWRSAA, is to create a network of lands and waters for the "conservation, management, and where appropriate, restoration of the fish, wildlife, and plant resources and their habitats."[128] The act directs FWS to provide for conservation by ensuring the biological integrity, diversity, and environmental health of the refuges.[129] The act creates a hierarchy of uses for the refuges: (1) conservation, (2) wildlife-dependent recreation, and (3) all other uses.[130] If a conflict exists between this hierarchy and the purpose for which a particular refuge was created by the president or Congress, the specific purpose of the refuge wins.

The NWRSAA defines "wildlife-dependent recreation" as activities such as hunting, fishing, wildlife observation and photography, and environmental education.[131] Wildlife-dependent recreation, though, must not "materially interfere with or detract from" the conservation mission of the refuge system or the purpose of a specific refuge.[132] In fact, the act prohibits any persons from disturbing or possessing any natural growth in a refuge, or taking or possessing any animal, unless permitted under the act.[133] All other human uses, such as livestock grazing, are given a lower priority than wildlife-dependent recreation. FWS regulations state that any economic activity on refuge lands must be compatible with and contribute to the purpose of the wildlife refuge system.[134] The act does specifically mention that mining and mineral leasing may continue on refuge lands.[135] Due to the hierarchy created by the NWRSAA, though, commercial use of wildlife refuge lands is much less of a priority than for BLM or Forest Service lands.

The 1997 amendments to the NWRSAA instructed FWS to create a "comprehensive conservation plan" for each refuge or complex of refuges by 2012.[136] Conservation plans must be revised every 15 years thereafter. In the plans, FWS must identify the problems that may adversely affect the populations and habitats of fish, wildlife, and plants in a given refuge. The plans must then describe the actions necessary to correct or mitigate such problems. The act also requires the conservation plans to outline opportunities for wildlife-dependent recreational uses.

The implementation of the conservation plans will be influenced by a relatively new overarching management program developed by FWS. In 2006 FWS created a program called Strategic Habitat Conservation (SHC), which applies to both the management of wildlife refuges and to the many other programs FWS administers.[137] For wildlife refuges, the idea behind SHC is to pick a handful of focal species in a particular refuge, and then increase or maintain the population sizes of each of those focal species. The belief is that managing the habitat to increase or maintain the populations of focal species will benefit all the species in the refuge. FWS states that it uses SHC because it creates measurable objectives, making the agency more accountable in reaching those objectives.

As with the use of focal species by the Forest Service, the use of focal species under the SHC program may help the population sizes of the chosen focal species, but do little to protect other species. More fundamentally, FWS explicitly states that "the conservation of populations is our mission. Habitat management is a tool."[138] The SHC program cares little for ecosystem management, instead being concerned only with the population sizes of a handful of focal species. Maintaining biodiversity or ecological integrity in a refuge does not matter as long as the focal species meet FWS objectives.[139]

Caring more about certain populations than overall diversity or ecological integrity may become especially problematic as the global climate changes. Managing refuge habitats for a few focal species instead of attempting to mitigate the effects of climate change on entire ecosystems may help the focal species while degrading the habitats for many other species. This makes wildlife refuges exist as refuges for a handful of chosen species, with other species protected only insofar as they can coexist with the focal species. This seems especially short-sighted, as many empirical studies show that increases in biodiversity tend to increase and stabilize ecosystem functioning, and tend to buffer ecosystems against stress.[140] Managing refuge ecosystems to maintain or increase biodiversity may do more to protect all the species in the wildlife refuges than simply trying to increase the populations of a few focal species.

National Wilderness Preservation System

The final federal land system is the National Wilderness Preservation System. Wilderness areas have the tightest statutory restrictions on human use of all the federal land systems. The Wilderness Act of 1964 created the National Wilderness Preservation System, and poetically defines wilderness as "an area where the earth and its community of life are untrammeled by man, where man himself is a visitor who does not remain."[141] The act continues the definition of wilderness as an area of undeveloped federal land "retaining its primeval character and influence, without permanent improvements or human habitation." To qualify as wilderness, an area should meet four requirements: (1) appear to have been "affected primarily by the forces of nature, with the imprint of man's work substantially unnoticeable"; (2) have "outstanding opportunities for solitude"; (3) be at least 2,023 hectares (5,000 acres) in size; and (4) contain important ecological, geological, or other features.[142]

Wilderness areas are lands that were formerly a part of another federal public land system but, because of their characteristics, have been designated as wilderness areas. The Wilderness Act designated 3.6 million hectares (9.1 million acres) as wilderness areas at the time of its passage. The act includes a process for designating more lands as wilderness areas. All the agencies discussed above that manage the federal public lands are required to assess their lands to determine which additional areas would be suitable for designation. They recommend those areas to the president, who then makes a recommendation to Congress. Congress may then pass an act adding those lands to the National Wilderness Preservation System. The federal agency that managed the land before it was designated continues to manage the wilderness area.[143] The passage of ANILCA resulted in the designation of 22.6 million hectares (56 million acres) of wilderness areas in Alaska. In 2009 President Obama signed the Omnibus Public Land Management Act, which added 810,000 hectares (2 million acres) of wilderness area. There are now nearly 44.5 million hectares (110 million acres) in the National Wilderness Preservation System. The Park Service is responsible for the largest share of those hectares, managing 17.8 million hectares (44 million acres).

The Wilderness Act decrees that the agencies managing wilderness areas are responsible for "preserving the wilderness character of the area."[144] The act goes on to state that wilderness areas shall be devoted to recreational, scenic, scientific, educational, conservation, and historical use.[145] The act is very strict in attempting to keep wilderness areas unmodified by humans. It prohibits all commercial enterprise and permanent roads.[146] The act goes on to prohibit, except as necessary to meet the "minimum requirements for the administration of the area," temporary roads, motor vehicles, motorized equipment, and any structure or installation.

There are exceptions in the Wilderness Act allowing for human modification of wilderness areas. For instance, livestock grazing on national forest land, where it occurred prior to passage of the act, is still allowed.[147] Commercial services, such as commercial tour guides, are allowed if they help realize the recreational or wilderness purposes of the area.[148] Power projects and transmission lines are also allowed if authorized by the president. Considerable amounts of private land are surrounded by wilderness areas. The managing agencies usually allow individuals to use vehicles to go through the wilderness areas to reach these inholdings.[149] Additionally, while commercial logging is not allowed in wilderness areas, the managing agencies may remove trees to control fire, insects, and disease.[150] Consequently, while the intent of the Wilderness Act is to preserve certain lands as untouched by humans, in reality many wilderness areas are heavily influenced by anthropogenic activity.

The creation of the National Wilderness Preservation System has long stoked debate about the purpose and meaning of preserves for protecting wildlife: should wilderness areas be actively managed to help return them to a state as close to their historic condition as possible, or should they be completely left alone, allowing nature to run its course?[151] Those who advocate active management argue that the wilderness areas may have faced human modification for many years before being designated, and continue to face damage coming from outside in the forms of pollution, adjacent habitat degradation, and global climate change. As a result, they contend that active management is necessary to counteract those human-caused impacts. Further, the Wilderness Act itself specifically allows temporary roads, motorized vehicles, and structures to be used in the management of wilderness areas.

Those on the other side of the debate respond that management of wilderness areas, even if well-meaning, is still human interference with nature. To actively manage an ecosystem is to make human value judgments on what that ecosystem is supposed to be. They also make the case that managing ecosystems is difficult and prone to unanticipated consequences that may ultimately harm the ecological functions of wilderness areas. Finally, they argue that the point of the Wilderness Act is to keep wilderness areas as free from human interference as possible, and any management of such areas should be the exception, not the rule. As proof of this, they point to the requirement of the act that temporary roads, motor vehicles, and structures may be used only to meet the "minimum requirements" of management.

As the introduction to this chapter suggests, the agencies controlling wilderness areas have been willing to actively manage those areas. Building water tanks for bighorn sheep in the Kofa Wilderness Area is one such example; other examples include attempting to eradicate invasive species, and reintroducing species that were native to a wilderness area but had been lost.[152]

NATIONAL ENVIRONMENTAL POLICY ACT

The National Environmental Policy Act (NEPA) is a fascinating statute—it does not have the power to stop a federal agency from undertaking an action that affects the environment, but it does require agencies to show their thinking in making such decisions, which may by itself be enough to stop many actions.

Congress passed NEPA in 1969. The act requires all agencies of the federal government to write environmental impact statements (EISs) for all "major Federal actions significantly affecting the quality of the human environment."[153] The EIS must include a detailed description of any adverse environmental effects that cannot be avoided if the proposed action is undertaken. Significantly, the EIS must also include alternatives to the proposed action. These alternatives are described in a regulation as "the heart of the environmental impact statement."[154] The EIS must explicitly compare the environmental impacts of the proposed activity and the alternatives, providing a clear basis for a choice among options.[155] One of the alternatives considered must be no action. No action does not necessarily mean that the government does not do anything; no action may simply mean that the government continues doing what it is already doing, thereby maintaining the status quo. Finally, the EIS must include the relationship between short-term uses of the environment and long-term productivity, and any irreversible commitment of resources if the proposed action is undertaken.

NEPA also created the Council on Environmental Quality (CEQ). CEQ promulgates the regulations that determine how NEPA is implemented by federal agencies. The agencies must follow the regulations CEQ promulgates, but CEQ does not have any enforcement authority over the agencies. Enforcement of NEPA comes from citizens suing the government, which is discussed below.

CEQ regulations state that an agency must begin preparing an EIS when an agency develops a proposal for action.[156] That way the EIS can be included in any recommendation on the proposal. As a consequence, agencies do not have to write an EIS for every action they are simply considering, but they must write an EIS before a decision to take action has been made. In writing an EIS, the regulations require agencies to obtain all information relevant to foreseeable significant adverse effects on the environment if the cost of obtaining the information is not exorbitant.[157] CEQ regulations also require an agency to write a supplemental EIS if significant new circumstances arise or information bearing on the proposed action becomes available.[158]

Perhaps most importantly, the CEQ regulations require that an EIS include discussion of both direct and indirect effects of the proposed activity on the environment, as well as consider the cumulative effects of the activity. Direct effects are caused by the action and occur at the same time and place.[159] Indirect effects are caused by the action later in time or farther removed in distance, but are still reasonably foreseeable.[160] Indirect effects include changes in land-use patterns, population density, growth rate, or related effects on other natural systems. Lastly, cumulative effects result from the incremental impact of an action when added to past, present, and reasonably foreseeable future actions, regardless of what agency or person undertakes the action.[161] The regulations specifically mention that direct, indirect, and cumulative effects can include ecological effects, such as the effects on the components, structure, and functioning of ecosystems.[162]

While determining the potential direct and indirect effects of a project is difficult, determining the potential cumulative effects is downright arduous. An analysis of cumulative effects is often extremely important, though, because the direct impacts of a single project may not be particularly large, but the cumulative impact of several small actions over time may drastically affect an ecosystem. In the past, federal agencies often handled the requirement for a cumulative-effects analysis by simply not doing one.[163] After intervention by CEQ and citizen lawsuits, agencies became much better about including an analysis of cumulative effects, although the quality of that analysis is frequently questionable. For instance, the effects of proposed actions on species are often considered in EISs. Because population abundance and distribution data are not available for many species, however, agencies often use habitat as a proxy for population size and distribution.[164] An effect is deemed insignificant if habitat is not extensively affected. Cumulative effects are then determined as the amount of habitat that has been destroyed by past actions and the amount that would be destroyed by the proposed action. As an example, for lynx (*Lynx canadensis*) the Forest Service has decided a project is not allowable if more than 30% of a "lynx analysis unit" has already been degraded by past actions.[165] A lynx analysis unit is not based on where lynx have their actual home ranges, but is instead best thought of as theoretical lynx home ranges. Consequently, as long as less than 30% of a theoretical home range is not destroyed, the Forest Service considers the cumulative effects of a proposed action to be acceptable. This means that the actual abundance and distribution of lynx receive much less consideration.

Perhaps the most important cumulative effect confronting all federal agencies is the release of greenhouse gases that contribute to global climate change. How an agency should determine the cumulative effect of emissions from one of its proposed activities is not at all obvious. So many sources emit greenhouse gases that it is nearly impossible to determine how the specific emissions from a proposed project will impact the environment. A draft guidance document issued by CEQ in 2010 attempted to provide some clarification.[166] The document states that when a proposed project will release more than 25,000 metric tons of carbon dioxide–equivalent greenhouse gases a year, the agency should consider that a direct effect in its EIS. The CEQ document, though, leaves an analysis of the cumulative effects of greenhouse gas emissions up to the agencies. Further, the document explicitly states that the guidance it provides is not meant to apply to agencies that manage federal lands. The ways in which land-management decisions influence climate change, both positively and negatively, are very complex, and CEQ seems to believe that they are beyond the ability of most land managers to forecast.

An EIS is not always required for a proposed agency action. If it is not obvious that a proposed agency action will significantly affect the environment, instead of preparing an EIS, the agency can prepare an environmental assessment (EA).[167] An EA is a concise document analyzing whether an EIS is necessary or not. If the EA concludes that an EIS is not necessary, then the agency will issue a finding of no significant impact.

Writing an EIS is a considerable amount of work, and federal agencies try to get out of writing them whenever possible. One way agencies have tried to get around NEPA is to divide up a proposed action into several smaller actions. Each "segmented" action by itself would not significantly affect the environment, and would therefore not need an EIS, although the entire action may very well require an EIS. Courts have stated that this is allowable only if the segmented actions each have "independent utility."[168] If one segmented action makes no sense without the other segmented actions, then the agency must consider them all as one larger action in an EIS.

All the land-management agencies discussed above are subject to the requirements of NEPA. Preparing land-use plans qualifies as a major federal action requiring an EIS. Not all actions that will significantly affect the environment require an EIS, though. Some statutes, such as the Clean Air Act (discussed in chapter 10), state that an EIS is not necessary before taking action under the statute.

A federal agency may also decide that certain types of actions do not individually or cumulatively have an effect on the environment.[169] The agency may then decide that all proposed activities of that type do not require an EA or EIS. This is called a categorical exclusion. The Forest Service has declared that vegetation management activities that improve timber stands or wildlife habitat fall within the class of categorical exclusions. Between 2003 and 2005 the Forest Service approved nearly 2,200 vegetation management projects on 2.9 million acres of national forest land by claiming categorical exclusion.[170] Some scholars argue that such categorical exclusions allow needed projects to be performed quickly and without wasting time on paperwork. Others argue that each project by itself may very well have little significant environmental impact, but by not writing an EA or EIS, there is no attempt to determine the cumulative effect of all of these projects. NEPA exists to force agencies to consider the wider consequences of their actions, and the Forest Service may not be fulfilling that requirement by assuming that many of its vegetation management projects have little or no cumulative environmental impact.

If a federal agency docs go through all of the work of writing an EIS, the agency is then free to completely ignore its conclusions. NEPA does not require agencies to choose the least environmentally damaging action. As the Supreme Court has made clear, NEPA is a procedural statute, not a substantive one.[171]

The idea behind NEPA is clever, though if agencies are forced to consider all the environmental ramifications of their proposed actions, along with the alternatives, they are likely to make better-informed decisions that do less damage to the environment. As an added benefit, writing an EIS gives the public valuable insight into the impacts of proposed actions and the decision-making process of federal agencies. Such openness may provide citizens and Congress the opportunity to pressure federal agencies into making less environmentally damaging decisions.

The public has two main opportunities to comment on and help shape an EIS. First, when an agency decides to prepare an EIS, it must publish a notice of intent in the *Federal Register*. The agency must then ask for comments by interested parties, including interested citizens, before beginning to write the draft EIS.[172] Second, after an agency has written a draft EIS and before it has prepared the final EIS, the agency must again solicit comments from people or organizations that may be interested in the proposed agency action.[173] The agency must respond to the comments it receives from the public in the final EIS.

Perhaps the most important way NEPA helps protect the environment is that it creates the ability for citizens to sue federal agencies for not properly following its procedural requirements. NEPA may not control the actions agencies ultimately undertake, but those agencies must follow the procedures NEPA sets out. If an agency does not follow those procedures, a citizen or environmental group can sue. The citizen may first have to file an administrative appeal with the agency, but if the citizen's concerns are still not met, he or she may sue in federal court under the Administrative Procedure Act. The court may then stop the agency from undertaking an action before completing an adequate EIS. The time and cost an agency must spend in litigation, combined with the time and cost of preparing an adequate EIS, may persuade an agency that it is simply not worthwhile to undertake a proposed action.

CHAPTER SIX

Private Lands

The city of Knoxville was simply trying to slow urban sprawl. Between 1982 and 1997, the developed area of Tennessee had increased from 607,000 to 970,000 hectares (1.5 million to 2.4 million acres).[1] Most of this development indiscriminately converted agricultural and forestland to residential use, resulting in urban sprawl.

Urban sprawl is low-density single-family housing that haphazardly occurs at the edges of an urban area. Because development is low density, it takes up much more land per dwelling than high-density urban centers. Often sprawl results in leapfrogging, where a section of land is developed, then adjacent land is skipped in favor of developing land farther out in the countryside. As discussed below in greater detail, urban development in general is horrible for natural ecosystems. The roads and buildings and people in urban areas destroy or fragment native habitats, and dramatically alter the ecology within the urban area.

To try to slow its urban sprawl, the state of Tennessee passed the Growth Policy Act of 1998, which required all cities in Tennessee to define an urban growth boundary. An urban growth boundary is a line drawn around a city within which high-density development is allowed, while areas outside of the boundary can still be developed, but only at much lower density. Developers prefer high-density developments because the more houses that can be built in a given area, the greater the profit. Urban growth boundaries are meant to stop urban sprawl by encouraging development to occur in a more compact manner within the growth boundary, thereby preventing the leapfrogging of development out into the countryside. The city of Knoxville set its urban growth boundary in 2001.

And then the sprawl got worse.[2] After Knoxville established its urban growth boundary, urban sprawl in the rural areas outside of the boundary greatly intensified. There were several reasons for the increased sprawl.[3] The urban growth boundary set the line where the city of Knoxville would likely annex (incorporate into the city) new residential developments. Annexation meant paying additional city property taxes, making a house outside of the growth boundary more attractive. Additionally, many people prefer to live in low-density areas, and the urban growth boundary indicated where high-density development would eventually occur, again making housing outside of the boundary attractive. Finally, the city of Knoxville controlled development within its city limits, but did not directly control development in surrounding Knox County. Although the city collaborated with the county in planning for development, the Knox County government had ultimate say over development in rural areas. The city of Knoxville could not by itself stop urban sprawl. The consequence of all this was that haphazard development of low-density residential housing—urban sprawl—increased in rural areas after the creation of the Knoxville urban growth boundary.

There do not appear to be any ecological studies of the impact of the Knoxville urban growth boundary on regional ecosystems (and very few ecological studies of urban growth boundaries in general). A study of the urban growth boundary around Seattle found that after the city set a boundary in 1992, urban sprawl also continued into rural areas.[4] From 1997 to 2001, only 14% of new residential building permits were for areas outside the urban growth boundary, but the land area developed outside of the boundary was 61% of the total developed land. Over a period of 20 years, this sprawl resulted in the loss of 41% of interior forest habitat in the rural areas surrounding Seattle. In the Knoxville region, it is very likely that the increased urban sprawl prompted by the creation of the Knoxville urban growth boundary resulted in greater regional ecological damage than if no boundary had been drawn.

This is obviously a lesson in unintended consequences, but the story points to a larger moral. Almost all land-use planning and decision making in the United States is done at the local level.[5] The city of Knoxville could control the use of land within its jurisdiction, but it had no direct control over nearby cities or the county within which it is located. These other local governments each made their own decisions about how to use the land in their jurisdictions. Knoxville could attempt to stop sprawl in the areas next to the city, but it could not prevent that sprawl from jumping beyond the urban growth boundary to other jurisdictions.

This fragmentation of jurisdictions directly leads to the fragmentation of ecosystems. States rarely do any regional land-use planning, and when they do, local governments are mostly free to ignore it. Furthermore, county governments usually have much less land-use authority than cities do, making them less able to prevent sprawl. On top of that, cities often compete against each other for new single-family housing developments because they bring high tax revenue. To win more residential development, cities may be willing to allow low-density development, again promoting sprawl. Finally, even if a local government decides that conservation is a top priority, there is no guarantee that an adjacent government will have the same priority. Differences in priorities make it very difficult for local governments to work together. As a result of this fragmentation of jurisdictions, there is little planning for, or actual implementation of, ecosystem protection at the regional level.

The chapter looks first at how land is used in the United States. Next, a discussion of the laws and ecology of agriculture is presented, followed by a discussion of urban areas and land-use planning. The fragmentation of jurisdictions, all the way down to individual landowners who have wide latitude in how they use their land, shows up in the laws regulating private lands, and in the resulting ecological effects of those laws. The chapter concludes with an examination of conservation easements, which are a popular tool for preserving ecological resources on private lands.

While not discussed in this chapter, it should be noted that governments use other means to regulate private lands. Recall from chapter 3 that the US Fish and Wildlife Service requires the submission of a habitat conservation plan from a private landowner before the agency will issue an incidental take permit for the threatened or endangered species on the landowner's property. Additionally, a private landowner usually must receive a permit from the US Army Corps of Engineers before she may fill in a wetland on her property. The filling of wetlands will be discussed in considerable detail in chapter 8. Finally, federal and state governments may simply take private land for public use. This is called eminent domain. Eminent domain is discussed in the context of wetlands, in chapter 8 (see box 8.1).

USES OF LAND

Every few years the Department of Agriculture releases a report on the uses of land in the United States. The most recent report found that

- The United States has a total land area of 915 million hectares.
- Forestland comprises 271 million hectares (29.7%).
- Grassland pasture and rangeland comprise 248 million hectares (27.1%).
- Cropland comprises 165 million hectares (18.0%).

- Special uses such as parks and wildlife areas comprise 127 million hectares (13.8%).
- Miscellaneous uses such as deserts, wetlands, and rural residential areas comprise 80 million hectares (8.7%).
- Urban land comprises 25 million hectares (2.7%).[6]

This final category needs a bit of explaining. Urban lands are defined as areas with at least 50,000 people, or urban clusters with 2,500 to 50,000 people.[7] Total rural residential area is more difficult to accurately calculate, so the Department of Agriculture does not include it as its own category. The department estimates, though, that rural residential areas cover 42 million hectares (103 million acres), or 4.6% of total US land area.[8]

Of all the land in the United States, 60% is privately owned. The federal government owns 28%, state and local governments own 9%, and 2% is held in trust by the Bureau of Indian Affairs.[9] Forestland and rangeland are the two largest categories of land in the United States. The forestland and rangeland numbers listed above include both private and federal lands. As a large percentage of US forestland and rangeland is owned by the federal government and is discussed in chapter 5, along with national parks and wildlife areas, this chapter instead focuses on cropland and urban land.

CROPLAND

Almost all cropland in the United States is privately owned.[10] Although cropland makes up only 18% of the land area of the United States, in some regions it covers a much higher percentage. In the northern plains, cropland covers 50.3% of the land.[11]

Naturally, any activity that involves half the land in a region is going to have significant environmental effects. Farming often involves plowing up native vegetation and planting a monoculture. Such activity destroys natural habitats and reduces biodiversity. Approximately half of threatened or endangered plant and animal species were listed in the Endangered Species Act in part because of agricultural activity.[12] Farming also causes soil erosion and loss of soil organic matter. Soil organic matter is mostly carbon, and soils act as the largest carbon pool on land.[13] This means that soils can become a source or sink for carbon dioxide, depending on the type of agricultural activity, with important consequences for global climate change. Planting crops influences not just the land, though; it also greatly affects water bodies. Farming results in soil, fertilizer, and other pollutants being washed into waterways; farming also uses vast amounts of water to irrigate crops. The combination of water use and pollutants severely degrades many aquatic ecosystems. The effects of agriculture on aquatic ecosystems will be discussed in greater detail in chapters 7 and 8.

Despite the environmental damage agricultural activity can inflict, federal environmental laws do very little to regulate farming. Some statutes, such as the Clean Water Act (discussed in chapters 7 and 8), specifically exclude most aspects of farming from regulation. The reasons for this include the power of the agricultural lobby in Congress, and the belief held by many that farming is a noble profession that does not require government regulation.[14]

The following sections describe a federal statute that directly affects how croplands are used, and two statutes that indirectly affect such land. The federal statute that directly affects cropland is the farm bill, and as will be explained next, the bill is intended to support agricultural activity rather than to mandate what farmers may or may not do with their lands. The two statutes that indirectly affect cropland are the Federal Insecticide, Fungicide, and Rodenticide Act and the Toxic Substances Control Act, which both help regulate what chemicals farmers may put on their crops.

Farm Bill

Congress passed the first farm bill in 1933, and passes a new farm bill approximately every five years. Each bill has a different name but is almost always referred to simply as the farm bill. The most recent farm bill is called the Agricultural Act, and passed Congress in 2014.

Each farm bill provides authorization for a number of different programs, and the US Department of Agriculture oversees those programs. The bill funds programs such as guaranteed loans to farmers, crop insurance, and agricultural research. The vast majority of support for farmers goes to those growing corn, rice, wheat, cotton, and soybeans. The farm bill also includes authorization for nutrition programs, such as the Supplemental Nutrition and Assistance Program (frequently referred to as the food stamp program).

Importantly for this chapter, the farm bill contains authorization for several conservation programs. All these programs are voluntary. Perhaps surprisingly, one of the more controversial provisions of the farm bill is a conservation program, the Conservation Reserve Program (CRP).

The CRP makes yearly "rental" payments to farmers who remove portions of their land from agricultural production. Farmers must also plant species on the land that improve environmental quality, such as native grass-

FIGURE 6.1. Restored wetlands on CRP land in northern Iowa. Photograph from USDA NRCS/Lynn Betts.

land species. CRP contracts last 10 to 15 years, and may be renewed for additional years. By removing land from agricultural production, the program aims to improve water quality, prevent soil erosion, and reduce loss of wildlife habitat (see figure 6.1). The CRP is the single largest program in the world designed to pay individuals to protect ecosystem services. In 2013 there were 10.9 million hectares (26.9 million acres) enrolled in the CRP under 700,000 contracts. For comparison, the state of Kentucky has a total area of 10.5 million hectares (26 million acres).

To qualify for the program, a farmer's land must meet one of several criteria, such as being suitable as a riparian barrier or wildlife habitat, or being highly erodible.[15] Lands that meet one of these criteria are then ranked against each other, based primarily on the environmental benefits that will result from removing the land from agricultural activity. Seven factors are considered: (1) soil erosion, (2) water quality, (3) wildlife benefits, (4) soil productivity, (5) likelihood land will remain in nonagricultural use after end of contract, (6) air quality, and (7) cost of enrolling land.[16] Lands that meet more of these factors are more likely to receive an offer of enrollment from the Department of Agriculture.

The CRP is controversial because it is often cast by detractors as simply a way to pay farmers for not planting crops. They argue that the CRP is less about promoting environmental health, and more about providing subsidies to the agricultural industry. Others argue that the Department of Agriculture is lax in enforcing the conservation requirements for land enrolled in the CRP.[17]

The best way to determine the environmental value of the CRP is of course to examine its environmental outcomes. The Department of Agriculture claims that in 2012 alone, the CRP helped reduce nitrogen losses from farms by 274 million kilograms (605 million pounds), and phosphorus losses by 55 million kilograms (121 million pounds), while reducing soil erosion by 272 billion kilograms (300 million tons).[18] The department also claims the CRP has helped restore 809,000 hectares (2 million acres) of wetlands and wetland buffers. Finally, it claims that the CRP results in the sequestration of more carbon than any other conservation program in the United States. Recall that the way in which land is used determines whether soil acts as a carbon source or sink. Plowing native vegetation and planting monocultures tend to reduce soil organic matter, turning the soil into a carbon source. CRP land removed from agricultural activity and planted with perennial vegetation, however, tends to increase soil organic matter, turning soil into a carbon sink.

Peer-reviewed ecological studies have found that CRP lands can have positive effects on the populations of species that use those lands. A study of CRP lands in the north-central United States discovered that several species of grassland birds that had seen population declines, such as bobolinks (*Dolichonyx oryzivorus*) and dickcissels (*Spiza americana*), had those declines lessened or reversed after implementation of the CRP in the region.[19] A different study found that in the Midwest and Great Plains states, grassland bird species with increasing populations tended to inhabit landscapes with more CRP land than those species that had stable or decreasing populations.[20] Other studies at smaller scales suggest that several species of mammals, reptiles, and insects use CRP lands, but information on how those lands affect population abundances is limited.[21] At least one study found, though, that it takes more than 20 years of enrollment in the CRP before former cropland begins to have vegetation cover and composition like that of nearby undisturbed land.[22]

While not a panacea, the environmental benefits of the CRP seem to be fairly evident. There are hints, though, that the CRP is becoming much less environmentally beneficial. The reason has to do with the planting of renewable fuels.

The Energy Independence and Security Act of 2007 sets the minimum volume of renewable fuel required to be in transportation fuel sold in the United States. The volume of renewable fuel is mandated to increase yearly until 2022.[23] This directly affects farming because most renewable fuel is ethanol made from plant matter, primarily corn. As the mandated volume of renewable fuel increases, so does the price of corn. To take advantage of the increase in corn prices, farmers have begun to convert their CRP lands back to agricultural production. Since 2009 farmers have removed 2 million hectares (5 million acres) of land from the CRP.[24]

The two most important sources of ethanol in the United States are corn and cellulosic plant matter, frequently from grassland species. Ideally, both corn and cellulosic ethanol will reduce net greenhouse gas emissions because as the plants grow, they remove carbon dioxide from the atmosphere. This carbon dioxide is released when the ethanol is burned as fuel, but the carbon dioxide is captured again when new plants are grown for more ethanol. This is in contrast to fossil fuels that simply release carbon dioxide into the atmosphere, where it remains. As noted above, however, CRP lands sequester large amounts of carbon in the soil. If CRP land is converted back to agricultural use to produce plants for ethanol, the carbon that had been sequestered in the soil may be released. Studies suggest that keeping land set aside in the CRP produces a greater net savings in greenhouse gases than converting the land to corn ethanol production.[25] Land previously in the CRP would have to be used for corn ethanol production for approximately 48 years before the net savings in greenhouse gases would outweigh converting the land.

The result is different for cellulosic ethanol production. Cellulosic ethanol made from grassland species would actually continue to sequester carbon in the soil similarly to CRP lands. Converting CRP lands for production of cellulosic ethanol would produce an immediate mitigating effect on climate change.[26] However, considering CRP lands solely for their ability to sequester carbon ignores all the other ecological values such lands provide. The ability of CRP land to provide such benefits as wildlife habitat and pollutant runoff prevention would be mostly lost. Weighing all the environmental benefits and costs of setting aside land in the CRP versus using it for ethanol production is very tricky.

The obvious solution to preventing farmers from converting their CRP lands back to corn production would be for the federal government to use CRP funds to actually buy land from farmers. While this would make sense ecologically, there appears to be no political support for such a program. Additionally, even farmers who support conservation are often reluctant to sell their land outright. This reluctance helps explain the rise in conservation easements, which are discussed in the final section of the chapter.

Several other conservation programs are authorized by the farm bill. They include such programs as the Conservation Stewardship Program and the Environmental Quality Incentives Program. The farm bill also authorizes the Swampbuster program and the Agricultural Conservation Easement Program, both of which help protect wetlands on agricultural lands (and are discussed in more detail in chapter 8). While the CRP has received a fair amount of ecological study, these other conservation programs have received at best a handful of peer-reviewed studies. More studies of these programs are needed to assess their ecological value.

Regulating Agricultural Chemicals

While the farm bill directly influences how a considerable amount of cropland is used, the Federal Insecticide, Fungicide, and Rodenticide Act (FIFRA), and to a lesser extent the Toxic Substances Control Act (TSCA), have an indirect influence on cropland. Both FIFRA and TSCA affect cropland by regulating the chemicals farmers put on their fields.

Many farmers use pesticides to help increase the productivity of their lands. Pesticides are designed to kill living organisms, so it is little surprise that they can have widespread environmental effects. Approximately 500 million kilograms (1.1 billion pounds) of pesticides is applied each year in the United States, with most of that being herbicides applied to agricultural lands.[27] These pesticides have a tendency to not stay where they are sprayed. The pesticides wash off the land and into the water, contaminating waterways (discussed in the next chapter), or drift through the air onto neighboring lands. Pesticides can kill nontarget wildlife several ways—from direct exposure, after wildlife eats contaminated food, or indirectly by eliminating resources or refuges.[28]

FIFRA regulates chemicals used as pesticides. Before manufacturing or selling a pesticide, the producer must perform tests to determine the toxicity of the potential pesticide.[29] The producer must then register the chemical with the EPA. The EPA will approve the registration of a chemical as a pesticide only if the chemical performs the tasks the producer claims it will perform, and if the chemical "will not generally cause unreasonable adverse effects on the environment."[30] FIFRA defines "unreasonable adverse effects on the environment" as an unreasonable risk to humans or the environment, "taking into account the economic, social, and environmental costs and benefits of the use of any pesticide."[31] This means that in deciding whether to register a potential pesticide, the EPA must weigh the benefits of using the pesticide to kill pests against the risks to humans and the environment.

Critics argue that the tests required under FIFRA do little to evaluate the risks of pesticides to wildlife or ecosystems.[32] The testing requirements consider only acute toxicity in a few representative species.[33] Additionally, the tests do not examine cumulative or synergistic effects of pesticides in the environment.

After the EPA registers a pesticide, the chemical receives a label describing how the pesticide must be used.[34] The EPA may also allow pesticide application by anybody, or may restrict application to certified applicators.[35] Certified applicators must maintain records of when, where, and how much pesticide they apply, but do not have to show those records to a federal agency unless asked.[36] FIFRA does little else to regulate the application of pesticides to farmland. There are no permits needed before application, and there is no direct monitoring of pesticide levels in soil or runoff.

Most other chemicals that are not used as pesticides are regulated by TSCA. TSCA requires the EPA to compile and publish a list of all chemicals manufactured or processed in the United States.[37] For instance, the chemical ingredients that are used to make fertilizers must be registered under TSCA. TSCA also gives the EPA authority to require testing of the toxicity of chemicals used or manufactured in the United States. This requirement, however, depends on the category a chemical compound falls into.

TSCA creates two categories of chemicals, "new" and "existing." Many of the chemicals that were in use before the passage of TSCA in 1976 are considered to be existing chemicals, and can continue to be used in commerce unless the EPA shows that they pose a hazard. Over 60,000 chemicals were thus grandfathered into the existing chemicals category when TSCA was implemented.

Conversely, to begin manufacturing a new chemical, a producer must file a premanufacture notification with the EPA. Unlike FIFRA, though, TSCA does not always require chemical producers to test the toxicity of the new chemicals they have created before manufacturing and selling them. If the EPA finds that a new chemical may present an unreasonable risk of injury to health or the environment, and there are insufficient data to make a reasonable prediction, then the EPA may require testing of the new chemical.[38] The EPA can require tests for carcinogenesis, mutagenesis, teratogenesis, behavioral disorders, cumulative or synergistic effects, or any other effect that may present an unreasonable risk of injury to health or the environment.[39] If the EPA finds after testing that a chemical presents an unreasonable risk of injury to health or the environment, TSCA allows the EPA to prohibit the manufacture or use of the chemical.[40]

The EPA normally requires testing, however, only when a new chemical is molecularly similar to another chemical known to be toxic. Usually, the chemical manufacturer need only give the EPA any available data that indicate the new chemical will "not present an unreasonable risk of injury to health or the environment."[41] This means that most chemicals regulated under TSCA have not undergone controlled testing of toxicity. There are roughly 84,000 chemicals in use in the United States. Critics estimate that only 500 to 1,000 of those chemicals have been adequately tested for safety.[42]

As already mentioned, the chemical ingredients that are used to make fertilizers must be registered under TSCA. Beyond that, though, federal statutes do not regulate the normal application of fertilizers. Despite the damage fertilizers can cause when they are washed into waterways, they are virtually unregulated. The ecological effects of fertilizers washed into waterways is taken up in the next chapter. This chapter now turns to urban land-use planning, where the federal government does even less regulating than in farming.

URBAN LAND

Although the percentage of urban land in the United States is relatively small, the changes urban areas make to ecosystems are extensive. Urban development tends to destroy native habitats, and locally eliminates most native species.[43] The pattern of streets and buildings in urban areas also creates widespread fragmentation of those habitats that remain. This habitat fragmentation helps to greatly reduce the diversity and abundance of animal species in urban areas relative to that found in nearby rural areas.[44] Studies show that animal species richness decreases as one moves from rural areas to the center of the urban area. Several other indicators of anthropogenic disturbance increase when moving from rural areas to the center of the urban area: the percentage of the land that is impervious surface (e.g., pavement and buildings); road density; average ambient air temperature (urban heat island effect); air and soil pollution; and soil alkalinity.[45]

Habitat fragmentation in urban areas also tends to create a great deal of habitat heterogeneity. As a result, and perhaps surprisingly, plant species richness is actually higher in urban areas than in surrounding rural areas.[46]

All the different habitats that are created by urban habitat fragmentation allow a wide array of plant species to grow. It should be noted that plant species richness is also high in urban areas because humans bring in and plant many nonnative species. Additionally, even though there is high species richness, most plant species occur at small population sizes. Bird species richness also tends to increase when moving from rural to urban areas, peaking at intermediate levels of urbanization.[47] This appears to be due to the heterogeneity of edge habitats created by urban fragmentation.

Urban areas influence not just the number of species that occur in those areas, but also the trophic dynamics of those species.[48] Where roads and buildings are placed, there are no plants, so net primary productivity is close to zero. However, green spaces within a city, such as parks and golf courses, often have higher net primary productivity than surrounding rural areas. The reason is that humans water and fertilize such green spaces, and the green spaces have a longer growing season in the spring and fall because of the urban heat island, which raises average temperatures. At the same time, native vertebrate predators tend to exist at a lower abundance, or not at all, in urban areas. High net primary productivity coupled with low predation means that the vertebrate species that are able to exist in urban areas, such as some bird species and mammals such as cottontail rabbits or raccoons, tend to have much higher abundances than in rural areas. This in turn greatly alters the trophic dynamics for those species. Competition for food resources becomes a much stronger driver of population dynamics relative to rural areas. At the same time, predator-prey interactions become much less important. More research in a larger sample of urban areas is necessary, though, to better understand the trophic dynamics that urban areas foster.

The following sections on urban land examine laws that directly influence the use of urban lands, and two federal statutes that indirectly influence such lands. The section begins by examining land-use planning by local governments, and the considerable power those governments have to directly determine the use of the private lands under their jurisdiction. This is followed by an explanation of the Resource Conservation and Recovery Act and the Comprehensive Environmental Response, Compensation and Liability Act. These two federal statutes indirectly influence urban lands by regulating the disposal of solid waste and the cleanup of hazardous sites, respectively.

Land-Use Planning

Although urban areas have extensive, and frequently devastating, effects on native habitats and species, the effects can be partially mitigated. The most important means to do so is through land-use planning. A land-use plan creates a coherent strategy for the development and use of land within a defined area. To protect native habitats and species, a plan may allow development in areas with limited ecological resources, while prohibiting development in more ecologically sensitive areas.

As the beginning of this chapter suggests, almost all land-use planning in the United States is done at the local level. Planning at the state or federal level would almost certainly be better at creating consistent plans that protect ecosystems spanning large areas. There is no comprehensive federal land-use planning, however. Similarly, there is relatively little comprehensive state planning. Federal law does directly influence land-use planning in one statute, the Coastal Zone Management Act. This act promotes the development of state-level planning for coastal zones. Participation in the act is voluntary, however, and planning is done only for coastal areas. The Coastal Zone Management Act is discussed in greater detail in chapter 9.

The individual states have the power to regulate the uses of private lands within their borders, but the states delegate that authority to the local governments within the state. As part of this delegation of authority, however, a state may require local governments to create comprehensive land-use plans, and then pass zoning laws and subdivision regulations that implement those plans.[49]

Comprehensive Land-Use Plans

At the local level, a comprehensive land-use plan guides where future development will occur, while balancing economic, social, and environmental considerations. Cities, as well as the surrounding counties, engage in land-use planning. A plan usually begins by stating the goals of the community, and then describes how development of land, transportation, utilities, and other factors will achieve those goals. Depending on the local government, a planning commission or board may write the comprehensive land-use plan, with the input of a professional land-use planner. The local legislative body then has to vote to adopt the plan before it takes effect. Local jurisdictions update their comprehensive plans on a regular basis, often every five years.

The comprehensive plan may seek to protect ecological resources by restricting where development can occur. Many state wildlife agencies provide site-specific maps of biodiversity that land-use planners can use to determine where areas of high biodiversity exist. The maps often include species distributions and habitats, and may include potential threats to those habitats. The maps may further suggest areas that local governments should avoid developing to preserve linkages between important conservation areas.[50] Local conservation groups also occasionally provide maps of environmentally sensitive areas where they believe development should be restricted. Planners may use all that information in deciding where to restrict development.

Critics have pointed out, though, that maps of biodiversity and environmentally sensitive areas are often not at a small enough scale to help make planning decisions at a local level. On top of this, state agencies and conservation groups often do not suggest areas where development would be least environmentally damaging.[51] Planners with little ecological expertise may understand which areas should be reserved from development, but not where development would produce the least amount of environmental damage. Additionally, if an area has been slated for development, there is often little guidance for planners to determine what kind of development, and at what density, would do the least ecological damage. While there are case studies examining how urban land use effects landscape structure, there have been few attempts to elucidate general principles for use by planners.[52] Finally, the maps created by state agencies showing where environmentally sensitive areas exist are usually not binding on local governments. Land-use planners may decide that developing such areas would be in the best interest of the community.

All of the above criticisms are minor if local governments care about conservation and environmental protection and reflect those concerns in their comprehensive land-use plans. However, surveys of cities show that the vast majority of local governments spend very little or no time considering ways to conserve biodiversity, including in their comprehensive land-use plans.[53] Local governments are usually much more interested in promoting development than in slowing it down to help conserve ecological resources. Even local governments that do devote time to planning for conservation rarely do so in consultation with adjacent jurisdictions.

This may be slowly beginning to change, however. Recall that the city of Knoxville collaborated with the surrounding county in creating its development plan. The result was not what Knoxville intended, but there was collaboration between local governments. Many states have statutes that authorize the creation of regional planning boards.[54] These planning boards create comprehensive land-use plans that cover a regional area encompassing two or more local jurisdictions. The decision on whether to create one of these boards, however, is often left to the local governments. If a local government does not particularly care about protecting ecological resources, it will likely be unwilling to give up its land-use planning authority to a regional board.

The state of New York in 1971 created a regional planning board that was not left to the discretion of local governments. The Adirondack Park Agency undertakes land-use planning for the 2.4 million hectares (6 million acres) of the Adirondack Park in New York. This is significant because more than half the land in the park is privately owned, and the park encompasses 12 counties and 105 towns and villages. If the creation of the Adirondack Park Agency had been left to the discretion of the local governments in the park, it almost certainly would not have been formed. The creation of a regional planning board by New York State helped prevent the land-use planning in Adirondack Park from becoming extremely fragmented.

Zoning Laws

Once a comprehensive land-use plan has been written and passed by a local government, it must be implemented. Zoning laws (also called zoning ordinances) are the primary means for realizing a land-use plan. Zoning laws set allowable uses of land, such as permitting only residences in one zone, commercial buildings in another, or open space. The laws also regulate factors such as the density of buildings in a zone, how tall they may be, and how much green space must be left on a lot. As for comprehensive land-use plans, the local legislative body must vote to adopt zoning laws. State statutes frequently require that zoning laws be consistent with the comprehensive land-use plan of the jurisdiction. A zoning law that is not consistent with the land-use plan may be challenged in court.

New York City in 1916 adopted the first citywide zoning law in the United States. The zoning law created different zones within the city, and then allowed only certain uses on private land in those zones. In 1926 the Supreme Court in the case *Village of Euclid, Ohio v. Ambler Realty Co.* held that creating zoning districts is constitutional.[55] The court stated that the due process clause of the Fourteenth Amendment was not violated by such zoning because it was a reasonable way to protect the public interest.

There are several ways in which zoning laws help implement a comprehensive land-use plan. As mentioned in the beginning of this chapter, a city may draw an urban growth boundary around itself. The city, in cooperation with the surrounding county, may set zones for high-density development within the growth boundary, while setting zones for low-density development outside of the boundary. Zones for agriculture, forestland, and open space may also be set outside of the growth boundary.

Another common use of zoning laws is the creation of overlay zones. An overlay zone is an additional set of zoning requirements laid over a pre-existing zone. As a result, land within the overlay district must meet the requirements of both the preexisting zone and the overlay zone. A city may use an overlay zone to protect historic areas or to promote development, but they are often used for conserving natural areas. For example, Brookfield, Wisconsin, created an overlay zone to protect wildlife habitat areas and scenic areas in a section of upland forest. The overlay prohibits building any structures, removing vegetation, and filling or excavating the land within the overlay zone.[56]

Studies of zoning laws are divided on whether they do much to slow urban sprawl into rural areas. Cities and counties may set zones to protect agricultural land, forestland, and open space. If urban growth pressure is high, though, local governments may be tempted to change zoning requirements to allow development as a means to increase the tax base. Additionally, zones for low-density residential development in rural areas may simply encourage urban sprawl by forcing development to occur over a wider area to accommodate all those who wish to move into the area. One study of county zoning laws in southern Indiana found that zoning weakly protected farmland, but did little to protect forestland.[57] Urban sprawl is a powerful force that local governments have difficulty controlling.

Subdivisions

Along with zoning laws, comprehensive land-use plans may be realized through the use of subdivision regulations (box 6.1). Subdivision regulations control the division of land into two or more lots for development. For instance, a developer may buy a large piece of land zoned for residential development, and then divide that land into several smaller lots upon which houses are built. Subdivision regulations control how the land is divided into lots, and the infrastructure, such as roads and sewer lines, that must be built within the subdivision.

Where development is slated to occur, subdivision regulations have been touted as one of the best ways to protect native habitats. The preeminent method to conserve native habitats is not to destroy those habitats in the first place.[58] To that end, many jurisdictions have subdivision regulations that encourage "cluster subdivisions." A cluster subdivision allows building the same number of houses as would be built in a conventional subdivision, but clustering the houses close together on small lots. The rest of the subdivision is then left as undeveloped open space. The result is that part of the

BOX 6.1. OTHER CONSERVATION
PLANNING APPROACHES

Besides conservation subdivisions, three other conservation planning approaches are widely used.

Transfer of development rights links an area where the government wishes to discourage development (sending district), to an area where the government wishes to encourage development (receiving district). An individual in the sending district may sell his right to develop his land to an individual in the receiving district. The sending district is usually in an ecologically sensitive rural area, and the receiving district is usually near other population centers. By buying development rights from the sending district, a developer may build houses at a higher density than would normally be allowed under the zoning laws of the receiving district. The same amount of development ultimately occurs, but it is at higher density in the receiving district than in the ecologically sensitive sending district.

In *conservation and limited development projects*, a developer or nonprofit organization uses revenue from development on a small portion of land to pay for conservation on the rest of the land.*

Planned developments require planning of an entire community before development begins. Plans include space for residential and commercial development, as well as space set aside for land conservation. Planned developments occur most frequently in the southern and western United States, where there is significant population growth and large tracts of land zoned for development.†

*Jeffrey C. Milder, "A Framework for Understanding Conservation Development and Its Ecological Implications," *BioScience* 57 (2007): 761.
†Milder, "Conservation Development," 763.

subdivision contains houses at a higher density than normally allowed under the zoning law, while upward of half of the subdivision may remain as open space where native habitats have not been directly touched.

If the developer attempts to maintain the ecological structure and function of the undeveloped land in the subdivision, then the cluster subdivision is often called a "conservation subdivision." In practice, the terms "cluster subdivision" and "conservation subdivision" are used interchangeably, especially if a developer can market a subdivision as being sensitive to the environment by calling it a conservation subdivision.[59] The rest of this section will use the term "conservation subdivision," as it is the one favored in the ecology literature.

Developers often prefer building conservation subdivisions to building conventional subdivisions: they get to sell the same number of lots as in a conventional subdivision, but by clustering the lots, they spend less money building infrastructure for the subdivision. Some jurisdictions also provide an incentive for developers to build a conservation subdivision by providing density bonuses. A density bonus permits one or two more houses to be built in a conservation subdivision than would be allowed in a conventional sub-division, further increasing the ability of the developer to make a profit on the conservation subdivision. On the other hand, many jurisdictions require developers to go through extra steps and pay extra fees to build a conserva-tion subdivision, making it easier and faster to simply build a conventional subdivision.[60]

Do conservation subdivisions actually help protect native habitats and species? Unfortunately, there are few empirical studies of the ecological effects of conservation subdivisions. Existing studies mainly examine how much open space has been preserved by the subdivision, without exploring ecological function in the open spaces or outside of the subdivision.[61]

Nonetheless, a handful of studies have suggested conservation subdivi-sions are better at protecting ecological resources than conventional subdi-visions. For example, a study of housing in Wisconsin found that clustering did help limit habitat loss.[62] A separate study in Wisconsin found that on average, 47% of conservation subdivisions were preserved as open space.[63] A different study concluded that because residential development occurs on such a large scale in the United States, if conservation subdivision develop-ment replaces conventional development it could protect several million hectares of land every year.[64]

Other studies, though, have found little difference between conservation subdivisions and conventional subdivisions. One study from Colorado found that the open spaces in conservation subdivisions contain many nonnative plant species.[65] As a consequence, the conservation subdivisions had wild-life communities much more similar to conventional subdivisions than to undeveloped areas.

There are several reasons why conservation subdivisions may not pro-vide a better ecological outcome than conventional subdivisions. The open spaces within a conservation subdivision are immediately adjacent to human residences. Studies have found that the zone of influence around houses, defined as the zone where human-adapted species outcompete human-sensitive species, extends at least 180 meters (197 yards).[66] In the study of Colorado conservation subdivisions, the average distance from open space to the nearest house was 254 meters (278 yards).[67] If the open area of a conservation subdivision is ringed by houses, then the area outside the zone of influence may be quite small.

There is also little effort to connect the open spaces in conservation subdivisions to adjacent ecologically significant areas.[68] Ideally, all the conservation subdivisions in a region would link their open spaces to create a regional network of preserved spaces.[69] As might be expected, this ideal has never been met. Land-use planners often give little thought to how the development that occurs next to the conservation subdivision affects habitats in the open spaces.[70] For instance, if the area next to a conservation subdivision is zoned such that it has considerable impervious surfaces, then the storm water that runs off that area may enter the open space of the subdivision, bringing large amounts of pollutants.

On top of that, the way in which houses are clustered in the conservation subdivision may negatively impact the ecology of the area. The land that is preserved as open space in a conservation subdivision may not be the land that is most ecologically sensitive. For example, the houses in the subdivision may be preferentially clustered around the most ecologically sensitive area in the subdivision, such as a lakefront. At regional levels, houses tend to be built in areas of high biodiversity and ecological value, suggesting that people and wildlife tend to agree on what makes an area desirable to live in.[71] The same appears to be true at the smaller scale of conservation subdivisions. There is a need for additional research to determine the best way to cluster the developed part of conservation subdivisions, and whether the best arrangement depends on the type of landscape where the subdivision occurs.[72]

Finally, the developer of a conservation subdivision may create a conservation easement (discussed in detail at the end of the chapter) to protect the open area in the subdivision. More often, though, developers leave the open area under the control of the homeowners in the subdivision. Homeowners may have little knowledge of ecology, and may be poor protectors of the open space. With no guidance, homeowners are likely to plant in their yards nonnative species that may spread to the open area, and to allow domesticated animals free run of the open area. Both actions would negatively affect the open area, likely increasing the zone of influence of the houses over the open area.

Future research may find that conservation subdivisions are able to overcome all the pitfalls listed above and help protect native habitats. However, to completely understand the effects of conservation subdivisions, researchers will likely need to study them for a long time. The reason is that conservation subdivisions, like almost all development, fragment landscapes. This fragmentation may produce different results for different species in the landscape. Fragmentation that is extensive enough may result in the immediate loss of many species from the landscape. This is called local extinction. Conversely, other species may be able to survive just fine in the new landscape. A third group of species, though, may continue to exist in the new landscape, but they are not fine. The landscape has changed enough that although the populations of these species are still there for now, they will go locally extinct over the course of several years or decades. This is known as an extinction debt. These "living dead" species did not disappear when the landscape was first changed, but their populations will slowly decline until they do disappear. Because conservation subdivisions are a relatively new phenomenon, the extinction debt that they create may not yet be fully paid. Maintaining large patch sizes is one way to reduce extinction debt, so conservation subdivisions may result in a much lower debt than conventional subdivisions.[73] Nevertheless, the extinction debt created by conservation subdivisions must be accounted for to truly determine the ecological effects of this form of land use.

Regulating Waste

Besides destroying habitats and fragmenting ecosystems, urban areas affect the environment by creating large amounts of waste. The Resource Conservation and Recovery Act (RCRA) regulates the generation and disposal of solid waste. The statute defines "solid waste" to include almost all discarded material, including liquid waste, some types of contained gaseous waste, and hazardous waste.[74] There are important exceptions to what RCRA defines as solid waste, though. It does not include solid or dissolved material in domestic sewage, irrigation return flows, industrial discharges regulated under the Clean Water Act, or nuclear waste.

While RCRA regulates solid waste that is currently being produced, waste created decades ago may be more important for land-use planning in many urban areas. Before Congress passed RCRA in 1976, there was little federal oversight of hazardous waste disposal. As a consequence, there are many sites in the United States with buried waste that has begun to leak, or sites where the owner went bankrupt but left hazardous waste in warehouses or in the ground.

To deal with this problem, in 1980 Congress passed the Comprehensive Environmental Response, Compensation and Liability Act (CERCLA). CERCLA is also known as the Superfund Act. CERCLA authorizes the EPA to clean up a site whenever "any hazardous substance is released or there is a substantial threat of such a release into the environment."[75] The act defines these terms broadly so that they cover many different releases and types of substances. CERCLA has an important exception in its definition of hazardous substances: it does not include petroleum, crude oil, natural gas, or synthetic gas usable for fuel.[76]

There are so many sites in need of cleanup that CERCLA directs the EPA to create a National Priority List of contaminated sites.[77] The act states that priority should be based on the relative risk or danger to the public health or the environment, and must take into account such criteria as the potential for destruction of sensitive ecosystems.[78] As of 2013, there were 1,314 sites on the National Priority List, and 54 sites proposed for listing. An additional 1,984 sites were eligible to be listed based on their level of contamination, but the EPA deferred those sites to cleanup approaches outside of CERCLA.[79] Illustrating the difficulty of cleaning up hazardous waste, the EPA considers only 371 sites to have been sufficiently cleaned up to warrant deletion from the National Priority List.

Cleanups are expensive, and a central part of CERCLA is trying to determine who should do the cleanup and who should pay. The EPA can do the cleanup itself using money from a trust fund created for such cleanups, called the Superfund. The Superfund was originally funded by crude oil and chemical taxes and an environmental tax assessed on corporations, but authority for those taxes expired in 1995. The Superfund also receives revenue after the EPA performs a cleanup and sues the responsible parties for cost recovery. Most revenue for the Superfund trust, though, now comes from direct appropriations by Congress.

Instead of performing the cleanup itself, the EPA has authority under CERCLA to order a potentially responsible party (PRP) to do the cleanup.[80] CERCLA names four types of PRPs: (1) the current owner or operator of a contaminated site, (2) any person who owned or operated a site at the time hazardous substances were disposed of, (3) any person who arranged for disposal or treatment of hazardous substances, and (4) any person who accepted hazardous substances for transport.[81]

The EPA's ability to order a PRP to do a cleanup is a powerful tool because CERCLA does not allow pre-enforcement review of such an order. This means that courts do not have jurisdiction to review an EPA order for a cleanup. After the cleanup, a PRP may sue to be reimbursed from the Superfund, arguing that it was not actually a PRP. A PRP can also sue other PRPs that contributed to the contamination for some of the costs of the cleanup. This is called a contribution action.

Courts have interpreted liability under CERCLA to be strict, joint and several, and retroactive. Strict liability means that if there is a release or threat of release of a hazardous substance, then the PRPs are liable, even if there is no proof that the PRPs were negligent. Joint and several liability means that a single PRP can be forced to pay for an entire cleanup, even if many other PRPs contributed. Consequently, joint and several liability ensures that the entire cost of a cleanup will be paid for by PRPs, even if only one or a few of them can be found. Finally, retroactive liability means that PRPs are liable for actions that took place before Congress enacted CERCLA.

A PRP may be able to avoid joint and several liability if it can prove that its waste is divisible. This means that if a PRP can show a reasonable basis for determining its contribution to the harm, then costs may be apportioned.[82] There is also an exemption in CERCLA for PRPs that contributed very little hazardous waste. If a PRP contributed less than 416 liters (110 gallons) of liquid waste, or 90.7 kilograms (200 pounds) of solid waste, to a site on the National Priorities List before April 1, 2001, then the PRP is not liable for cleanup costs.[83] Additionally, there is an exemption for owners of residential property, small businesses, and tax-exempt organizations that sent municipal waste to a site on the National Priorities List.[84]

In the list of PRPs that CERCLA names, owners of a contaminated site are listed first. This is perfectly fair if the owner knew about the hazardous waste at the site. If, however, the owner bought the site not knowing there was any hazardous waste at the site, then holding the owner liable seems less fair. CERCLA deals with this by not holding owners liable if when they acquired the site, they did not know, and had no reason to know, that any hazardous substances were disposed of on, in, or at the site.[85] To establish that they had no reason to know, owners must demonstrate that before purchasing the site they "carried out all appropriate inquiries," including hiring an environmental professional, interviewing past and present owners, and reviewing historical sources.[86]

Although CERCLA provides an exemption for innocent owners of contaminated sites, many potential buyers are reluctant to buy land if there is even the remote possibility that hazardous substances are at the site. Why risk a court's deciding that you did not carry out all appropriate inquiries, and thus become a PRP, when you can instead buy land that has never been developed before and thus has no risk of containing hazardous waste? As a consequence, many sites that previously had an industrial or commercial use have difficulty finding new owners. Without being able to find a buyer, many of these sites are simply abandoned.

These abandoned sites are called brownfields. Many cities have to deal with a considerable number of brownfield sites in their urban centers. More importantly, because developers do not want to buy these brownfield sites, they instead buy and develop greenfield sites on the edges of cities that have not been developed before. Avoiding brownfields and developing greenfields encourages urban sprawl and destroys open spaces. To counteract this, many states have passed statutes that provide financial incentives for developers to buy and develop brownfield sites. The EPA also offers grants to help in redeveloping brownfield sites. Nevertheless, CERCLA indirectly creates an incentive for urban sprawl.

To make matters more complex, though, brownfield sites may not always be so troublesome—they may in fact be beneficial. Brownfields that have been abandoned for a long time can support considerable biodiversity.[87] The plants and animals on the brownfield site may also contribute to the ecosystem services of the urban area. For example, trees growing on a brownfield site may help capture air pollutants, reduce and filter storm-water runoff, and mitigate flooding.[88] Trees also help reduce the urban heat-island effect, thereby reducing the amount of energy used for air conditioning. It might seem obvious that brownfield sites should be redeveloped before greenfield sites face the backhoe. If a greenfield site is of little ecological importance, though, while a brownfield site offers habitat for a wide range of species and helps provide ecosystem services, then the answer is not so obvious. More research is needed on the ecological value of brownfield sites, and if and when they should be preserved in favor of developing greenfield sites.

CONSERVATION EASEMENTS

Local governments can place limits on the uses of private lands, but landowners still retain a great deal of discretion in deciding how they use their land. Even if land is in a zone that allows development, the landowner may decide not to develop and to leave the land as open space. Conversely, if land is in a zone that does not allow development, the landowner may simply leave the land as it is, or may restore and manage the land for native species and habitats. As 60% of the land in the United States is privately owned, this means that landowners have enormous influence on the ecology of the United States.

Many conservation organizations spend most of their budgets trying to positively affect the ecology of private lands. One way to do this is to buy ecologically important land outright, called owning the land in fee simple. By owning the land, the conservation organization can ensure that the land is properly managed and never developed. There are two sizeable difficulties for conservation organizations buying land in fee simple, though: the person who owns the land may not want to sell it, and buying the land outright may be very expensive.

One solution is called a conservation easement. A conservation easement is created when a conservation organization or government agency purchases the development rights for a piece of land from the landowner. The landowner still owns the land, but the conservation easement restricts the activities of the landowner, to protect the land's conservation resources. Often the landowner may not build new buildings or roads on the land, but may continue activities such as farming or ranching on the land. The holder of the conservation easement may enter the land to verify that the landowner is not undertaking any activity in violation of the easement. The conservation easement may also include a management plan for the land, requiring the landowner to do such things as remove introduced species. A conservation easement can be thought of as a negative easement (see box 6.2) because it prohibits certain actions. If land burdened by a conservation easement is sold, the holder of the easement may enforce the easement against the new landowner. Conservation easements are usually written to conserve the land in perpetuity (i.e., forever).

Conservation easements have become very popular with conservation organizations because they are much less expensive to acquire than buying land, and landowners are more willing to sell conservation easements than to sell their lands outright. In fact, most conservation easements are actually donated by landowners, because they are interested in protecting the ecological value of their land and because there is a monetary incentive. If a conservation easement is donated, the landowner is eligible for a federal tax deduction in the year of the donation. The landowner may also pay reduced state property and estate taxes. To qualify for the federal tax deduction, the easement must fulfill the Internal Revenue Service's (IRS) requirements for a conservation easement. As most landowners are interested in taking the deduction, most conservation easements are consistent with the IRS requirements. The requirements are that the easement (1) is perpetual, (2) is held by a conservation organization such as a nonprofit conservation group or a government entity, and (3) serves a "conservation purpose."[89] Conservation

BOX 6.2. EASEMENTS, COVENANTS, AND
EQUITABLE SERVITUDES

Land-use terminology has built up over many years and can be rather complex; moreover, it is not always the same from state to state. Nonetheless, this box attempts to explain some of the most common terms.

An *easement* is an interest in a piece of land, giving a person the right to use someone else's land for a specific purpose. For example, an easement may give a person the right to cross someone else's private land to get to a nearby lake. *Easements in gross* benefit the owner of the easement personally, and usually cannot be transferred to another person. *Commercial easements in gross*, frequently owned by utility companies, are transferrable.

An *appurtenant easement* is a benefit that is attached to a piece of land, and runs with the land. The land that benefits from the easement is called the dominant estate, while the land that is burdened by the easement is called the servient estate. This means that a person has a right to use the servient estate for a specific purpose, but only when that person owns the dominant estate. When the person sells the dominant estate, the easement goes to the next owner of the dominant estate. So in the example above, the person may cross the servient estate to get to the lake, but only when she owns the dominant estate.

A *real covenant* is similar to an easement, but may be affirmative or restrictive. An *affirmative covenant* obligates the owner of the burdened piece of land to undertake certain acts on that piece of land. For instance, the owner of the burdened land may promise to maintain a drainage ditch so that water does not drain onto someone else's land. A *restrictive covenant*, on the other hand, restricts the owner of the burdened land from doing something on that land. For example, the owner of the land may promise not to build a second story on his house that would block the view of a mountain from an adjacent piece of land. Restrictive covenants are also called negative easements.

An *equitable servitude* is similar to a real covenant, the main difference being in how they are usually enforced by the courts. If the owner of the burdened piece of land breaks the terms of a real covenant, courts will usually grant monetary damages. To enforce an equitable servitude, courts will issue an injunction preventing the owner of the burdened property from undertaking the proscribed action.

purposes are defined as preservation of land for recreation by the public, protection of a relatively natural habitat, preservation of open space (including farmland and forestland), and preservation of a historically important land area or structure.[90]

There are few empirical studies examining the effectiveness of conservation easements in protecting ecological resources. One study looked at lands with conservation easements in Wyoming, and compared them to nearby lands without easements.[91] The study found that in areas with high pressure for residential development, lands with conservation easements had fewer structures and fewer roads than lands without easements. Similarly, lands with conservation easements in high-pressure areas had higher use by wildlife than nearby lands without easements. On the other hand, in high-pressure areas, the canopy cover of introduced plant species was similar on lands regardless of whether they had a conservation easement or not. In areas with low pressure for residential development there was little difference in structures, roads, introduced plants, or wildlife occurrence between lands with and without conservation easements. The results of this study show that conservation easements in areas of high pressure are able to slow development of land and provide habitat for wildlife. Easements, though, are not able to prevent the spread of exotic plant species from nearby developed land. This means that land protected by a conservation easement but surrounded by developed land may need intensive management to remove the exotic species constantly being introduced from nearby areas. Finally, although conservation easements seem to add little value to the ecology of areas with low development pressure, such pressure may increase in the future, making the ecological value of those easements potentially much greater. Additional research exploring how biodiversity, landscape connectivity, and ecosystem functioning are influenced by conservation easements is needed. A useful resource for conducting research is the National Conservation Easement Database, which attempts to record and map all the conservation easements in the United States.[92]

There are several criticisms of conservation easements as conservation tools. The first two criticisms are polar opposites: some scholars argue conservation easements are not permanent enough, while others argue they are too permanent.[93]

Those who argue conservation easements are not permanent enough point to the doctrine of changed conditions. This doctrine allows a court to terminate a conservation easement if conditions on the land or surrounding areas have changed enough that it becomes impossible or impractical to continue using the land for a conservation purpose.[94] For example, if a conservation easement is created specifically to protect an endangered species, but the species then goes extinct, a court may decide to terminate the easement. Scholars who worry conservation easements are not permanent enough are concerned that when the price being offered to a landowner to develop his land is high enough, he will ask a court to terminate the conservation easement. A change in economic conditions is usually not enough to terminate a conservation easement, but a landowner may point to transformations caused by climate change or development surrounding the land. A judge may then use the doctrine of changed conditions as an excuse to give the landowner what he wants. Courts have the discretion to make perpetuity last for a lot less than forever.

On the other side are those who argue that conservation easements are too permanent. They point out that the above scenario is likely to be very rare. Many conservation easements are written to protect open space and habitats, even if those habitats change due to climate change or some other cause. Consequently, a court will be unlikely to terminate a conservation easement under the doctrine of changed conditions because the easement will have been written in such a way that the land will still serve a conservation purpose even if conditions change.[95]

The argument that conservation easements are too permanent rests on the idea that an easement may actually prevent the protection of ecological resources as conditions change. This could happen in two ways. First, conservation easements may prevent adaptive management of lands as conditions change. For example, a conservation easement may stop development on a piece of land, but still allow a fixed level of grazing on that land. If changing environmental conditions make that grazing more damaging to the land than it was before, the holder of the conservation easement will not be able to force the landowner to reduce the level of grazing. The landowner will be able to point to the easement as guaranteeing her a certain level of grazing. Over time the language of conservation easements has become more detailed and complex, often allowing the holder of the easement to alter how the landowner may use the land in response to changing conditions. However, it is difficult to foresee all potential changes in condition in an easement meant to last forever.

Second, changing conditions may make land burdened by a conservation easement today of little ecological value tomorrow. Land that contains a great deal of biodiversity, or a population of a rare species, may see those species move off the land as the climate changes or as development surrounds the land. Although the land may be of little ecological value, the conservation organization that holds the easement must still expend money and manpower to administer the easement and ensure that the landowner is in compliance with its provisions. The resources of the conservation organization would be better spent by terminating the easement and acquiring a conservation easement on a more ecologically valuable piece of land. Some scholars have proposed amending the federal tax code to encourage landowners to donate conservation easements that last for only a fixed number of years.[96] These "term" easements would be similar to farmers' removing land from agricultural activity under the CRP. The fixed term would allow conservation organizations to better use their resources to protect land that becomes ecologically important as conditions change. Moreover, if a conservation organization and landowner decide to renew a term easement, they could negotiate new management requirements for the land at the end of each term.

The argument against such fixed-term conservation easements is that they do not supply real conservation. As soon as the easement has expired, the landowner is free to develop the land. The landowner who is least likely to renew a term easement will often be the one who is receiving the largest cash offer to develop the land. One of the important characteristics of a conservation easement is that once signed, no amount of development pressure will result in the loss of open space on the protected land.

There are two other criticisms of conservation easements. One study of conservation easements in Wisconsin found that the majority of the easements examined did not include a right for the easement holder to conduct ecological monitoring of the land.[97] The holders of the easements were allowed to inspect the lands to make sure there was no development, but were not allowed to do tasks such as enter the land to quantify environmental indicators. If a conservation easement includes a management plan for the land, but the holder does not have the right to perform ecological monitoring of the land, then management decisions will be made essentially blind.

The final criticism of conservation easements is that while they protect the conserved land from development, development is simply shifted elsewhere.[98] If land burdened by a conservation easement is located near the advancing suburban edge of a city, the conservation easement may force development to leapfrog onto land farther out into rural areas. A conservation easement may thereby increase the range of urban sprawl. Perversely, land protected by a conservation easement may even be viewed as an amenity, encouraging greater development near that land.

As these criticisms suggest, conservation easements are not the perfect instrument for protecting ecological resources on private lands. Local land-use planning and the conservation provisions of the farm bill also tend to do a less than stellar job of protecting environmentally sensitive lands. Disjointed though they may be, conservation easements, local land-use planning, and the farm bill are three primary means for protecting the ecological value of private lands in the United States. Hopefully future research will help illuminate ways in which these three approaches can be better combined to protect terrestrial ecosystems.

PART III

Water

CHAPTER SEVEN

Polluting Lakes, Streams, and Rivers

On June 22, 1969, the Cuyahoga River caught on fire. The river, flowing through Cleveland and emptying into Lake Erie, had been the dumping site for industrial waste from facilities in the Cleveland area for over a century.[1] Among the facilities were several oil refineries that would dump the unusable fraction of refined crude oil directly into the river. Fires on the river were actually nothing new—the Cuyahoga had been catching fire at regular intervals since 1868. A fire on the river in 1952 was the largest blaze, resulting in a five-alarm inferno. Surveys in the 1960s showed that there were long stretches of the river where no life, other than algae, could exist.

Federal statutes aimed at regulating water pollution existed by the 1960s, such as the 1956 Federal Water Pollution Control Act. The statutes were weakly enforced, though, and did little to stop the discharge of pollutants into water bodies. In response to the 1952 fire, the city of Cleveland had actually begun cleaning up the Cuyahoga River, managing to make significant progress in reducing pollution in the river.

The cleanup, though, was not enough to stop the Cuyahoga River from burning one more time in 1969. An oil slick and debris under a railroad trestle caught on fire, possibly from a spark thrown by a passing train. The fire was relatively small and was under control within 30 minutes. Indeed, there was no damage from the fire except for a few railroad ties warped by the heat of the flames.

This fire, though, was different from the earlier ones. The environmental movement in the United States was becoming increasingly mainstream by the 1960s. The 1969 fire received national media attention from Time magazine and National Geographic. Other environmental disasters earlier that year, such as a huge oil spill in Santa Barbara Channel and a fish kill caused by the discharge of pollutants in Lake Thonotosassa in Florida, which had killed 26 million fish, also received national media attention. These calamities, along with the fire on the Cuyahoga River, became potent symbols of the dismal state of water quality in the United States. The Cuyahoga River fire helped create public pressure for a major revision of the Federal Water Pollution Control Act, creating in 1972 the statute we now know as the Clean Water Act (CWA). The CWA is the primary federal statute regulating surface waters in the United States.

Despite the passage of the CWA more than 40 years ago, the Cuyahoga River is still not clean. The river is much cleaner than in the past, certainly, supporting 44 species of fish in places where no fish used to exist. Under the CWA, though, states must identify "impaired waters" that are too polluted to meet water-quality standards. Several stretches of the Cuyahoga River are classified as impaired. Industrial discharges of pollutants continue, while rainwater runoff carrying nutrients from agricultural land adds additional pollutants. As this chapter discusses, the CWA regulates the discharge of pollutants from industrial sources, but not runoff from agricultural land. The CWA has increased water quality in the United States, but the waters are far from clean.

In a remarkable parallel to the fire on the Cuyahoga River, hydraulic fracturing (often called fracking) is generating great public concern over its environmental impacts since the news media began showing videos of tap water burning in houses near fracking wells. The image of water burning seems to be so unnatural that it forces many people to conclude that the environment must be deeply degraded. Just as the Cuyahoga River fire helped promote passage of the CWA, many environmentalists hope that video of tap water burning will help promote new regulations for fracking. As the ecological implications of fracking are still under investigation, and the law on fracking is also in flux, this chapter only briefly considers hydraulic fracturing.

Many different pollutants find their way into US water bodies. The three that cause the most widespread problems are nutrients, bacteria, and sediment. Nutrients, such as nitrogen and phosphorus, are put on crops as fertilizer or are present in animal waste stored on farms. When it rains, those nutrients are often washed into the nearest water body. Nutrients entering water bodies can have several effects on aquatic ecosystems, including the promotion of algae blooms and eutrophication (discussed below and in box 7.1). Bacteria enter water bodies from sewers that overflow during heavy rains and pour untreated sewage into waterways. Bacteria also enter water bodies from animal waste stored on farms. Bacteria in water can directly impact human health and the health of wildlife. Finally, sediment (e.g., dirt) washes into water bodies after humans undertake activities that disturb land, such as logging, plowing fields, or bulldozing construction sites. Sediment in waterways can clog the gills of fish, make the water so murky that aquatic plants cannot perform photosynthesis, and kill fish eggs.

While these are the three most widespread pollutants, many other pollutants regularly enter water bodies in the United States. Industrial facilities may dump chemicals directly into water bodies. Pollutants other than nutrients that are sprayed onto land, such as insecticides and herbicides, may get washed into water bodies. There is also considerable thermal pollution in US water bodies. Thermal pollution occurs when a power plant puts water directly into a water body after the water was used to produce electricity and while it still has an elevated temperature. Thermal pollution may also occur when a dam releases water that is warmer or cooler than the river it is going into. Thermal pollution can kill fish and other aquatic organisms. In addition to all the pollutants that are directly dumped into water bodies or washed off land, pollutants also enter water bodies from the atmosphere. Pollutants that have been emitted into the air, such as from power plants, eventually fall back to the ground and may be deposited in water bodies. In fact, one of the leading causes of pollution in lakes is mercury deposited from the atmosphere. Air pollutants and their deposition are discussed in considerable detail in chapter 10.

The CWA controls the polluters and pollutants that impact US water bodies—although as will be seen, the CWA has greater authority to control some polluters and pollutants more than others. This chapter has an overview of the CWA, and describes under what circumstances polluters may discharge pollutants into bodies of water.[2] Next, the chapter examines water-quality standards set by the states. The next sections deal with water quantity: they delve into the law and ecology of water withdrawals from streams and lakes, such as for power plant cooling and irrigation of crops. This is followed by a discussion of the effects of dams on aquatic ecosystems. The chapter concludes with a brief look at the laws and ecology of fracking. The CWA also regulates the physical alteration of water bodies and wetlands, but this will be discussed in detail in the next chapter.

THE CLEAN WATER ACT AND WATER QUALITY

The CWA regulates water quality through provisions that fall into three categories.[3] (1) The first category consists of pollution from point sources. A point source can be thought of as a factory or sewage treatment facility that discharges pollution directly into a waterway. These are called point sources because pollution flows from a source such as a pipe, ditch, or channel and enters the water at one particular point.[4] (2) The second category consists of pollution from nonpoint sources. Nonpoint sources are land uses such as farms, abandoned mines, and parking lots. Rain or snow falling on such land picks up pollutants and washes them into nearby water bodies. These sources are called nonpoint because the pollutants do not enter a water body at one particular point. (3) The third category consists of water-quality standards. The CWA requires that each state set water-quality standards for their waterways. The states are then required to limit discharges of pollutants into those waterways to achieve the standards. To do this, states can limit pollutants entering waters from both point and nonpoint sources. These three categories will be discussed in turn.

Point Sources

The CWA states that without a permit, the "discharge of any pollutant by any person shall be unlawful."[5] The act defines "discharge of a pollutant" to mean the "addition" of any pollutant to "navigable waters" of the United States from any point source.[6] The act defines "person" broadly to include, among other things, individuals, corporations, and states.[7] "Pollutant" is also defined broadly to mean "dredged spoil, solid waste, incinerator residue, sewage, garbage, sewage sludge, munitions, chemical wastes, biological materials, radioactive materials, heat, wrecked or discarded equipment, rock, sand, cellar dirt and industrial, municipal, and agricultural waste discharged into water."[8] The CWA specifically states that water, gas, or other materials injected into a well for oil or gas production do not count as pollutants.

There has been considerable debate and litigation over exactly which water bodies are regulated by the CWA. The CWA states that it covers "navigable waters" of the United States. What constitutes a navigable water is not as clear as it may seem, and has been the subject of several Supreme Court decisions. As the need to clearly define "navigable water" usually arises in the context of developers wanting to fill in wetlands, the definition will be taken up in the next chapter. For now, it is enough to know that a navigable water does not have to be one that is traditionally navigable by a boat; very small streams and wetlands are covered under the CWA. On the other hand, a water body usually has to have some connection to a traditionally navigable water to fall under the CWA. Even though pollutants in groundwater may migrate to surface waters, most courts have held that groundwater is not regulated under the CWA.[9]

Pollution Permits

A point source may discharge pollutants into a water body under certain provisions of the CWA. To legally discharge pollutants, usually a point source must first receive a National Pollutant Discharge Elimination System (NPDES) permit.[10] (These permits are also frequently referred to as Section 402 permits, for the section of the CWA that authorizes them.) NPDES permits usually last for five years, and are issued by the EPA or by a state if it can show it has the expertise to issue the permits. A point source given a permit must stay below the effluent limitations in the permit. The point source must also submit reports on its discharges, often on a monthly basis. The reports are public records and can be checked by anyone. Additionally, the CWA allows citizens to bring suits for alleged violations of NPDES permits.[11]

The amount of pollutants a point source may discharge depends on the type of point source. The next five sections consider discharges of pollutants from publicly owned treatment works, industrial sources, indirect dischargers, concentrated animal feeding operations, and storm-water runoff.

Publicly Owned Treatment Works. A significant source of pollution is sewage treatment plants. There are 14,770 sewage treatment plants in the United States, treating more than 121 billion liters (32 billion gallons) of wastewater each day.[12] The CWA requires these publicly owned treatment works (POTW) to use a specific level of technology before discharging pollution into waterways. There are three levels of treatment that POTWs may use. Primary treatment consists of separating solid waste from liquid waste through the use of filters, screens, and settlement tanks. Secondary treatment uses microorganisms to break down organic material. Tertiary treatment consists of nitrification, denitrification, disinfection to kill bacteria, and physical and chemical means to remove dissolved metals. When treatment is complete, the liquid that remains is often discharged into a body of water. Any remaining solid sludge is used on land as fertilizer or simply placed in a landfill. Under the CWA, most POTWs are only required to use primary and secondary treatment on their sewage.[13] The EPA may require tertiary treatment where the body of water that the POTW discharges into already has poor water quality.

Industrial Sources. If the point source is something other than a POTW, such as an industrial source, the CWA requires the source to meet a particular numeric effluent limitation. The effluent limitation is based on what a specific technology has been shown to accomplish in reducing pollution from that type of point source.[14] A point source is not required to use that specific type of technology, though, as long as it is able to meet the effluent limitations in its NPDES permit. This allows point sources the flexibility to experiment with technology to find the cheapest way to meet effluent limitations.

The EPA does not set effluent limitations on a facility-by-facility basis. As the Supreme Court ruled in 1977, the EPA can set industry-wide guidelines for effluent limitations.[15] If this were not allowed, the EPA would have to set limitations for tens of thousands of facilities on an individual basis, taking up considerable time and resources. The EPA has instead set standards for more than 50 industry categories. A particular facility may petition the EPA for a variance from the standard for its industry category if it can show that unique characteristics of that facility make the standard inappropriate.

The technological standard the EPA uses to set effluent limitations depends on the category of the pollutants being discharged by the industry: toxic, conventional, or nonconventional.[16] The CWA defines a "toxic pollutant" as one that after discharge and exposure or assimilation into any organism causes "death, disease, behavioral abnormalities, cancer, genetic mutations, physiological malfunctions (including malfunctions in reproduction) or physical deformations."[17] Assimilation into the organism can come either directly from the environment or indirectly by ingestion through the food chain. Additionally, the effects of the pollutant can be felt either by the exposed organism or in its offspring. The EPA lists 65 toxic pollutants, although some of the entries refer to groups of several chemicals.[18] To determine the technological standard for toxic pollutants, the EPA must use the "best available technology economically achievable" (BAT). The EPA can consider the costs of the technology in choosing the BAT, but may not directly compare costs with benefits.

Conventional pollutants are common pollutants of waterways, and are defined as biochemical oxygen demand, total suspended solids, pH, fecal coliform, and oil and grease.[19] Under the CWA, conventional pollutants are controlled through either the "best practicable control technology currently available" or the "best conventional pollutant control technology." Best practicable technology is a less stringent standard than BAT. In setting the standard for best practicable technology, EPA must perform a limited cost-benefit analysis, weighing the cost of the technology against the effluent reduction benefits. When Congress first enacted the CWA, it intended NPDES permits to be based on best practicable technology by 1977, and on BAT by 1983. Congress amended the CWA in 1977, though, fearing that effluent limits for conventional pollutants based on BAT would cost too much. Instead, Congress created best conventional technology as a new category. Best conventional technology is often a more stringent standard than best practicable technology, but less stringent than BAT. Best conventional technology limits are imposed on an industry if the EPA considers them to be cost-effective.

Finally, nonconventional pollutants are those that do not fall under the first two categories. They include such pollutants as chloride, nitrate, and iron. Effluent limitations for nonconventional pollutants typically must meet the BAT standard.

Newly built facilities often have tighter effluent limitations than existing facilities. The CWA requires that new facilities meet "new source performance standards" (NSPS).[20] Regardless of whether the facility is releasing toxic, conventional, or nonconventional pollutants, NSPS are often even stricter than BAT. The reasoning is that older facilities must retrofit their plant to meet the limitations, but new facilities can build pollution-control technology into the design of their plant. Consequently, new facilities can meet the tighter effluent limitations more easily than existing facilities. The tighter limits of NSPS tend to create perverse incentives, however. By creating stricter standards for new facilities, some companies find it less expensive to keep their existing facilities operating for as long as possible to avoid the stricter standards that would come with building a new facility. As a result, the requirement for stricter standards for new facilities may ultimately result in more pollution being discharged.

Indirect Dischargers. Some point sources do not discharge their pollutants directly into a body of water, but instead into a POTW. These point sources are called indirect dischargers.[21] Most POTWs are not capable of adequately treating the chemicals and other pollutants discharged by industrial facilities. In response, the EPA prohibits indirect dischargers from discharging pollutants to a POTW that interfere with the proper operation of the POTW, or will pass through the POTW untreated.[22] The EPA has also created categorical pretreatment standards that specify the concentrations of pollutants that an indirect discharger may release to a POTW. The standards often require indirect dischargers to meet BAT standards for toxic and nonconventional pollutants. It is assumed that POTWs can adequately handle conventional pollutants released by indirect dischargers. POTWs may also set their own more stringent limits for which pollutants may be discharged into the POTW.

Concentrated Animal Feeding Operations. So far this chapter has assumed that point sources are discrete facilities discharging pollutants literally through a pipe into water. The CWA includes other types of discharges as point sources, however. Although they do not seem to fit the definition of a point source, the CWA treats concentrated animal feeding operations and certain instances of storm-water runoff as point sources.

Animal feeding operations in the United States create roughly five million tons of manure annually. That is three times the amount of human sanitary waste produced each year in the United States.[23] Animal waste is high in nutrients such as nitrogen and phosphorus; if these nutrients are released into water bodies, they may lead to eutrophication (see box 7.1 for a description of eutrophication). Ammonia in animal waste may also be toxic to aquatic species if it reaches a high concentration in the water. The EPA found that between 1981 and 1999, at least four million fish were killed by runoff from large feeding operations.

BOX 7.1. EUTROPHICATION

Eutrophication occurs when a water body receives high levels of nutrients, such as nitrogen and phosphorus, leading to excessive growth of phytoplankton and aquatic plants. The growth may turn the water green and limit sunlight from reaching the bottom of the water body. When the phytoplankton and plants die, they fall to the bottom of the water body, where bacteria decompose the organic matter. Because so much organic matter is falling to the bottom, the bacteria quickly deplete the oxygen supply in the deeper water. As a result, the deeper parts of the water body can no longer support aerobic species, such as fish. The ecosystem of the water body may change dramatically, often losing considerable biodiversity. In shallow eutrophic lakes, the buildup of organic matter at the bottom may eventually fill in the lake, resulting in a marsh or swamp, or even dry land.

Eutrophication can occur naturally, as rain carrying nutrients from the land flows into a water body. However, if the runoff carries high concentrations of nutrients, such as occurs when rain washes over agricultural land where fertilizers have been applied, then eutrophication can happen at a much faster rate than would occur naturally. Sewage containing high levels of nutrients may also cause eutrophication if released into a water body.

The CWA specifically defines point sources to include "concentrated animal feeding operation[s]" (CAFOs).[24] CAFOs include operations such as cattle feedlots and poultry farms. The EPA defines a CAFO as an animal feeding operation that confines a specified number of animals; for example, a feeding operation with 1,000 head of cattle, 2,500 swine, or 125,000 chickens would be classified as a "[l]arge CAFO."[25] An animal feeding operation may also be designated a CAFO if it is a significant contributor of pollutants to US waters.

Because they are point sources, CAFOs must have NPDES permits to discharge pollutants into the waters of the United States. The permits use technology-based effluent limits that frequently prohibit the discharge of all manure, litter, or wastewater from the CAFO. There is an exception, though, for high rainfall events causing an overflow of manure, litter, or wastewater, which may then be discharged into water bodies.[26] The EPA estimates that more than one-third of the roughly 15,000 CAFOs in the United States that require NPDES permits have not applied for one.

There is another way that pollutants from CAFOs may reach water. CAFOs may legally spread manure on agricultural land for use as fertilizer. To get rid of all the manure that the animals produce, CAFOs frequently overapply manure on land.[27] The application of manure on phosphorus-saturated soils may result in large amounts of nutrients being carried to local water bodies by storm-water runoff. Storm-water runoff occurs when rain falls and the water flows over land areas, picking up pollutants from the land and carrying them back to nearby water bodies. In the past, the EPA did not regulate the land application of manure from CAFOs, but that changed in 2003 when the EPA issued rules to regulate it.[28] If manure is applied to land in a way that ensures "appropriate agricultural utilization of the nutrients in the manure," then the EPA considers runoff from that land to be agricultural runoff and not regulated under the CWA. If a CAFO applies excessive manure, though, the runoff is considered a point-source discharge regulated under the CWA.

Storm-Water Runoff. The CWA also regulates storm-water runoff under circumstances other than overapplication of manure.[29] In the early history of the CWA, all storm-water runoff was treated as unregulated nonpoint-source pollution. The EPA quickly began to require NPDES permits for some types of storm-water runoff, but it resulted in little actual control of storm-water discharge. Eventually, Congress in 1987 added a section to the CWA to establish a system for regulating storm-water runoff.[30] The system regulates storm-water discharges from three sources: industrial facilities, city storm-sewer systems, and construction sites.

Storm water often comes into contact with materials or equipment used at industrial facilities, thereby picking up pollutants that may flow into nearby water bodies. The CWA requires an NPDES permit for storm water discharged from any conveyance at an industrial facility that collects such storm water.[31] If the facility can prevent storm water from being exposed to any industrial material, then the EPA does not require a permit.[32]

Many cities have separate storm sewers for storm-water runoff. Such runoff often contains pollutants picked up from roads, lawns, and other urban areas. Cities that collect storm water separately from their POTWs must obtain an NPDES permit before discharging the storm water into a water body.[33]

Finally, storm water running off a construction site may carry pollutants such as sediment and debris to nearby water bodies. The owner of a construction site that disturbs more than one acre of land must receive a NPDES permit for storm-water runoff.[34]

Pharmaceuticals and Endocrine Disrupters

The CWA has been relatively good at reducing the discharge of pollutants that the EPA lists as toxic. There are many chemicals, however, that scientists believe harm organisms, but that are not listed as pollutants by the EPA and are therefore not regulated under the CWA. Of increasing concern is the impact that pharmaceuticals and endocrine disrupters are having on aquatic environments.[35]

Pharmaceuticals are chemicals that are manufactured for use as medicinal drugs. These chemicals were developed because they influence biological activity, so it should come as little surprise that when released into water bodies, they influence the biological activity of aquatic organisms. Pharmaceuticals as diverse as antibiotics, antidepressants, and beta-blockers may affect aquatic organisms.

Endocrine disrupters include pharmaceuticals such as estrogen and many pesticides, polychlorinated biphenyls, and bisphenol A. Endocrine-disrupting chemicals interfere with the normal endocrine function of organisms (the endocrine system secretes hormones into the circulatory system). Disrupting the endocrine system can lead to a wide range of effects, such as developmental and reproductive problems.

Pharmaceuticals and endocrine disrupters find their way into water bodies through several pathways. When a human uses a pharmaceutical, a large portion of the pharmaceutical passes through the body unmetabolized and ends up at the local POTW. The POTW waste treatment process does a poor job of breaking down pharmaceuticals before discharging them into water bodies. Industrial facilities may directly discharge pharmaceuticals into water, or indirectly via POTWs. Pharmaceuticals are also widely used in animal feeding operations and are found in the resulting animal waste. The animal waste may then find its way into water bodies, such as through rainwater runoff carrying manure spread on agricultural land. Finally, pesticides that act as endocrine disrupters are frequently used in agriculture, and enter water bodies when rainwater carrying the pesticides runs off the land into nearby water bodies.

All these different pathways mean that many waters in the United States contain pharmaceuticals and endocrine disrupters. In 1999–2000 the US Geological Survey looked for 95 different pharmaceutical and endocrine-disrupting chemicals in US waterways. The agency took samples from 135 streams in 30 states. They found that 80% of the samples contained at least one of the chemicals, with 54% containing more than five.[36]

The EPA currently lists very few pharmaceuticals or endocrine-disrupting chemicals as toxic pollutants. One reason is that the cost of requiring POTWs to remove all pharmaceuticals from the treated water they discharge would be enormous.[37] Additionally, as is discussed next, the runoff of pollutants from agricultural lands is not regulated under the CWA. As a consequence, even if these chemicals were no longer released by POTWs, many would still find their way into water bodies from agricultural runoff.

Once discharged into water bodies, these pharmaceuticals and endocrine disrupters can have a wide range of effects on aquatic ecosystems. Antibiotics that are discharged from POTWs may interfere with the natural bacterial community in a water body, evolutionarily selecting for bacteria that are resistant to those antibiotics.[38] Antibiotics may also decrease the growth of aquatic plants. At the same time, antidepressants cause premature spawning in shellfish, and influence egg-laying in snails and slugs.[39]

Estrogen compounds in water have several effects on aquatic vertebrates. The estrogen tends to result in feminization of males, abnormal sexual development, and impaired reproduction. In a study of streams in Nebraska, laboratory-raised fathead minnows (*Pimephales promelas*) were placed in streams containing POTW discharge of estrogen compounds.[40] After only seven days, the fish showed signs of feminization. In a separate study, white suckers (*Catostomus commersoni*) were collected upstream and downstream of a POTW in Colorado.[41] The discharge from the POTW contained several endocrine-disrupting chemicals. The frequency of males downstream of the POTW was half that of the upstream site. Moreover, intersex fish were not found upstream of the POTW, but intersex fish made up 18%–22% of the downstream sample.

The species potentially affected by pharmaceuticals and endocrine disrupters occur at all trophic levels in aquatic ecosystems. These chemicals may consequently have considerable impacts on the dynamics of aquatic communities. Unfortunately, there has been little empirical research to examine whether pharmaceuticals and endocrine disrupters can alter the dynamics of entire aquatic communities.

As an aside, the drinking water in many locations contains trace amounts of pharmaceuticals and endocrine disrupters. The Safe Drinking Water Act, the 1974 federal statute that sets standards for drinking water, regulates almost none of these chemicals.[42] Many water utilities do not test their drinking water for pharmaceuticals, and conventional methods for treating drinking water do not remove many organic contaminants.

Nonpoint Sources

Unlike point sources, the CWA does not clearly define what constitutes a nonpoint source. As a consequence, anything that conveys pollutants into waters of the United States and is not a point source is categorized as a nonpoint source. Most often, nonpoint sources of pollution are storm-water runoffs. As the section on animal feeding operations discusses, some forms of storm-water runoff are regulated as point sources under the CWA. All other types of runoff are nonpoint sources. In addition, the CWA explicitly states that storm-water runoff from agricultural land and irrigation water returning to a water body are not included in the definition of point sources, and are therefore nonpoint sources.[43]

Pollution entering waterways from nonpoint sources is considerably greater than pollution from point sources. For example, nonpoint sources account for 76% of nitrogen and 56% of phosphorus reaching water bodies in the United States. Agricultural runoff is the greatest nonpoint source of pollutants, and thus the greatest single source of water pollution.

Despite this, the CWA does very little to regulate pollution from nonpoint sources. The act essentially leaves the regulation of nonpoint sources up to the states. The CWA requires the states to create a management program for nonpoint sources. Under the state programs, nonpoint sources must use the "best management practices" to reduce pollution runoff at the "earliest practicable date."[44] Best management practices include actions such as terracing land or better timing in the application of fertilizer. Because runoff is diffuse and enters water bodies from many different places, the states are encouraged to take a watershed-by-watershed approach in creating their management programs.[45] The EPA may authorize annual grants to states to help implement their management programs, although the grants must be less than 60% of the cost of the state's management program.

Other than the EPA's refusing to give a state a grant, though, the CWA does not include any provision for the federal government to enforce state nonpoint-source management programs. If a state does not implement its management program, the federal government may not impose a program on the state. The inability of the CWA to enforce management of nonpoint sources is likely a result of political pressure, as the agricultural lobby has been very resistant to regulation of nonpoint pollution.[46]

Three main pollutants enter water bodies from nonpoint sources: sediment, poisonous substances, and nutrients.[47] Sediment is soil particles that have been carried into a water body. When these sediment particles become suspended in the water column, they can reduce photosynthesis by algae, thereby reducing the productivity of the ecosystem. Suspended sediments can also reduce the ability of sight-feeding fish to feed. Additionally, sediment can act as a transporter, bringing attached pollutants into the water. Sediment carried into water is approximately five times greater from cropland than from forestland. Urban areas also tend to increase the sediment in water, as does livestock grazing and severe wildfires.

Poisonous substances coming from nonpoint sources include herbicides from agricultural land and pesticides from urban land. Heavy metals from the tailings of abandoned mines and from traffic exhaust are an additional source of poisonous substances in water. Poisonous substances may slow the growth of, or kill, aquatic organisms.

Finally, excess nutrients, most often phosphorus and nitrogen, can lead to rampant algal growth. Such growth can deplete the dissolved oxygen in the water body, suffocating fish and other aquatic organisms. Excess nutrients may lead to the eutrophication of water bodies. Fertilizer for crops and manure from livestock are the main sources of nutrients from agricultural lands. Streams draining watersheds that contain predominantly agricultural land had nutrient loads that were nine times higher than streams draining forested watersheds.[48]

In a study examining over 15,000 watersheds, Brown and Froemke found a dramatic divide between watersheds on either side of the 100th meridian.[49] The 100th meridian roughly cuts the contiguous United States into an eastern half and a western half. Watersheds to the east of the meridian had much higher pollution levels than watersheds to the west of the meridian. The reason is that the eastern half of the contiguous United States has a higher human population, less severe topography, and more precipitation. All of these factors have encouraged greater agricultural development, which leads to considerable pollution of water bodies. Conversely, the western half of the country has fewer people, less precipitation, and more severe topography, such as the Rocky Mountains; and much of the land is owned by the federal government, meaning it is forestland and rangeland. All of these factors reduce agricultural development and pollution levels. Additional research examining how the more polluted watersheds of the eastern half of the United States affect species and ecosystems that have ranges stretching across both halves of the contiguous United States would be interesting.

An example of several ways that nonpoint-source pollution can affect the ecosystem of a water body is found in Lake Okeechobee.[50] Located in Florida, Lake Okeechobee has faced unrelenting ecological change from human activity. Runoff from agricultural land and urban development contributes nitrogen and large amounts of phosphorus to the lake. Several canals empty into Lake Okeechobee; being canals, they lack substantial vegetation. As a result, the canals allow phosphorus in agricultural runoff to be discharged to the lake unimpeded. Tributaries also discharge into the lake, often carrying fecal coliform bacteria from nearby animal feeding operations. In addition, multitudes of nonnative species have been introduced into the lake.

The pollution in Lake Okeechobee has caused considerable ecological change. The high levels of phosphorus allow noxious cyanobacteria to bloom. Pollution has resulted in large increases in the populations of pollution-tolerant macroinvertebrates such as aquatic worms (oligochaetes). Unlike many aquatic insect species, aquatic worms do not have an adult stage that emerges from the water. Many species, such as migratory waterfowl and some fish species, rely on emergent adult insects for food. As aquatic worms have come to dominate the lake, the decrease in aquatic insects has meant a reduction in an important food resource for many other species.

There is evidence that endocrine disrupting chemicals in Lake Okeechobee are affecting the growth and development of American alligators (*Alligator mississippiensis*). Studies have found that male alligators in the lake have small phalluses and reduced androgen hormones (hormones that control the development of male characteristics), likely due to the presence of endocrine disrupters in the water.[51] The endocrine-disrupting chemicals in the water are almost certainly affecting many aquatic species in Lake Okeechobee.

Finally, the aquatic vascular vegetation of Lake Okeechobee has also changed. The high level of phosphorus in the lake has stimulated growth of nonnative cattails (*Typha* spp.). Cattails have outcompeted and displaced the native spikesedges (*Eleocharis* spp.). Spikesedges create excellent habitat for fish to forage in; cattails, on the other hand, grow so densely that they provide very poor habitat for fish. The dense growth of cattails may also adversely affect birds such as the endangered snail kite (*Rostrhamus sociabilis*), which needs relatively open habitat to see its prey.

As the example of Lake Okeechobee suggests, pollution from nonpoint sources can affect the ecology of a water body in multiple ways. Enough pollution entering a water body may alter the interactions within, and the composition of, the entire aquatic community.

Water-Quality Standards

As the previous section suggests, the primary way that the CWA protects the waters of the United States is not by regulating nonpoint sources, but by setting effluent limitations for point sources in NPDES permits. These effluent limitations are based on what is achievable technologically, not on how the pollution released by the point source is affecting a water body. As a consequence, the pollution entering a water body from point sources under the NPDES, combined with the pollution coming from nonpoint sources, may be high enough to severely degrade a water body. There is a secondary way that the CWA attempts to protect the water quality of individual water bodies, though: the CWA regulates ambient water quality.

Ambient water quality is the concentration of pollutants in a specific water body. To regulate ambient water quality, the CWA first requires that every state designate appropriate uses for the water bodies in their borders.[52] The uses include public water supply; protection of fish, shellfish, and wildlife; recreation; agriculture; industry; and navigation.[53] States must review the designations of their water bodies every three years.[54]

After the designation of water bodies, states must then determine the water-quality criteria necessary to achieve that designation. Criteria may include the chemical, physical, and biological condition a water body must attain to reach its designation. The quality of the aquatic ecosystem that must be met for a water body to achieve its designation is called the biological criteria. Biological criteria and the tests performed to assess the quality of aquatic ecosystems will be discussed in the next section.

Once the water-quality standards have been set, the CWA requires each state to also identify the water bodies in its borders that are "quality limited."[55] These water bodies are unable to attain their water-quality standards even with the effluent limitations of the NPDES. For a quality-limited water body, a state must determine the "total maximum daily load" (TMDL) of each pollutant that could be discharged into the water body to allow it to meet its water-quality standard.[56] To achieve the TMDL, point sources are required under the CWA to reduce their discharge of pollutants below what the technology-based effluent limitations for that type of point source would normally be.[57] As a result, a point source discharging into a water body that is quality limited and has a TMDL will have lower effluent limitations than other facilities of that type discharging into waters that are not quality limited. If a state does not create a list of its quality-limited waterways and TMDLs, the EPA must prepare its own list for the state.[58] However, the CWA does not give the EPA authority to actually implement those TMDLs.

Often a water body will be quality limited not from pollution from point sources, but from pollution from nonpoint sources. The TMDL standards apply to waters even if they are polluted primarily by nonpoint sources.[59] As discussed previously, the EPA lacks the authority to regulate nonpoint sources under the CWA. Consequently, point sources may be forced to lower their discharge of pollutants to meet the TMDL even though nonpoint sources continue to be the primary polluters of a water body. Any attempt to meet TMDLs by regulating nonpoint sources must be done by the individual states.

As it is often difficult and costly to develop TMDLs for each body of water, states have begun creating TMDLs for entire watersheds.[60] The largest watershed program is the TMDL the EPA created for the Chesapeake Bay. Several states are in the Chesapeake Bay watershed, and a watershed-based TMDL coordinated by the EPA is much more efficient than each state creating its own TMDL. Under the Chesapeake Bay TMDL, each state in the watershed is given a pollution budget, thereby ideally reducing the pollutants entering the bay from each state.

Despite the goal of the CWA to protect ambient water quality, many water bodies in the United States are still highly polluted. In a 2009 report, the EPA stated that in the US waters it had assessed, 44% of rivers and streams, 64% of lakes, and 30% of bays and estuaries were impaired, meaning they were not clean enough to support designated uses such as fishing and swimming.[61]

Bioassessment

As scientists began to realize that pollution in water bodies was negatively affecting aquatic ecosystems, they started to create tests to determine how organisms and ecosystems were responding to pollution. There is a long history of such bioassessments, dating back to the early twentieth century.[62]

Bioassessments provide a relatively quick means of determining the health of a water body.

Recall that biological criteria are the qualities that must be present in a body of water for it to meet the designation made by the state for that water body. A bioassessment defines the present state of the water body, with the biological criteria being the goal for the water body. Bioassessments ultimately allow a state to determine when it has reached the biological criteria for a given water body.

The most common bioassessment performed by the EPA and the states surveys a water body to determine the presence, condition, and number of fish, invertebrate, amphibian, algae, and plant species. The list of observed species is then compared to the list of species expected to be present if the water quality in the water body had not been degraded. This type of bioassessment measures only ecosystem structure. Ecosystem structure consists of those characteristics of an ecosystem that can be measured at a specific point in time. Measurements of ecosystem structure are often preferred in bioassessments because they are relatively easy and inexpensive to perform. The structure of an ecosystem is assumed to indicate the overall condition of that ecosystem.

Measurements of ecosystem structure alone, however, may not adequately indicate the true condition of an aquatic ecosystem. The dynamic properties of an ecosystem are equally important in determining its condition. In an analogy from Palmer and Febria, a doctor measures not only the structural attributes of a patient, such as height and weight, but also the dynamic properties, such as heart rate and blood pressure, to determine if a patient is healthy or not.[63] The dynamic properties of a water body are determined through measurements of ecosystem function. Ecosystem function includes such properties as productivity, reproduction and migration of species, pollutant removal rates, decomposition rates, and nutrient-cycling rates. Measurements of ecosystem function must be taken repeatedly through time, and are thus more difficult to obtain.

Combining measurements of ecosystem function with measurements of ecosystem structure should provide the best indication of the condition of an aquatic ecosystem.[64] Structural measurements alone may be good indicators of when something is wrong in an ecosystem. For instance, the disappearance of a keystone species in a river would clearly suggest that the river ecosystem is in trouble. However, functional measurements will often be necessary to determine what precisely is wrong with a water body, and how to fix its ecosystems.

There is little scientific agreement on what is the best combination of structural and functional measurements to indicate the true condition of aquatic ecosystems. Part of the reason is that very few studies have examined both structural and functional measurements of the same ecosystem. While more research is necessary to determine the best combination of measurements, the EPA and states would certainly gain a better understanding of the true condition of water bodies by measuring both the structure and function of waters of the United States.

Water-Quality Trading

As mentioned earlier, a TMDL for a water body may require that point sources reduce their discharge of pollutants below that required by technology-based effluent limitations. Reducing discharges will often be very costly for a facility if it has to install new technology or operate below its full capacity. As one way to help point sources meet effluent limitations in a cost-efficient manner, the EPA has begun to allow water-quality trading.

Under the water-quality trading program, a point source that wants to discharge pollutants above its effluent limitation may do so by first buying a credit from a different point source that has reduced its discharge of pollutants below its limitation.[65] Trading programs can also allow a point source to buy credits from a nonpoint source. Trades between point sources and nonpoint sources have the potential to be very cost efficient. Point sources have been regulated under the CWA for over 40 years, and many have already made all the inexpensive changes to reduce pollutant discharges that can be made. Further reductions in discharges, such as would be required if a new TMDL were established, would likely require an expensive investment in control technology. Conversely, nonpoint sources are not regulated under the CWA and may be able to reduce the runoff of pollutants by making small, inexpensive changes to their practices. For example, a farmer may install a riparian buffer between agricultural land and a stream, thereby preventing a significant percentage of the pollutants carried off the land from reaching the stream. Water-quality trading programs do not require a TMDL for their existence, but TMDLs are often the motivation for the creation of such programs.

Given the complexity of aquatic ecosystems, however, water-quality trading may not be ecologically neutral. If a trade reduces the discharge of pollutants into a water body that is not particularly ecologically sensitive, but increases discharges into a sensitive water body, then the net ecological outcome will be negative. For example, a trade that results in more pollutants being discharged into a stream where a particular species of fish spawns may have a large ecological impact. The EPA has created policies for the water-quality trading program that attempt to account for this complexity. EPA policy states that water-quality trades should take place within the same watershed. This should help ensure that nonpoint sources in watersheds located in rural areas do not become suppliers of credits while point sources in watersheds located in urban areas become buyers of credits, thereby creating highly polluted urban water bodies. EPA policy also states that if a downstream source buys credits from an upstream source, the upstream source must usually reduce its discharge of pollutants below the amount of credit sold to the downstream source. The reason is that the concentration of many pollutants, such as phosphorus, decreases as it moves downstream. Phosphorus is taken up by aquatic plants, settles out of the water, and is diverted with water used for agriculture. As a result, an upstream source may have to reduce its discharge of a pollutant by several pounds to be equivalent to a pound of pollutant discharged by a downstream source.

The difficulty in creating equivalency in trades increases when the trade is between point sources and nonpoint sources. While it is possible to monitor the pollution discharged from the end of a pipe by a point source, monitoring the pollution released by a nonpoint source is extremely difficult. Water-quality trading programs assume that when a nonpoint source undertakes a particular activity, such as installing a riparian buffer, it will result in a certain percentage reduction in pollutants reaching the water. The amount of pollution that is actually prevented from reaching the water is rarely if ever measured. Some trading programs introduce trading ratios in an attempt to compensate for the uncertainty in the amount that nonpoint sources reduce pollution. A trading ratio of 2:1 would require a point source to purchase two pounds of pollutant reduction from a nonpoint source for every pound of discharge by the point source.[66] Trading ratios act as insurance against the uncertainty inherent in the ability of nonpoint sources to reduce pollution by a given amount.

Despite the uncertainty, trades between point sources and nonpoint sources may have an unintended benefit for the environment. A trade between a point source and a nonpoint source will often occur for only a single pollutant. Actions taken by the nonpoint source to reduce that single pollutant, though, may end up reducing the runoff of several pollutants. For example, a nonpoint source may sell credits for reducing phosphorus after putting in place a riparian buffer. The buffer would not only reduce the amount of phosphorus reaching the water; it would also reduce the amount of nitrogen and other nutrients reaching the water. As a result, trades between point sources and nonpoint sources may prevent more pollutants from reaching water bodies than simple reductions in discharges by point sources.

Two additional difficulties in water-quality trading programs should be considered. First, depending on how particular pollutants act when discharged into water, trades may result in "hot spots" where pollutants have a high local impact. This may be especially true for point sources located close to each other in urban areas making trades with sources located far away in the watershed. Second, there may be differences in the timing of pollutant discharges between sources making trades. For example, a farm will likely release few pollutants during the winter when fields are fallow. An industrial facility, on the other hand, may discharge at the same rate throughout the year. If the farmer sells credits to the facility, all the reduction in pollutants from the farm will have to come during the summer months. As a result, pollution levels in the watershed during the summer will be lower after the trade, but in the winter, pollution levels will be higher after the trade. The reason pollution levels will be higher in winter is because the farm will still be fallow during the winter, but the facility will be discharging at a higher rate throughout the year, including the winter months. The EPA has policies that attempt to compensate for these possibilities. However, the equivalency of all of these different types of trades is determined by modeling the behavior of pollutants in individual watersheds. The quality of the models created for the watersheds play a large role in determining the true equivalency of water-quality trades.

Despite the potential for cost efficiency, water-quality trading has been relatively rare. By 2008 only 100 facilities had engaged in trading, and 80% of the trades had occurred in only one trading program, the Long Island Sound program. Water-quality trading may become more prevalent, though, as regulators continue to push market-based mechanisms to achieve environmental goals.

WATER WITHDRAWAL

The federal government has broad authority to manage water as a resource under constitutional sources such as the commerce clause, which allows federal regulation of water involved in or affecting interstate commerce.[67] The federal government, however, often defers to the states in determining how to allocate water resources. For example, the CWA states that nothing in the act is meant to supersede the authority of the states to allocate water within their jurisdictions.[68] As a result, the states have split into two camps on how they allocate water: the eastern states generally follow the riparian doctrine, while the western states follow the prior appropriation doctrine.

The riparian doctrine has its origins in English common law. Under the riparian doctrine, owners of land next to a waterway have the right to use the water for any reasonable purpose. If there is a water shortage under this doctrine, all water users must face the shortage in proportion to their water right.

The prior appropriation doctrine has its origins in the western states in the 1800s. Gold rush miners needed water to operate their mines, so they applied mineral ownership rights to water. Under the prior appropriation doctrine, water rights do not arise out of landownership. Instead, individuals who obtain water rights first have seniority over those who obtain rights later. Additionally, those who obtain water must put the water to beneficial use or they lose their water rights. If there is a water shortage under the prior appropriation doctrine, those who obtained their water rights last in time have their water share reduced or eliminated, while those with seniority receive their normal share of water. Some states have passed legislation that allows power plants and other water users that fulfill essential public needs to have preference over more senior water rights in times of shortage.[69]

The two biggest users of surface water in the United States are power plants and agriculture. In 2005 power plants accounted for 53% of fresh surface-water withdrawal in the United States. Trillions of gallons of water are withdrawn every year for generating electricity with steam-driven turbine generators. Agricultural irrigation accounted for 28% of fresh surface-water withdrawal. Accounting for considerably less surface-water withdrawal were industrial use at 5% and domestic use at 0.03%.[70]

For the most part, power plants take water from a water body for cooling, and then return that water to the water body. Agricultural activities tend to consume most of the water that is withdrawn. The CWA regulates the way in which power plants withdraw water, but the decision on how much water power plants may withdraw is left up to the states. The states also decide how much water may be withdrawn for agricultural activities. The withdrawal of water by both of these users has enormous implications for aquatic ecosystems.

Power Plants

Power plants withdraw water from water bodies such as rivers, estuaries, lakes, and the ocean to cool machinery and drive turbines. There are two types of power plants: open-loop and closed-loop facilities. Open-loop power plants withdraw water from a water body, use the water to drive and cool turbines, and then discharge the heated water back into the water body. Closed-loop power plants withdraw and use water, but then use cooling towers or cooling ponds to allow the water to be reused. Closed-loop power plants withdraw less water than open-loop plants, as they need to withdraw water only to replace that lost to evaporation and leakage. Closed-loop plants, however, consume more water than open-loop plants because open-loop plants return almost all the water withdrawn back into the water body from which it was taken.

Water withdrawal by power plants kills a vast number of aquatic organisms. The EPA estimated in 2004 that large power plants kill 3.4 billion fish and shellfish annually. The withdrawal traps organisms against the water intake structure (impingement), or pulls them inside the facility (entrainment). If the power plant is open-loop, the discharged heated water may also kill organisms.[71] For fish species, water withdrawal often means the death of eggs, larvae, and juveniles that are less capable of escaping the water intake structure than adult fish. For example, striped bass (*Morone saxatilis*), delta smelt (*Hypomesus transpacificus*), Sacramento splittail (*Pogonichthys macrolepidotus*), and several other fish species in California are killed in enormous numbers by the cooling-water intakes of the Contra Costa Power Plant in the San Joaquin River and the Pittsburgh Power Plant in Suisun Bay. Each plant has the capacity to withdraw more than 1.1 million liters (300,000 gallons) of water per minute.

The CWA regulates cooling-water intake structures by requiring the "best technology available for minimizing adverse environmental impact."[72] The EPA is in the process of issuing new rules for cooling-water intake structures after an appeals court stopped the EPA from applying the old rules.[73] Even after the new rules are finalized, however, water withdrawals will continue to kill large numbers of aquatic organisms.

Unfortunately there have been relatively few recent empirical studies of the effects of cooling-water withdrawal on aquatic populations, and even fewer examining the effects on ecosystems. Deaths from cooling-water withdrawals are often density independent. Density-independent mortality means that the probability of any individual in a population dying is unrelated to the size of the population. This is in contrast to density-dependent mortality, where the probability of an individual dying is related to the size of the population, often because of increased competition for resources.

A series of models by Newbold and Iovanna predict how the density-independent mortality caused by power plant withdrawals may influence aquatic populations and ecosystems.[74] Unsurprisingly, the models predict that cooling-water withdrawals will reduce the populations of species that are killed by the withdrawals. Using life-history parameters and mortality rates from water-cooling withdrawals, a separate study suggested that the striped bass population in California mentioned earlier is 30% smaller than it would be if there were no cooling-water withdrawal.[75] If a population is already small and has a relatively low growth rate, density-independent mortality from water withdrawals may drive the population locally extinct. Mortality from cooling-water withdrawals also tends to reduce the resilience of aquatic populations, so that such populations take longer to bounce back from random shocks.

Intuitively, one would assume that as cooling-water withdrawals increase, more individuals in an affected population would die. This may not always be the case. If cooling-water withdrawal is large enough, a population will shrink to a low number. Any additional withdrawals will continue to reduce the size of the population, but given the now small population size, the absolute number of deaths will be small. As a result, a regulator monitoring the effect of cooling-water withdrawals on aquatic populations may see relatively few individuals killed and assume it is because the withdrawals are not affecting aquatic species. The reverse may be true, though, as the water withdrawals may be inducing so much mortality that affected populations have been pushed to very low levels. Conversely, attempts to reduce mortality from withdrawals may increase the number of individuals killed as aquatic populations increase. Regulators must monitor both the number of individuals killed by withdrawals and the population sizes of affected species to know the true effect of the withdrawals.

Modeling suggests that the ecological ramifications of cooling-water withdrawals become even more complicated when considering an ecosystem of several interacting trophic levels. The Newbold and Iovanna study assumes four trophic levels, with the lowest representing primary producers such as algae, the next representing primary consumers such as forage fish, above that secondary consumers that prey on the primary consumers, and finally top predators such as marine mammals. The model also assumes that all but the highest trophic level faces mortality from cooling-water withdrawals. As mortality from cooling-water withdrawals increases, the abundance of lower trophic levels may actually increase due to reductions in the abundance of higher trophic levels that prey on the lower levels. Cooling-water withdrawals may alter the structure of aquatic ecosystems, resulting in less biomass in the aquatic ecosystem overall, but relatively more biomass in lower trophic levels. As mortality from withdrawal increases further, though, each trophic level is pushed to local extinction one by one, starting with the highest trophic level. More empirical research is needed to determine how generally the results of these studies apply to aquatic ecosystems.

As already mentioned, open-loop power plants withdraw water to use in generating electricity, and then discharge the heated water back into the water body. Under the CWA, the definition of pollutant includes "heat" discharged into water.[76] This means that the EPA has authority to regulate thermal pollution discharged by power plants. Because power plants are point sources, they must obtain an NPDES permit before discharging thermal pollution into US water bodies. However, a different section of the CWA allows the EPA or an authorized state to grant less stringent effluent limitations for thermal pollution.[77] This is called a thermal variance. The EPA or state may grant a thermal variance if the power plant demonstrates that the more relaxed limitations will protect a "balanced, indigenous population of shellfish, fish, and wildlife" in the water body. In part because of thermal variances, power plants across the country release billions of liters of heated water into water bodies every day.

This thermal pollution can have extensive effects on an aquatic ecosystems. For instance, cold-water species, such as trout, may be replaced by warm-water species, such as carp, within the thermal plume. Warmer water may make some aquatic species more vulnerable to pollutants or pathogens already in the water. Thermal pollution results in less dissolved oxygen because warm water holds less oxygen than cold water. Aquatic species that are sensitive to oxygen concentration may move away from the warm water. Thermal pollution also reduces dissolved oxygen levels in an indirect way. Warm water increases the growth of aquatic plants. When the plants die, they are decomposed by bacteria, which further reduces the level of dissolved oxygen. All the ecological effects arising from thermal pollution may result in reduced biodiversity and altered structure in aquatic communities.

While cooling-water withdrawals and subsequent return to water bodies appear to have large effects on aquatic ecosystems, there may be considerable drawbacks in requiring power plants to reduce their intake of water. If a power plant lessens its cooling-water withdrawals, the efficiency of the plant will decrease. This drop in efficiency may result in greater emissions of particulate matter or other air pollutants. The benefit to aquatic ecosystems of reducing cooling-water withdrawals must be weighed against the environmental harm caused by increased emissions of pollutants into the air.

Consumptive Withdrawal

While open-loop power plants withdraw huge amounts of fresh surface water, most of that water is returned to the water body from which it was taken. The water is not consumed. Agricultural irrigation and other forms of water withdrawal, on the other hand, do consume water. This often affects aquatic ecosystems differently than the nonconsumptive withdrawal of many power plants.

As water is consumed from a stream, the width of the stream tends to decrease, as does the depth and velocity of the water in the stream. Low stream flow reduces riffle habitats, which are the shallow areas of streams with faster water flow relative to pool areas that are deeper. The loss of riffle habitats frequently results in the loss of species that depend on such habitats for survival. In general, as water flow in a stream decreases, species richness also decreases.[78] Low stream flow has other ecological effects, such as inducing individual fish and insect organisms to have smaller body sizes.

In a study of streams in Georgia, water consumption for municipal use altered fish assemblages. Water withdrawal led to the loss of specialist fish species that depend on fast-flowing water, and resulted in dominance by habitat generalist species, such as sunfish (*Lepomis* spp.), largemouth bass (*Micropterus salmoides*), and crappie (*Pomoxis* spp.).[79] In a study that experimentally altered stream flow, the total biomass of aquatic insects decreased significantly.[80] The loss of riffle habitats also led to a particularly low density of insects in the remaining riffle habitats.

As these studies show, water consumption alters aquatic ecosystems mainly by reducing habitats and decreasing species richness. As the flow of water in a stream decreases even further, though, the stream becomes a series of disconnected pools, resulting in an ecosystem that only vaguely resembles a stream ecosystem. In polluted waterways the reduced flow of water may mean that there is less water to dilute pollutants, thereby causing higher concentrations of downstream pollution. All these outcomes from reducing stream flow will almost certainly interact in complicated ways in aquatic ecosystems. As mentioned earlier in the chapter, it is almost entirely up to the individual states to decide how much water people may consumptively withdraw. It is therefore up to each state to determine how consumptive withdrawal is impacting the aquatic ecosystems within the state and in downstream states.

DAMMING RIVERS AND STREAMS

There are a surprisingly large number of dams on rivers and streams in the United States. In 2010 there were approximately 84,000 dams in the United States. Almost all of those dams are owned by individuals, municipalities, or states. The federal government owns only 4% of the dams in the United States.

Building dams can have a variety of negative impacts on ecosystems.[81] For instance, if the dam creates a storage reservoir, the reservoir floods a substantial amount of land. This flooding results in a large loss of habitat, and will kill an enormous number of terrestrial plants. The creation of a reservoir may even have global implications. When a reservoir floods large amounts of vegetation, that vegetation begins to decay. Dams also stop the downstream flow of organic carbon. As all this flooded and trapped organic matter decays, there are large releases of greenhouse gases, such as methane and carbon dioxide. As a consequence, dams can act as significant contributors to global climate change.

Dams may also negatively affect the aquatic ecosystem downstream of the dam. A dam will likely alter the natural flow regime of the river. Naturally occurring pulses in river flow trigger migration or spawning for many aquatic species, and scour away sediments, helping to maintain aquatic habitats. If the dam alters this flow regime, there may be losses of habitat and disruption to migration and spawning. Additionally, if water is released from the reservoir to the downstream river, it may be a different temperature than the water that would naturally be in the stream. Water released from the bottom of the reservoir will be cooler in summer and warmer in winter than natural river water, while water released from the top of the reservoir will be warmer than natural river water throughout the year. These differences in water temperature may disrupt the life cycles of aquatic species, reducing the diversity of species for several kilometers downstream. The water flowing out of the dam may also become supersaturated with gases. This may cause fish to suffer a condition similar to the bends, often killing many downstream fish. On top of all this, dams may block anadromous species such as salmon, or catadromous species such as eels, from migrating upriver or downriver, respectively, to spawn. Finally, dams frequently block sediment from moving downstream. This may eventually result in the elimination of beaches and coastal deltas that rely on upstream sediment for their existence.

Many major rivers contain multiple dams. These dams fragment the aquatic ecosystem, in much the same way that roads and farmland fragment terrestrial ecosystems. Additionally, water quality continually worsens as water passes through each successive dam. For example, the Grand Coulee Dam on the Columbia River receives water that has come through dams upstream in Canada. When the river water reaches the Grand Coulee Dam, it is already high in total dissolved gases from the upstream dams. As the water passes through the dam, the dissolved gas levels increase even further. The ability of successive dams to progressively impair water quality results in increasingly altered aquatic communities, and often results in environments more conducive to nonnative species than to native species.

Some of the largest and most ecologically disruptive dams are those that produce hydroelectric power. Roughly 1,940 dams in the United States produce hydroelectric power; of these, the federal government owns 240. Hydroelectric power accounts for approximately 7% of electricity generation in the United States, but 80% of renewable electricity generation. Dams that produce hydroelectric power, and are not owned by the federal government, are regulated by the Federal Energy Regulatory Commission (FERC) under the Federal Power Act.[82] Nonfederal dams that do not produce hydroelectric power are regulated by the states.

The primary way in which FERC regulates hydropower dams is through issuing licenses for their construction. FERC also relicenses hydropower dams every 30 to 50 years. Before filing a license application with FERC, the applicant must conduct studies to determine the environmental impacts associated with the proposed dam.[83] The applicant must also consult with other agencies, such as the EPA and US Fish and Wildlife Service, who may have concerns about the proposed project. This information is then used in submitting a license application to FERC. Once FERC receives the application, it evaluates the project using the NEPA framework (see chapter 5). FERC must give equal consideration to how the project will influence power production, energy conservation, recreational opportunities, the "protection, mitigation of damage to, and enhancement of, fish and wildlife," and other aspects of environmental quality.[84] Finally, FERC must consider whether the project will "serve the public interest."[85]

In issuing a license or a relicense, FERC may require changes be made to a dam to benefit aquatic ecosystems. A FERC license may require a dam to continually release a certain amount of water from its reservoir to maintain a minimum water flow in the downstream river. This minimum flow may reduce the peak power that the hydroelectric dam can achieve, but will often be necessary to maintain downstream habitats for aquatic species.

There may also be a requirement for the release of pulses of water from the dam. As mentioned above, natural pulses in a river help maintain habitat and provide cues for spawning, hatching, and migrating.[86] FERC may require that the pulses released by the dam replicate natural pulses. Dam owners often want to produce pulsed flows at other times, though, to produce peak energy when required by electricity customers. Pulsed flows that occur at the wrong time or that are too large, however, may have adverse consequences. Such pulsed flows can result in reduced spawning, downstream displacement of organisms, and reduced rearing survival. Unfortunately, the scientific research on the ecological effects of pulsed flows from dams is limited. Additional research is needed on how large a pulsed flow from a dam can be before it begins to harm the aquatic ecosystem.[87] Other areas for research are how pulsed flows affect water quality, the cumulative effect of many pulsed flows throughout a year, and how to mitigate the negative effects of pulsed flows. Research on these topics will help FERC when licensing hydropower dams to decide when to require pulsed flows and when to prohibit them.

In licensing or relicensing, FERC may also require the installation of technology to allow fish passage beyond a dam. The most common technology is a fish ladder that allows fish migrating upstream to jump up several steps and eventually into the water beyond the dam (figure 7.1). Fish ladders, however, are quite poor at facilitating fish passage. More than half the fish attempting to migrate past a dam to spawn may not make it up the fish ladder.[88]

As might be expected, the inability of so many fish to make it to spawning grounds reduces the fish population. Perhaps less expected, fish ladders also reduce the flow of nutrients upstream. Anadromous fish species that migrate upstream to spawn and then die move considerable amounts of nutrients from the oceans upstream to rivers. Reducing the number of fish that are able to make it upstream has the result of reducing nutrients in headwaters.

In another unexpected consequence, installing fish ladders acts as an evolutionary force on fish populations.[89] The fish that are able to make it up the fish ladder and reproduce may be different in size or have more white muscle, which increases their ability to jump up the steps, than fish that do not make it up the ladder. The fish ladders are therefore selecting for certain traits in the species that use the ladders. Whether this selection influences the dynamics of fish populations and their larger aquatic communities needs further investigation.

Finally, most fish-passage technology is based on salmon. Fish ladders that somewhat work for salmon may not work at all for other anadromous

FIGURE 7.1. Fish ladder on Columbia River, Washington. Photograph from FWS.

fish species. Even worse, there is often no fish-passage technology at dams for other types of migratory fish, such as catadromous fish species that are headed downstream to spawn.

While FERC has the greatest control over the construction and function of hydropower dams, other federal statutes have a large influence on where dams may be located. One such statute is the Wild and Scenic Rivers Act, which prevents dams from being constructed on certain rivers.[90] The act creates a system for designating rivers that are to be protected from development. Rivers may be designated if they possess "outstandingly remarkable scenic, recreational, geologic, fish and wildlife, historic, cultural, or other similar values."[91] A river may be designated either by an act of Congress, or by the secretary of the interior after an act of a state legislature and request of the governor. Under the Wild and Scenic Rivers Act, no federal agency is allowed to assist in any water resource project that would have a direct and adverse effect on a designated river. The statute specifically states that FERC may not license the construction of a dam on a designated river. The statute protects 203 rivers, covering more than 20,200 kilometers (12,600 miles);

however, this is less than 0.25% of the rivers in the United States. Many of the river sections that are designated under the act would be excellent sites for hydropower dams. Consequently, the Wild and Scenic Rivers Act is an important statute for precluding the construction of dams.

Similarly, as discussed in chapter 5, the Wilderness Act prohibits any commercial enterprise, structure, or installation within a national wilderness area.[92] This prevents the construction of dams within such areas. Additionally, the Energy Policy Act of 1992 prohibits FERC from issuing a license for a hydropower dam within the boundaries of the National Park System if it would have a direct adverse effect on the land within the park.[93]

Finally, the CWA plays an important role in determining where hydropower dams may be built. One could be forgiven for thinking that the power to control where dams are built comes from the NPDES. After all, a dam seems to clearly meet the definition of a point source. Most dams release water from pipes into navigable waterways of the United States. The water a dam releases may be a different temperature or contain less oxygen than the water it flows into, both of which may negatively affect fish and other aquatic populations. Surprisingly, the courts and EPA have interpreted the NPDES as not applying to dams.

As previously mentioned, the CWA defines "discharge of a pollutant" as "any addition of any pollutant."[94] As is common in the law, the precise meaning of a single word, in this case "addition," can determine whether a statute applies to a broad range of actions. What does it mean to add pollution to a waterway? The CWA does not define "addition," and courts have found some activities to be additions while others are not. Several courts have held that water moving from one side of a dam to another is not an addition of pollutants.[95] The EPA finalized a regulation in 2008 codifying the exclusion of water transfers from one water body to another from the NPDES.[96] This includes the movement of water from one side of a dam to the other side. In fact, the EPA does not even consider the movement of water from one side of a dam to the other to be a "transfer" of water because the water is all part of the same river.[97] As a result, an NPDES permit is not necessary for the operation of a dam.

The CWA, though, has a different provision that does help control where a hydropower dam may be built. Under the CWA, any applicant for a federal license for a project that may result in a discharge into navigable waters must receive a certificate from the state where the water is located, indicating that the project will comply with state water-quality standards (often called Section 401 certification).[98] Given the interpretation of the NPDES, it would seem logical that dams would not qualify as dischargers under this provision of the CWA, and Section 401 certification would not be necessary for dams. The Supreme Court, however, has held that all activities, including the release of water from a dam, must comply with state water-quality standards.[99] Consequently, the state may consider the potential activities of a dam, such as providing a minimum water flow from the dam, as part of the Section 401 certification process. As a result, applicants to FERC for a license to build a dam must also receive the approval of the state where the dam is to be located. States therefore have the ability to prevent a dam from being built in a certain area, or to require changes in the dam to protect water quality, before approving the project.

There has recently been movement away from building new dams, and there has even been increasing pressure to remove old dams. Part of the cause for this is the increasing recognition of the ecological damage done by dams. In 1999 FERC refused to renew the license for the Edwards Dam on the Kennebec River in Maine, thereby leading to its removal. The rationale was that the negative environmental impacts of the dam outweighed the benefits of the dam. However, part of the reason that there have been fewer dams built recently than in the past is because so many of the best sites already have dams.

FRACKING

The law involving fracking, and the scientific understanding of fracking's impact on ecosystems are both unsettled. Fracking uses fracturing fluid, consisting of large volumes of water mixed with chemicals and sand, at high pressure to create fissures in rock where oil or gas is trapped. The water and chemicals are then pumped out, leaving the sand to keep the fissures propped open. The fissures allow the oil or gas trapped in the rock to escape to the surface well. The water and chemicals that are pumped out are wastewater. Wastewater may be recycled for another fracking project, or it may be disposed of at a POTW, injected into a disposal well, or left in an evaporation pond. Fracking is increasingly being used to extract natural gas from shale, such as in the Marcellus Shale formation that lies underneath six states in the mid-Atlantic region. Annual production of natural gas from shale increased nearly fivefold between 2006 and 2010.

Federal statutes are currently circumscribed in their regulation of fracking. In general, the CWA does not regulate the discharge of pollutants into groundwater. The Safe Drinking Water Act is the primary federal statute regulating groundwater. The act requires that substances injected underground do not contaminate groundwater, and similar to the CWA, authorized injections must first receive a permit.[100] In 2005, however, Congress passed another Energy Policy Act, which specifically exempts fracking from regulation by the Safe Drinking Water Act (unless diesel fuels are used).[101] As a consequence, the fracturing fluids injected into a well during the fracking process are not regulated under the act. On the other hand, wastewater from fracking injected into a disposal well is regulated under the act. In 2015 the Bureau of Land Management finalized regulations for fracking on federal lands.[102] The regulations require a cement barrier between the wellbore and any water zones the wellbore passes through. It also requires that companies publicly disclose the chemicals they use in fracking, and that wastewater be held in aboveground storage tanks.

The federal laws regulating fracking may continue to evolve. Bills have been introduced in Congress to regulate fracking under the Safe Drinking Water Act, and the EPA continues to study the environmental effects of fracking. Currently, though, the regulation of fracking is mostly left up to each state.

Some studies have examined the potential ecological and environmental consequences of fracking. A 2011 study of water wells in Pennsylvania and New York found levels of dissolved methane on average 17 times higher in groundwater within one kilometer of a fracking site than in groundwater farther away.[103] The high level of dissolved methane may help explain the videos of tap water catching on fire. The study did not find any contamination of the groundwater from fracking fluids, however. A subsequent study by the US Department of Energy also found no evidence of gas or fluids migrating from the Marcellus Shale into drinking-water aquifers after fracking.[104]

Fracking, though, does use a huge volume of water. The water is frequently diverted from surface waters, reducing stream flow and thereby reducing habitats for aquatic organisms.[105] Moreover, many of the chemicals used in the fracturing fluid are highly toxic or known carcinogens. The wastewater from fracking contains not only these toxic chemicals, but also metals, organics, and salts from the drilling process. If the wastewater is left in an evaporation pond, there is the distinct possibility of the wastewater spilling and running into streams or lakes, or leaching into groundwater. Even wastewater that has been treated may still have negative effects on aquatic ecosystems. One study found that wastewater treated at a commercially operated treatment plant and discharged into surface waters contained high levels of ammonium and iodide.[106] More research is needed to determine whether and in what ways fracking may affect aquatic ecosystems.

Despite the potential that fracking has for negatively affecting ecosystems, there is another consideration that should be weighed. Natural gas recovered from fracking is cleaner burning than coal or oil. The potential environmental degradation that results from fracking should be balanced against the degradation that accompanies burning coal or oil in place of natural gas.

CHAPTER EIGHT

Filling In Streams and Wetlands

"No net loss" of wetlands has been the goal and catchphrase of wetlands policy for the last 20 years. "No net loss" is defined to mean that land developers are supposed to avoid destroying wetlands when developing land, but if they do destroy wetlands, they must compensate by restoring or creating new wetlands somewhere else. The idea is that destroying wetlands in some places but then restoring them in other places will result in no net overall loss of wetlands in the United States.

Such an important national environmental policy would ideally be the result of considered scientific and political calculations. Instead, the no net loss policy came about because George H. W. Bush was running for president in 1988.[1] Behind in the polls to his challenger, Michael Dukakis, Bush wanted to change perceptions that he was hostile to the environment. He did so in part by giving a policy speech in which he proposed the goal of no net loss of US wetlands. After Bush won the presidency, the EPA and the US Army Corps of Engineers wrote a memorandum of agreement in 1990 that officially created the no net loss policy.[2] The no net loss policy was enforced through the powers the EPA and the Corps were granted by the Clean Water Act (CWA). The no net loss policy continued in subsequent presidencies, and is still in place today. Examining wetland area, the federal government has claimed that the policy has been a huge success, and that there has been no net loss of wetlands since 1990. There are currently 44.6 million hectares (110.1 million acres) of wetlands in the United States.[3]

The no net loss policy certainly seems likes a sensible idea, but like many federal environmental goals, its arbitrariness undercuts much of the purported ecological benefit. By the time George H. W. Bush created the no net loss policy, the United States had already lost half its wetlands. In the 1780s the coterminous United States had approximately 89.4 million hectares (221 million acres) of wetlands. By 1990 the Fish and Wildlife Service estimated that the United States had lost roughly 53% of those wetlands. Despite the large extent of wetlands already lost, the no net loss policy sets the baseline for no losses of wetlands at 1990 levels. Why set the baseline at 1990 levels, thereby creating a goal that is less than half the wetland area the United States previously had? Why not set a goal of increasing wetlands to the amount in 1780, or at some other point in the past? The answer appears to be simply that George H. W. Bush was running for president in 1988 and happened to win.

There are additional areas of arbitrariness in the no net loss policy. The memorandum of agreement specifically states that the no net loss policy applies only to wetlands—not to streams, lakes, or any other water bodies. Again, this appears to be the case because George H. W. Bush happened to focus on wetlands in his policy speech, not other types of water bodies. Finally, there had to be a way to quantify whether there was no net loss of wetlands or not. The easiest way to do that was to determine the number of hectares of wetlands destroyed every year, and compare that to the number of hectares restored or created. However, the reason for the no net loss policy is to maintain the functions and services that wetlands provide. As will be discussed in greater detail later in the chapter, research suggests restored wetlands may never reach the same level of function as the pristine wetlands they replace. Simply comparing hectares lost versus hectares restored says nothing about the functions and services of those wetlands. No net loss of wetlands may mean no overall loss of wetland hectares, but a huge overall loss of wetland functioning.

Despite the arbitrary nature of the no net loss policy, the biggest, and potentially most arbitrary, decision that must be made is which bodies of water federal environmental laws apply to. This chapter discusses the Supreme Court cases that have decided when the CWA applies to a water body. The chapter then turns to an examination of how the CWA regulates the dredging and filling of those water bodies. As for wetlands, when a developer destroys any type of water body, the Army Corps of Engineers may require the developer to compensate for that destruction by restoring a different water body. This compensatory mitigation for streams and wetlands is taken up in the next part of the chapter. The chapter concludes with a look at the Ramsar Convention, which is an international treaty with the goal of preserving wetlands around the world.

NAVIGABLE WATERS

As mentioned in the last chapter, the CWA prohibits the discharge of pollutants to the "navigable waters" of the United States.[4] The question of what is or is not a navigable water has received a great deal of attention from industry and environmental groups, and has been the subject of numerous court cases. If a waterway is not considered a navigable water, then any individual or corporation can discharge pollution into that water body, or completely fill in and destroy that water body, without violating the CWA. There may be state laws that prohibit such an action, but the federal government will have no say in the matter.

The term "navigable" would seem to imply that the waterway must be wide enough and deep enough for a vessel to pass through. Perhaps not surprisingly, given the importance and complexity of the CWA, this simple definition is not the one used to determine whether a waterway is navigable under the act. The problem with defining a navigable waterway as one a ship could pass through is fairly clear—small streams and lakes too small for a ship often feed directly into larger waterways. If a navigable waterway under the CWA included only waterways large enough for ships to pass through, industries would simply discharge pollutants into smaller streams not covered by the CWA, and the act would do little to reduce the level of pollution in water bodies.

The CWA defines "navigable waters" as "waters of the United States, including the territorial seas."[5] If this definition does absolutely nothing to clear up your confusion over what is or is not a navigable water, you are not alone.

To more clearly indicate which water bodies fall under the CWA, the EPA and Army Corps of Engineers devised their own definition of "waters of the United States."[6] The definition includes waters subject to the ebb and flow of the tide; the territorial seas; all interstate waters and wetlands; and all (1) intrastate waters such as lakes, rivers, streams, mudflats, sand flats, wetlands, sloughs, prairie potholes, wet meadows, playa lakes, or natural ponds, which could (2) affect interstate commerce if they were destroyed. The definition also includes the tributaries of any of the above waters. Finally, the definition includes wetlands adjacent to the above waters. The regulation then defines "wetlands" as areas that are inundated or saturated by surface or groundwater frequently enough to support a prevalence of vegetation adapted for life in water-saturated soils.[7] Wetlands therefore include swamps, marshes, bogs, and similar areas (figure 8.1).

FIGURE 8.1. Wetland in Okefenokee National Wildlife Refuge, Georgia. Photograph from FWS/Ryan Hagerty.

The regulatory definition of waters of the United States is obviously quite broad, and would seem to include almost every body of water in the United States. Three important Supreme Court cases have clarified what the EPA and the Corps can regulate as waters of the United States.

First, in the 1985 case *United States v. Riverside Bayview Homes,* the court held that the regulatory definition of waters of the United States could indeed include wetlands adjacent to traditionally navigable waters and their tributaries.[8] The court stated that because water moves in hydrologic cycles, Congress intended the CWA to give broad federal authority to control pollution throughout the entire aquatic system. The court specifically noted in its decision, though, that it was not deciding whether isolated wetlands that are not adjacent to traditionally navigable waters fall under the CWA.

The second Supreme Court case considered whether human-made isolated ponds that were to be filled in for use as a solid waste site fell under the CWA. The Corps claimed to have jurisdiction over the ponds because migratory birds used the ponds. Recall that the regulation defining "waters of the United States" includes intrastate waters not part of a tributary system whose destruction could affect interstate commerce. The EPA and the Corps gave as examples waters used by migratory birds that are protected by treaty or that cross state lines, waters used as habitat for endangered species, and waters used to irrigate crops sold in commerce. This became known as the Migratory Bird Rule. The Supreme Court held in *Solid Waste Agency of Northern Cook County v. U.S. Army Corps of Engineers (SWANCC)* that the Migratory Bird Rule exceeded the authority granted to the Corps under the CWA.[9] The court stated that Congress intended the CWA to apply only to waters with some connection to traditionally navigable waters. The court therefore ruled that isolated ponds with no nexus to traditionally navigable waters do not fall under the CWA.

The third case by the Supreme Court further restricted the jurisdiction of the EPA and the Corps, but did so in such a way that what qualifies as waters of the United States is even less clear than it was before. *Rapanos v. United States*[10] concerned two separate cases involving wetlands: the filling of wetlands that were connected to Saginaw Bay and Lake Huron through eleven miles of man-made ditches and natural streams;[11] and a permit to fill a wetland physically separated from a man-made ditch that flowed one mile into Lake St. Clair.[12] The appeals court had found in both cases that the wetlands fell under the CWA. The Supreme Court considered the cases together, and in the *Rapanos* decision vacated the appellate court ruling. The Supreme Court sent the cases back to the lower courts with a new definition of "waters of the United States" to use. However, a majority of the Supreme Court could not actually agree on what that new definition should be, splitting 4–4–1. As a result, two definitions of waters of the United States have come out of the *Rapanos* decision.

Justice Antonin Scalia authored the plurality opinion. He wrote that waters of the United States include only "relatively permanent, standing or flowing bodies of water" described in "ordinary parlance" as streams, oceans, rivers, and lakes. Scalia wrote that this definition does not necessarily exclude bodies of water that dry up in extraordinary circumstances, or seasonal rivers with continuous flow during some months of the year but no flow during dry months. According to Scalia, wetlands fall under the CWA only if they have a continuous surface connection to traditionally navigable waters.

Justice Anthony Kennedy wrote a concurring opinion containing a different definition of waters of the United States. He wrote that wetlands fall under the CWA if they have a "significant nexus" to a traditionally navigable water. A wetland has such a nexus if it affects the "chemical, physical, and biological integrity of other covered waters more readily understood as 'navigable.'" Kennedy's definition of water of the United States depends on how a water body performs important functions for a traditionally navigable water, such as by filtering water or slowing the flow of surface runoff that enters a lake or river. As a consequence, one means of showing that there is a significant nexus between a wetland and a traditionally navigable water is through ecological considerations such as trapping and filtering nutrients that would otherwise reach the navigable water.

Justice Scalia tends to prefer "bright-line" rules so that the law on a subject is clear and individuals know precisely when the law applies to their activity and when it does not apply. His test for when wetlands are part of the waters of the United States is an example of such a bright-line rule. It should be relatively easy to determine whether the surface waters of a wetland connect to a traditionally navigable water, thereby making it clear when the CWA applies to a wetland. A bright-line rule is almost always an exercise in arbitrary line drawing, but the clarity it brings to the law may outweigh the arbitrariness. Conversely, the significant nexus test described by Justice Kennedy is not so clear. It may often be quite difficult to determine when a significant nexus exists between wetlands and other waters. However, the significant nexus test better reflects the scientific understanding of how wetlands interact with traditionally navigable waters.

Wetlands that do not have a continuous surface connection to traditionally navigable waters may nonetheless have hydrologic and ecological connections to those waters. There are several ways for hydrologic connections to exist. A wetland may have an intermittent surface connection to a stream through spillage or overbank flooding. Wetlands may also be connected to streams and other navigable waters belowground. Of particular importance may be a connection through the hyporheic zone. The hyporheic zone is the area below and along the sides of a streambed where stream water and groundwater mix. Hyporheic flow occurs when water from a stream enters the hyporheic zone and later returns to the stream. A wetland may have no surface connection to a stream, but if the wetland is located above the hyporheic zone of a stream, hyporheic exchange will allow nutrients or pollutants to move between the wetland and the stream. If a wetland is more isolated and is not connected to the hyporheic zone of a stream, it may still be hydrologically connected to the stream through groundwater. Water from the wetland may flow through the groundwater, which in turn may flow through the hyporheic zone of nearby streams. In such groundwater flows, material can move from wetlands to streams, but material does not move from streams to wetlands.

There may also be ecological connections between wetlands and traditionally navigable waters. Organisms may migrate between wetlands and other water bodies by water, land, or air. Migration is an important pathway for the movement of nutrients between wetlands and traditionally navigable waters. Wetlands may also serve as spawning sites for migratory fish, or as refuges from predators, competitors, or invasive species. All these connections may be important for the ecological functioning of traditionally navigable waters.

As there was no majority in the *Rapanos* decision, the new definition of waters of the United States is not entirely clear. As a consequence, the EPA and the Corps decided to use both definitions arising from the decision. For the EPA and the Corps, a water body is a water of the United States if it meets either the "relatively permanent" standard of Justice Scalia or the "significant nexus" standard of Justice Kennedy.

To help clarify how they define waters of the United States, the EPA and the Corps jointly proposed new regulations in 2014. The proposed regulations define waters of the United States as traditionally navigable waters, interstate waters, the territorial seas, and tributaries of those waters. The definition also includes wetlands adjacent to one of these waters. The proposed regulation differs most from the current regulation in the other types of wetlands it includes. It deletes from the definition wetlands whose degradation or destruction could affect interstate commerce. Instead, the proposed regulation includes wetlands that have a "significant nexus" to a traditionally navigable water. The proposed regulation then defines "significant nexus" as significantly affecting "the chemical, physical, or biological integrity of a water" that already falls under the CWA. The proposed regulation thus codifies the definition of "significant nexus" that Justice Kennedy supplied in the *Rapanos* decision.

As the significant nexus test will likely become part of the definition of waters of the United States in the new regulation, additional research needs to be conducted on the ways in which upstream components of stream networks influence downstream waters.[13] This is particularly true for research exploring the many ways in which traditionally nonnavigable waters influence traditionally navigable waters. Scientific research will hopefully influence regulators in deciding when a significant nexus to a traditionally navigable water exists, and thus in determining which water bodies fall under the CWA.

DREDGING AND FILLING

Once the EPA and the Corps determine that the CWA applies to a particular body of water, they can then enforce the requirements of the CWA for that water body. As discussed in the previous chapter, the CWA prohibits the discharge of pollutants to the waters of the United States. This prohibition includes filling wetlands or other water bodies. Section 404 of the CWA, however, creates an exception that allows the filling of water bodies if a person first receives a permit. For instance, a developer near Bigfork, Montana, received a permit to fill 1.1 hectares (2.8 acres) of wetland to build nine homes and a road. Filling 1.1 hectares might not sound like much, but tens of thousands of permits are issued by the federal government every year. Once a wetland or stream is filled, the habitat for organisms living in that wetland or stream is destroyed, and the ecosystem services provided by that wetland or stream are lost. This section examines the conditions under which such permits are granted.[14]

Under the CWA, the Corps has authority to issue permits for the discharge of "dredged or fill material" into waters of the United States. These permits are often called Section 404 permits for the section of the CWA establishing them. At the same time, the CWA gives the EPA the power to create the environmental rules the Corps must follow in granting permits.[15]

In EPA and Corps regulations, fill material is defined as material placed into water that has the effect of replacing any portion of a water body with dry land, or changing the bottom elevation of any portion of a water body.[16] Examples of fill material include rock, sand, soil, overburden from mining, and material to create a structure. Fill material therefore comes from outside of a water body and, when placed in the water body, raises its bottom. The "discharge" of fill material is further defined as the addition of fill material to water bodies, and includes adding fill for construction, placing intake pipes for power plants, building artificial reefs, and placing overburden or tailings from mining activities.

Regulations define "dredged material" as material that is excavated or dredged from a water body.[17] There is no requirement that the dredged material change the bottom elevation of a water body when placed in that water body. The "discharge" of dredged material simply means the addition of dredged material to a water body, even if the dredged material is being redeposited in the water body from which it was taken. Incidental fallback of dredged material during the dredging process is not considered to be a discharge of material. For example, incidental fallback occurs if a stream is being dredged and a bucket containing soil from the bottom of the stream is raised, allowing some soil to fall from the bucket back into the stream.[18] As a consequence, dredging where there is only incidental fallback of dredged material is not regulated under the CWA.

It should be noted that the CWA and the associated regulations do not explicitly prevent the draining of wetlands. The process of draining a wetland, however, may result in the discharge of dredged or fill material, such as through the construction of a drainage ditch, thereby requiring a Section 404 permit. Exactly when drainage of a wetland falls under the CWA is still an open question.

In deciding whether to issue a permit for the discharge of dredged or fill material, the Corps first considers whether the proposed activity complies with the environmental guidelines issued by the EPA (often called the 404(b)(1) Guidelines).[19] Although called guidelines, they are actually regulations and therefore have the force of law. The Guidelines state that the discharge of dredged or fill material will not be permitted if there is a practicable alternative that would have less adverse impacts on the aquatic ecosystem.[20] The Guidelines presume that if an activity does not require access or proximity to a water body to fulfill its basic purpose, and is therefore not "water dependent," then a practicable alternative exists.[21] While this part of the Guidelines reads as though it would stop many development projects, in practice the water-dependency requirement rarely stops projects from receiving permits.[22]

The EPA Guidelines further state that no discharge of dredge or fill material is allowed if it violates state water-quality standards or toxic-effluent standards (see chapter 7), jeopardizes the continued existence of threatened or endangered species or results in the likely destruction or adverse modification of critical habitat (see chapter 3), or violates protections to marine sanctuaries (see chapter 9).[23] Additionally, no discharge is allowed if it causes or contributes to "significant degradation" of the waters of the United States.[24] Discharges that contribute to significant degradation cause significant adverse effects on human health, plankton, fish, shellfish, wildlife, or special aquatic sites. They may affect life stages of aquatic organisms and wildlife dependent on aquatic ecosystems. They may affect aquatic ecosystem diversity, productivity, and stability, such as through loss of habitat or through loss of capacity for wetlands to assimilate nutrients, purify water, or reduce wave energy. Lastly, they may adversely affect recreational, aesthetic, and economic values. Discharges of dredge or fill material are not allowed under the Guidelines unless appropriate and practicable steps are taken to minimize potential adverse effects of the discharge on aquatic ecosystems.[25] Ecologically relevant steps that may minimize adverse effects include avoiding changes in water currents and circulation patterns that would interfere with animal movement, selecting sites to avoid creating habitat favored by invasive species, and timing discharges to avoid spawning or migration seasons.[26]

If the proposed discharge of dredge or fill material meets the requirements of the EPA Guidelines, the Corps next considers the public interest in issuing the permit.[27] To do so, the Corps weighs all the positive and negative impacts the proposed activity will have on the public interest. The Corps considers the cumulative effect of such factors as conservation, economics, aesthetics, general environmental concerns, fish and wildlife values, recreation, water quality, energy needs, food and fiber production, and the needs and welfare of the people. Courts have split on how far the Corps may look when determining the impacts of a proposed activity. Some courts have held the Corps may not look outside of the waters of the United States.[28] Conversely, other courts have held the Corps may consider impacts on an entire ecosystem, not just on waters of the United States, when considering a proposed activity.[29] Although the public-interest review seems very comprehensive, it will almost never decide whether a Section 404 permit is issued or not. In practice, the EPA Guidelines determine whether the Corps issues a permit. There are very few if any Section 404 permits that are denied based solely on the public-interest review.[30]

In writing the CWA, Congress did not fully trust the Corps to protect environmental resources. When Congress passed the CWA in 1972 (and perhaps to this day), the Corps had a reputation for being more interested in undertaking civil projects than in protecting the environment. As a result, after the Corps has approved a Section 404 permit, the CWA gives the EPA authority to reject the permit.[31] The CWA states that the EPA should reject a permit if it has "unacceptable adverse effect[s]" on shellfish beds, spawning and breeding areas for fish, wildlife, recreational areas, or municipal water supplies. Despite this power, the EPA has rejected only 13 Section 404 permits since Congress passed the CWA. (Section 404 permits authorize roughly 40,000 activities every year.)[32] This number is slightly misleading, as the EPA and the Corps have ongoing negotiations over permits, which may frequently result in permits being withdrawn without the EPA's being forced to reject them. Nonetheless, given the number of permits issued every year, the number of permits rejected by the EPA is surprisingly minuscule.

Wetlands and Farming

Under the CWA, a Section 404 permit is not required for the discharge of dredged or fill material when it occurs during normal farming, silviculture, or ranching activities.[33] A permit is also not required for discharges that occur during the construction and maintenance of farm or forest roads, as long as construction or maintenance is done according to "best management practices."[34] These normally exempt activities do need a Section 404 permit under the CWA, though, if the activity changes the use of the land, and either impairs the flow and circulation of water or reduces the reach of the water.[35] This "recapture provision" is meant to prevent farmers from converting wetland into farmland.

Converting wetland to farmland has historically been the greatest cause of wetland losses. As mentioned in the beginning of this chapter, the United States has lost more than half its wetlands since the late 1700s; conversion of wetland to farmland was the primary reason for this loss. Between the 1950s and 1970s, approximately 87% of wetland losses were from agricultural conversion. Much of this conversion was encouraged by state and federal governments as an attempt to put "worthless" swampland to use.

Today, the federal government has two programs designed to prevent farmers from destroying wetlands, and to encourage them to restore previously degraded wetlands. The Swampbuster program, a provision of the Food Security Act of 1985 (frequently referred to as the 1985 farm bill), withholds federal farm-program benefits from farmers who convert wetlands into farmland. The program currently protects at least 2.4 million hectares (6 million acres) of wetlands.

The second program is the Agricultural Conservation Easement Program. The program used to be called the Wetlands Reserve Program, but the 2014 farm bill combined the Wetlands Reserve Program with other conservation programs into the newly named Agricultural Conservation Easement Program. Regardless of its name, the program pays farmers who volunteer to protect and restore wetlands on their lands. If a farmer agrees to a permanent conservation easement on an area of wetland, the Department of Agriculture pays for the value of the easement and 100% of the wetland restoration costs. A farmer may conversely opt for a 30-year easement or a cost-share agreement, with the Department of Agriculture paying less to the farmer. The program has protected over 930,000 hectares (2.3 million acres) of wetlands.

These two programs appear to be slowing the conversion of wetlands for agricultural use. Today, urban and rural development projects are the leading causes of wetland conversion.

Section 404 Permits and NPDES Permits

What happens when a developer wants to put fill material into a water body, and that fill material may constitute a pollutant under the National Pollutant Discharge Elimination System (NPDES)? Does the Corps issue a Section

404 permit, or does the EPA issue an NPDES permit? In 2009 the Supreme Court took up this question in *Coeur Alaska v. Southeast Alaska Conservation Council*.[36] The case involved a mining company that planned to discharge gold mine tailings into Lower Slate Lake in Alaska. The tailings contained heavy metals and thus could qualify as pollutants under the NPDES. In fact, the Corps expected the tailings to kill all life in the lake. The tailings also fit the definition of fill material, because the discharge would raise the elevation of the lakebed by 15 meters (50 feet). Additionally, recall that the regulatory definition for "fill material" includes overburden from mining, and the definition for "discharge of fill material" includes placing mine tailings in water.[37] Indeed, the mining company had already received a Section 404 permit from the Corps for the discharge, but an environmental group argued that the company needed an NPDES permit. The Supreme Court held that under the CWA, if the Corps has authority to issue a Section 404 permit, then the EPA does not have authority to issue an NPDES permit. Because the discharge of fill material includes placing mine tailings in water, the Corps had authority to issue a Section 404 permit, and the mining company could discharge the mine tailings into the lake. As it is often easier to receive a Section 404 permit from the Corps than an NPDES permit from the EPA, it remains to be seen whether the decision by the court will be used by individuals to skirt the requirements of the NPDES.

Mountaintop Removal Mining

One of the most controversial types of filling activities occurs during mountaintop removal mining, which removes the top of a mountain to get to the coal seams underneath. The overburden removed during the process is then placed into the valleys adjacent to the mountain. This often fills up a great portion of the valley, destroying its streams and wetlands. This type of mining is most common in the Appalachian region of the United States, with as many as 500 mountain peaks removed for mountaintop mining.

As might be expected, filling in entire valleys has significant environmental effects. Wetlands and other water bodies may be completely covered and destroyed by the fill material. Approximately 4,000 kilometers (2,500 miles) of waterways have been buried by mountaintop overburden. As the newly exposed rocks weather, they release minerals into water bodies. As a result, downstream waters have higher ion levels, such as selenium, calcium, potassium, sodium, magnesium, sulfate, bicarbonate, and chloride.[38] As ion levels increase, particularly sulfate, electrical conductivity of waterways also increases. High conductivity may negatively affect fish and macroinvertebrate species, often disrupting osmoregulation. This disruption can result in the almost complete loss of mayfly (Ephemeroptera) taxa in a waterway, which is significant because mayflies often account for 25%–50% of total macroinvertebrate abundance in undisturbed Appalachian streams in the spring.[39]

High levels of selenium can also result in teratogenic deformities in larval fish.[40] Teratogenic deformities are congenital malformations caused by excessive selenium in fish eggs. The selenium replaces sulfur in proteins, deforming the proteins. This leads to deformities in the hard and soft tissues of the larval fish. Often this produces fish with vertebral columns that look like sine waves. Additionally, some body parts of the developing fish may not form at all. Birds who eat fish with high levels of selenium may in turn risk reproductive failure.

Valley fills also produce increased sediment in downstream waterways. All the pollutants in the downstream water often lead to lower aquatic biodiversity, and result in a shift to more pollution-tolerant species.[41] The fill material may also create fragmentation of forestland, and form edge forests where interior forests once stood.[42] Despite what is known, more research is needed to examine how mountaintop removal mining affects the entire community structure of both aquatic and forest ecosystems.

Unlike in the *Coeur Alaska* case, mountaintop removal mining operations must apply for both a Section 404 permit and an NPDES permit. The NPDES permit is necessary for any point-source discharges that occur during the mining process, while the Section 404 permit is required for filling the valley with overburden. Although two permits are required, mountaintop removal mining has historically been poorly regulated. The states in the Appalachian region have authority to issue NPDES permits, and have encouraged mountaintop removal mining.

The Corps has also historically been lax in oversight because of its ability to issue general Section 404 permits. Under the CWA, the Corps can issue individual Section 404 permits, which are reviewed on a case-by-case basis. The CWA also allows the Corps to issue general permits that act as a blanket form of authorization for a category of activities. Under some general permits, individuals must notify the Corps before beginning the activity allowed under the permit, while other general permits do not even require prior notification. The CWA limits the issuance of general permits to categories that cause "only minimal adverse environmental effects" separately or cumulatively.[43] The Corps approved hundreds of mountaintop removal mining operations under a nationwide general permit known as NWP 21. As already noted, mountaintop removal mining certainly has had more than a "minimal adverse environmental effect," both separately and cumulatively.

Due in part to a greater focus by environmental groups and a greater understanding of the environmental effects of mountaintop removal mining, the Corps and EPA have recently begun greater oversight of mountaintop removal mining permits. In 2010 the Corps announced that it was suspending NWP 21 in the Appalachian region (although not in other parts of the country). The Corps then issued a new version of NWP 21 in 2012, with greater restrictions on the allowable loss of waterways, and with no authorization for valley fills. Mining operations that do not fit under the new NWP 21 permit may still apply through the individual permit process.

In 2011 the EPA used its veto power under the CWA to stop a Section 404 permit approved by the Corps from going to Arch Coal's Spruce No. 1 mine in West Virginia. The mountaintop removal mining operation would have buried approximately 12 kilometers (7.5 miles) of streams. Also in 2011, the EPA issued a final guidance document creating a stream pollution standard for permit requests for mountaintop removal mining. However, in 2012 a district court vacated the new EPA guidance, holding that the EPA had overstepped the authority given to it by the CWA.[44] As the court decision suggests, the laws regarding mountaintop removal mining will continue to evolve for the next several years.

COMPENSATORY MITIGATION

Earlier in the chapter, it was explained that under the EPA Guidelines for Section 404 permits, the Corps may not grant a permit for discharging dredge or fill material unless appropriate and practicable steps are taken to minimize potential adverse effects of the discharge on aquatic ecosystems.[45] Despite steps to minimize adverse effects, the discharge of dredge or fill material into a water body will almost inevitably produce unavoidable impacts. EPA and Corps regulations require that unavoidable impacts be compensated for, a process known as compensatory mitigation.[46] As described in the introduction to this chapter, one of the goals of compensatory mitigation is "no net loss" of wetland acreage and function.[47] Compensatory mitigation applies not only to wetlands, though; it also applies to adverse impacts to streams, lakes, and other aquatic resources. Most compensatory mitigation projects, however, are compensating for impacts to wetlands and streams.

The EPA and Corps regulations lay out the sequence developers must use when developing a site containing a water body: (1) developers must use all appropriate and practicable steps to avoid adverse impacts to waters of the United States, (2) developers must use all appropriate and practicable steps to minimize unavoidable impacts to waters, and (3) developers must compensate for the unavoidable adverse impacts they do cause.[48] The regulations define "practicable" to mean what is capable of being done after considering cost, technology, and logistics, given the purpose of the project. If compensatory mitigation is not possible, the Corps may refuse to issue a permit.[49] While the first two steps of the sequence, avoiding and minimizing adverse impacts, are supposed to come before compensatory mitigation, the Corps and EPA have placed a greater emphasis on compensation than on forcing developers to avoid or minimize damage.[50]

Restoration, enhancement, establishment, and preservation of water bodies are acceptable forms of compensatory mitigation.[51] Of these, restoration is preferred, because it has a higher probability of recreating a functioning aquatic ecosystem than does establishing a new water body and ecosystem. Additionally, restoration creates more new aquatic resources than does enhancement or preservation of already existing water bodies. Restoration ideally results in no net loss of aquatic resources: although one site is destroyed during development, a new site is restored to replace the old site. Preservation does not satisfy the no net loss goal: a site is destroyed, and the compensation is merely to protect a site that already exists. However, the Corps may allow preservation as compensatory mitigation if the preservation site is ecologically important and much larger than the site that is to be destroyed. The preservation site typically must be permanently protected, such as by a conservation easement.

The ability of the government to regulate water bodies and require compensatory mitigation raises constitutional questions. Approximately 75% of wetlands are privately owned.[52] Does the government's requiring a permit to fill in a wetland represent an unconstitutional taking of private property? The Fifth Amendment to the Constitution states that private property shall not be taken for public use without just compensation (see box 8.1). The government does not need to physically occupy an individual's private land for a taking to occur—government regulations that limit an individual's use of his property to such an extent that it is as though a physical taking has occurred also require just compensation. This is called a regulatory taking. The Supreme Court has held that requiring a permit to develop land or fill in a wetland is not by itself a regulatory taking.[53] In addition, the granting of such a permit may have conditions attached, such as requiring compensatory mitigation, without being a taking.

However, there are limits to the conditions that may be attached to a permit. The Supreme Court has held that there must be an "essential nexus" between the permit conditions and legitimate government interests, and there must be a "rough proportionality" between the conditions and the im-

BOX 8.1. TAKINGS CLAIMS

Governments in the United States have the inherent right to take private property and convert it to public use (this is called eminent domain). When a government does take private property, though, the landowner has a takings claim, and the government must pay the landowner for the land. This right is enshrined in the Fifth Amendment to the Constitution, which ends with the clause "nor shall private property be taken for public use, without just compensation." The Fifth Amendment has been extended to the states through the due process clause of the Fourteenth Amendment.

When the government physically occupies private property for its own uses, it is an obvious taking and will almost always require just compensation. Things become trickier when the occupation is not physical. The government may pass a law or promulgate a regulation that limits the use of a piece of private property to such an extent that it is as though a physical taking has occurred. Such a taking is called a regulatory taking, and is also subject to just compensation. Claims of regulatory taking often arise in the context of wetlands regulation.

How can one determine if a regulation limits the use of property enough to be a taking? In *Penn Central Co. v. New York City*,* the Supreme Court laid out three factors courts must weigh in deciding whether a regulatory taking has occurred: (1) the economic impact of the regulation on the landowner, (2) the extent to which the regulation interferes with distinct investment-backed expectations, and (3) the character of the government action (such as whether the government action creates a general regulatory regime or singles out a particular landowner).

There is no need for a court to balance the *Penn Central* factors, however, for a certain type of taking. If a regulation "denies all economically beneficial or productive use" of a piece of property, then the Supreme Court has held that a regulatory taking has occurred per se. The case establishing this principle is *Lucas v. South Carolina Coastal Council*.† Lucas had purchased two beachfront lots in South Carolina. The state of South Carolina then passed a law prohibiting construction within an erosion zone, meaning that Lucas could no longer build on the lots he had purchased. Lucas filed a takings claim against South Carolina, arguing that the property had been rendered valueless by the new law. The Supreme Court agreed, and Lucas received just compensation from the state of South Carolina.

It is rare for a law or regulation to render a piece of property valueless, though. This is especially true for takings claims involving wetland regulations. A regulation may prevent a landowner from developing on a wetland that occupies a section of a parcel of property. However, courts usually consider the economic value of the entire property. A landowner will often be able to develop upland portions of the property in question, meaning that the wetland regulation has not rendered the entire parcel of land valueless. For example, in *Palazzolo v. Rhode Island*,‡ the Supreme Court held that a decision by Rhode Island to prevent a landowner from building on waterfront property partially consisting of protected tidal marshes was not a taking. The court reasoned that the landowner could still build on the upland portion of his parcel, and therefore a total taking had not occurred. If a total taking has not occurred, the *Penn Central* factors again become applicable.

With the growth of mitigation banking, it will likely become increasingly difficult for landowners to argue that wetlands that cannot be developed have no economic value. A wetland that cannot be developed may be capable of being ecologically enhanced, which could then generate credits for sale to other developers.

*438 U.S. 104 (1978).
†505 U.S. 1003 (1992).
‡533 U.S. 606 (2001).

pact of the proposed development.[54] This is known as the *Nollan/Dolan* test after the two Supreme Court cases that established it. If the *Nollan/Dolan* test is not passed, then a taking has occurred, and the government must pay just compensation.

The *Nollan/Dolan* test raises two questions: Does the test apply only when the permit condition is to force a landowner to set aside land, or does it also apply to a condition to pay money? Does the test apply when the government simply denies a permit? In 2013 the Supreme Court answered these questions in *Koontz v. St. Johns River Water Management District*.[55] The court held that a condition in a permit to pay money must meet the *Nollan/Dolan* test. The court reasoned that if a condition to pay money did not fall under the test, then governments could simply require fees be paid for a permit and avoid the test. The Supreme Court then held that the *Nollan/Dolan* test applies if a permit is denied because the landowner failed to agree to the conditions. The reasoning behind this second holding is that if a government denying a permit does not need to pass the *Nollan/Dolan* test, then a government could get around the test by repeatedly denying a permit until the landowner gave in and agreed to the conditions. The aftermath of the *Koontz* case will likely be a limit to the conditions that governments try to place on permits, so that they do not face a deluge of takings claims. This may include the Corps requiring less stringent conditions for compensatory mitigation when issuing Section 404 permits.

The specific requirements for compensatory mitigation were issued jointly by the EPA and the Corps in 2008.[56] Unlike other compensatory mitigation schemes (see conservation banking in chapter 3), the compensatory mitigation rules for aquatic resources have the force of law. One of the more important requirements for compensatory mitigation projects is that mitigation sites should be located in the same watersheds as the sites being developed.[57] The reason for this watershed approach is the belief that the mitigation site will replace the functions and services lost when the development site is destroyed. This belief will be examined more closely in the following sections.

The regulations also state that in-kind compensatory mitigation is preferred to out-of-kind mitigation.[58] This means that if a particular type of water body, such as a perennial stream, is developed, then restoration of a perennial stream is preferred to restoration of something else, such as a wetland. Another requirement is that compensatory mitigation plans contain ecological performance standards that provide objective and verifiable criteria to measure the success of the mitigation.[59] Performance standards may include measurements of hydrology, measurements of ecosystem function, and comparisons to reference aquatic resources occurring at sites like the mitigation site. The person responsible for the compensatory mitigation must submit regular monitoring reports to the Corps for at least five years to determine whether performance standards are being met.[60] Finally, the mitigation site must receive long-term protection, such as through a conservation easement.[61] A compensatory mitigation project may require long-term management and long-term financing. If long-term financing is required, a financial instrument such as an endowment or trust must be created.

There are three types of compensatory mitigation: (1) permittee-responsible mitigation, (2) in-lieu fee mitigation, and (3) mitigation banking.[62] The following three sections briefly describe the different types of mitigation.

Permittee-Responsible Mitigation

Permittee-responsible mitigation occurs when the developer himself or herself undertakes a mitigation project to compensate for damage to an aquatic resource. The developer retains the responsibility for ensuring that the mitigation project meets its ecological goals. Because the developer is compensating for the damage caused by only one development project, the mitigation site is usually quite small. As a result, the mitigation site does not necessarily have to be located away from the development site.

Most developers do not have expertise in restoring stream or wetland ecosystems. Consequently, many permittee-responsible mitigation projects have, at best, questionable ecological benefit. Nonetheless, in 2008 permittee-responsible mitigation provided 59.1% of wetland and stream mitigation.[63]

In-Lieu Fee Mitigation

An in-lieu fee mitigation program requires developers to pay a fee to a local government or nonprofit organization to fulfill the compensatory mitigation requirement. The government or nonprofit organization collects fees from several developers, then uses the fees to restore streams or wetlands.[64] Because fees from several developers have been collected, the mitigation site may be relatively large. In 2008 in-lieu fee programs provided 5.6% of stream and wetland mitigation.[65]

To provide oversight to the in-lieu fee process, regulations require that the Corps district engineer assemble an interagency review team to review the establishment of in-lieu fee programs.[66] The interagency review team may include representatives from the EPA, the Fish and Wildlife Service, and other federal, state, and local agencies. The district engineer, however, retains final authority in allowing the creation of an in-lieu fee program. Because in-lieu fee mitigation requires developers to pay a fee, the *Koontz* decision means that in-lieu fee programs must pass the *Nollan/Dolan* test.

The primary criticism of in-lieu fee programs is that mitigation occurs after developers have already destroyed several stream or wetland sites. There may be a considerable time lag between when the functions of several water bodies have been destroyed by development projects, and when the in-lieu fee program begins to replace those functions. During this lag, there is a net loss of aquatic resources. Additionally, the nonprofit organization or government agency may not collect enough in fees to undertake a mitigation project that compensates for all the aquatic resources that were destroyed. There is also the possibility of a poorly managed nonprofit organization going bankrupt before any compensatory mitigation even begins. Consequently, the lag between development and mitigation may mean that compensation is ultimately inadequate or nonexistent.

Mitigation Banking

Mitigation banking is similar to conservation banking for endangered species habitat (see chapter 3). A stream or wetland is restored, and then credits for that restoration are sold to developers who develop a different stream or wetland site. Once the credits are sold, liability for the performance of the mitigation site is transferred from the developer to the mitigation bank owner. As of 2009 there were 797 mitigation banks, of which 431 were active.[67] In 2008 mitigation banks accounted for 35.3% of wetland and stream compensatory mitigation.

An interagency review team oversees the establishment of mitigation banks.[68] Like in-lieu fee programs, mitigation bank programs must pass the *Nollan/Dolan* test. The EPA and Corps regulations require that the Corps set mitigation ratios for mitigation banks.[69] A mitigation ratio determines the amount of stream or wetland that must be restored to compensate for the amount of stream or wetland that has been developed. For example, a 2:1 ratio would require that two feet of stream be restored for one foot of developed stream. The Corps may require a ratio greater than 1:1 for several reasons, such as when the site being developed contains high-quality habitat, when there are differences between functions lost at the development site and functions restored at the mitigation site, or when the mitigation site is simply preserving an already functioning stream or wetland. The unit of measurement for wetlands is usually acres, while the unit of measurement for streams is usually linear feet.

Ideally, a stream or wetland has been completely restored by the mitigation banker before credits are released by the bank. In practice, though, the Corps often allows mitigation banks to sell a portion of the credits they hope to eventually create as a way to help generate funds to pay for the restoration project. The Corps usually requires that a large share of the credits becomes available only when the mitigation bank has achieved its ecological performance standards.

The regulation establishing the rules for compensatory mitigation formally states that the EPA and the Corps prefer mitigation banking to in-lieu fee mitigation, and in-lieu fee mitigation to permittee-responsible mitigation.[70] Mitigation banks are preferred because they must generally have a plan to restore aquatic resources before being allowed to sell credits. As a result, there is less chance that a developer will fill a stream or wetland and an inadequate compensation program will be implemented after the damage. Additionally, most of the mitigation bank credits may be sold only after physical development of the bank site has begun. Mitigation banks also tend to be larger than the other forms of compensation, which may provide greater ecological benefits than several small disconnected compensation projects could provide.

In-lieu fee mitigation is preferred secondarily because it occurs after streams and wetlands have been filled, increasing the possibility that compensation will not occur or will be inadequate. In-lieu fee mitigation projects are preferred to permittee-responsible mitigation because in-lieu fee sites tend to be larger and more ecologically beneficial than permittee-responsible sites. That leaves permittee-responsible mitigation as the least preferred option.

Watersheds and Mitigation

As mentioned above, the regulations for compensatory mitigation state that mitigation sites should be located in the same watershed as the developed sites they are meant to replace. The EPA and the Corps define a watershed as "a land area that drains to a common waterway, such as a stream, lake, estuary, wetland, or ultimately the ocean."[71] Setting limits on where mitigation sites may be placed relative to development sites makes sense. If there were no restraints, developers and mitigation bankers would place mitigation sites where land was cheapest to acquire. Mitigation sites would often be located very far from the developed site, possibly in a different state or even in a different part of the country. The functions and services that a developed water body provided to an ecosystem would in no way be compensated for by the mitigation site.

To counteract this possibility, the Corps generally requires that permittee-responsible mitigation occur within the same watershed as the developed site.[72] For in-lieu fee programs and mitigation banks, the Corps creates a geographical service area. This is the area where a mitigation bank or in-lieu fee program may sell credits to mitigate for impacted sites. Geographical service areas are usually just the watershed that includes the mitigation bank or proposed in-lieu fee program. The interagency review team assigns geographical service areas for mitigation banks and in-lieu fee programs.

Geographical service areas are often determined using the US Geological Survey (USGS) classification system of hydrologic unit codes (HUCs). The HUC system divides the United States into increasingly smaller drainage areas. A two-digit HUC, or HUC-2, divides the United States into 21 drainage regions. The HUC-6 classification divides the United States into 378 drainage basins, while HUC-8 divides the United States into 2,264 drainage sub-basins. To help put this in perspective, a HUC-8 watershed may be 60 kilometers (37 miles) wide and 100 kilometers (62 miles) long. The regulation controlling compensatory mitigation states that for projects in urban areas, a HUC-8 or smaller watershed may be the appropriate service area, while in rural areas a HUC-6 watershed may be the appropriate service area.[73] As a result, in rural areas a much larger geographical area may be the geographical service area for a mitigation bank or in-lieu fee program, while in urban areas a smaller geographical area must be used.

Unfortunately, the HUC system potentially suffers from the arbitrariness that afflicts other environmental rules in this chapter. For example, many of the drainage units defined using the HUC system are not true topographic watersheds.[74] Water in a drainage unit may drain into different streams, or a unit may only be a small part of a larger topographic watershed. More importantly, the boundaries of a HUC-defined drainage unit do not necessarily overlap with the boundaries of an ecosystem. A single ecosystem may overlap several HUC watersheds, or more than one relatively distinct ecosystem may exist in a single HUC watershed.

In recognition of this, ecological characteristics may also form the basis for defining a geographical service area. Similar to the HUC classification, the EPA has developed the Omernik ecoregion classification system. Ecoregions are defined as areas where ecosystems are relatively homogeneous. The system has four ecoregion levels: level I divides North America into 15 ecoregions; level II creates 50 ecoregions; level III lists 182 ecological ecoregions; and level IV divides the United States into 940 ecoregions. Ecoregions are delineated using both biotic and abiotic factors, including hydrology, soils, climate, vegetation, geology, and animals.

In practice, the HUC classification system is much more commonly used in setting geographical service areas. Ecological characteristics are often considered when setting geographical service areas, but water-drainage characteristics are usually given greater consideration than ecoregions. By giving less weight to ecosystems, however, geographical service areas may reduce the likelihood of ecosystem functions' being compensated for after development. If two or more distinct ecosystems exist in a single geographical service area, an impacted site may be ecologically quite different from the mitigation site meant to replace it. Additional research would be helpful in assessing when and how using both the HUC and Omernik ecoregion classification systems together to set geographical service areas could result in better mitigation for aquatic impacts.

Stream Compensatory Mitigation

Once the watershed or geographical service area has been established for a compensatory mitigation project, the restoration work must begin. This section and the next explain the processes and likelihood of success for stream restoration and wetland restoration, respectively.

Stream restoration often includes activities such as riparian-zone revegetation and other forms of stream-bank stabilization. Restoration may also include the addition of large pieces of wood or boulders to streams, to provide habitat for aquatic species. Additional activities may include fencing to exclude livestock from grazing on riparian vegetation, removing forest roads to reduce sediments washed into streams, removing dams, or reconnecting streams with floodplains. Finally, restoration may include adding organic and inorganic nutrients to a stream.[75]

Unfortunately, there is growing evidence that stream restoration frequently does little to benefit stream ecosystems, or even to improve the physical or chemical conditions of streams.[76] One problem with many stream restoration projects is that the length of stream restored does little to affect the condition of the stream. Many restoration projects are less than 1 kilometer (0.6 miles) in stream length, while modeling suggests that projects should be approximately 10 kilometers (6 miles) in length to measurably benefit stream conditions.[77] Additionally, the condition of the terrestrial portion of the watershed seems to have as large a role in determining the condition of the stream as the stream itself. Merely restoring a stretch of stream or riparian zone has relatively little effect on stream condition. If the watershed for a stream is heavily urbanized, stabilizing a stream's banks will be of little help.

Another problem with stream compensatory mitigation is the spatial arrangement of the compensation. A study of mitigation in North Carolina found that the distance between impacted streams and their mitigation sites averaged 43.53 kilometers (27.05 miles).[78] Of the mitigation transactions examined, 23% of impacts were offset in a different HUC-8 watershed, and 89% were offset in a different HUC-11 watershed.[79] The study also found that impacted streams tended to occur in urban areas, while mitigation sites tended to occur in rural areas where restoration costs were lower. Restored streams also occurred farther upstream in the watershed than where the developed streams were located. This means that smaller streams were restored as compensation for damage to larger streams.

The results of the study in North Carolina suggest that compensatory mitigation often does not restore the same ecological communities or ecological functions that were lost at the development site. The study did find some potential ecological benefits to compensatory mitigation. First, there appeared to be no net loss of stream resources at the HUC-8 level. Second, the mitigation resulted in spatial defragmentation. Due to mitigation banking, several small impacted stream sites were compensated for by large restoration sites. As already noted, larger restoration sites are more likely to create functioning stream ecosystems than smaller, fragmented sites.

Relatively little ecological research, though, has been performed on the landscape-scale effects of compensatory mitigation.[80] There have been few attempts to quantify how moving aquatic resources from one site to a different site several kilometers upstream affects species and populations at the watershed or larger scale. Even the defragmentation brought about by mitigation banks or in-lieu fee programs may not be entirely beneficial. A large restoration site may be favorable for that particular stream, but it is not clear if it is advantageous for the aquatic ecosystem in the watershed as a whole. Defragmentation may result in the loss of metacommunity dynamics or unique spawning grounds. Several small sites may be relatively independent of each other ecologically, but still linked through migration or the movement of nutrients. If the small sites are developed and replaced by one large site, the ecological dynamics at the landscape scale may suffer.

Finally, in practice, the Corps primarily monitors the physical characteristics of a stream restoration project, and gives little attention to ecological characteristics. For instance, for a stream mitigation bank to begin selling credits, stream-channel width, slope, and extent of riparian vegetation on the restored site may be monitored. However, ecological aspects such as the composition of the aquatic community or the ability of the stream to retain nutrients are rarely measured.[81] Consequently, the Corps may never determine whether the ecological functions and services of the developed stream were actually replaced by the mitigation site.

Wetland Compensatory Mitigation

Wetlands are among the most biologically diverse ecosystems on the planet. While wetlands cover only about 5% of land in the contiguous United States, they provide habitat for 31% of plant species. More than one-third of threatened or endangered species in the United States live only in wetlands. Wetlands also provide several ecosystem services for humans. Approximately 75% of commercially caught fish and shellfish in the United States are dependent on wetlands for at least part of their life cycle. Wetlands are also important in protecting humans from some impacts of storms. Wetlands act like sponges, absorbing storm-water runoff, which helps prevent flooding. The wetlands then release the excess water slowly, helping to maintain a more constant stream flow throughout the year. Coastal wetlands also act as storm-surge protectors during hurricanes or tropical storms. Finally, wetlands are excellent at removing nutrients and pollutants from the water that flows through them, improving the water quality of nearby streams and lakes.

Wetland restoration overwhelmingly focuses on manipulating the hydrologic regime to replicate what occurred before the wetland was degraded. The hydrologic regime includes such factors as the depth and frequency of flooding. Establishing the hydrologic regime is important because it helps control seed germination and establishment in wetlands.[82] Restoration activities may also include manipulating the chemistry of the soil and water, removing nonnative species, and reintroducing native species that were lost from the site.

Similar to compensatory mitigation for streams, mitigation for wetlands moves wetland resources away from the development site, and results in the loss of wetlands in urban areas. For example, a study in Florida found that mitigation sites for wetlands were on average 24 kilometers (15 miles) away from the development site, and in more rural locations.[83] A study in North Carolina found that the distance between wetland impacts and wetland mitigation sites averaged 50.3 kilometers (31 miles).[84] Movement of the ecosystem services that wetlands provide away from urban areas may be very costly. For instance, as wetlands are important in preventing flooding, moving this ecosystem service away from highly populated areas to less populated areas may result in more flooding affecting more people.

To help combat this problem, some scholars have suggested unbundling the different ecosystem functions and services provided by wetlands during the compensatory mitigation process.[85] Currently, when a wetland is developed, a single mitigation site is expected to replace all the functions and services formerly provided by that wetland. Unbundling would allow mitigation at one site for one set of ecosystem functions and services, and mitigation at another site for a different set of functions and services. For example, a wetland near an urban area that is to be developed may provide flood protection for nearby residents, and provide habitat for several species of birds. Unbundling would require that the developer create a mitigation site near the developed wetland, and thus near the urban area, that is capable of replacing the flood protection service of the developed wetland. The developer would also have to create a different mitigation site, likely in a more rural area, that would provide high-quality habitat for the species of birds that were at the developed wetland.

A potential benefit of unbundling would be that it would allow a single mitigation bank or in-lieu fee program to have two or more geographical service areas. Wetland services related to hydrology could be constrained to a service area defined using the HUC classification, while wetland functions related to ecology could be constrained to a service area defined using an ecoregion classification. Creation of different geographical service areas for a single bank or in-lieu fee program may thus allow compensatory mitigation to more faithfully replace the lost ecosystem functions and services of the developed wetland.

The downside of unbundling is that dividing up the functions and services of a wetland may be extremely difficult. The characteristics of a wetland that allow it to protect against flooding, or to filter out pollutants, are also the characteristics that allow it to provide unique habitats for huge numbers of terrestrial and aquatic species. Unbundling these functions and services may be nearly impossible. Unbundling would also add to the complexity and administrative difficulty of compensatory mitigation. The ability of federal regulators to track the different credits being sold for different wetland functions and services at different mitigation sites may be limited. The administrative and ecological benefits and drawbacks of unbundling require more research.

While unbundling may ultimately be helpful in improving compensatory mitigation, restored wetlands may never truly replace all the functions and services of wetlands that have been lost to development. Moreno-Mateos et al. performed a meta-analysis on research papers examining wetland restoration sites around the world.[86] They found that hydrologic features of wetlands recover immediately after restoration. Hydrologic features include water level, flooding regime, and water storage. Biological structure, on the other hand, never fully recovered. Biological structure includes species abundance, richness, and diversity. Restored wetlands recovered on average only 77% of the biological structure of reference wetlands, even 100 years after restoration took place. Vertebrate and macroinvertebrate assemblages were quickest to recover, although never fully resembling reference wetlands. Plant assemblages took much longer, taking on average 30 years to converge with reference wetlands, although again, never fully recovering. Difficulty in dispersing to restoration sites and the presence of nonnative species may be some of the reasons plant assemblages take so long to recover. The inability of plant and animal species to disperse to a restored wetland site may be especially acute if the restoration site is surrounded by land that has been altered for human use. The species that are native to the restored wetland may be unable to make it through an inhospitable human landscape to arrive at the restoration site.

As mentioned above, one of the important ecosystem services of wetlands is nutrient storage. The study by Moreno-Mateos et al. found that phosphorus storage recovered quickly in restored wetlands, but carbon and nitrogen storage were much lower in restored wetlands than in reference wetlands. Even after 100 years, biogeochemical functioning in restored wetlands was only 74% of that in reference wetlands.

Finally, the study found that large restored wetlands, defined as greater than 100 hectares (247 acres), recovered more quickly than smaller restored wetlands. Smaller sites could not support all the species expected to reside in a high-quality wetland. Smaller wetlands also tended to be more isolated from other wetlands, making it difficult for species to disperse to the restored site. This would suggest that larger wetland mitigation bank sites and in-lieu fee sites should recover faster than smaller permittee-responsible mitigation sites.

The research shows that wetlands will only rarely, if ever, be restored to their condition prior to degradation. As a result, when a wetland is restored as part of a compensatory mitigation requirement, the restored wetland will not replace the functions and services of a wetland that was destroyed for development. While compensatory mitigation may result in no net loss of wetland area, there may be a continual loss of wetland functions and services. Mitigation ratios greater than 1:1 may help blunt this loss in some instances. For example, a restored wetland may not store as much nitrogen as a pristine wetland. The mitigation ratio may be made large enough that the total area of restored wetland can store the same amount of nitrogen as the developed wetland previously could. However, no mitigation ratio may be large enough to replace every wetland function. As the above research suggests, even relatively large restored wetlands may not be able to replicate all the ecological interactions and dynamics found in pristine wetlands, or the regional biodiversity found across several pristine wetland sites.[87] Consequently, compensatory mitigation is often complicit in allowing the continual loss of wetland functions and services.

Many of the studies examined here took place before the new 2008 regulations that were promulgated to control compensatory mitigation. While the regulations simply codified guidelines that the Corps and the EPA had used for years, the regulations may be subtly changing how compensatory mitigation is currently being implemented. Continuing research is needed to determine if the 2008 regulations have improved compensatory mitigation projects, or simply maintained the status quo.

RAMSAR CONVENTION

Wetlands have been lost on a global scale at a similar rate to losses in the United States. Roughly half of global wetland area has been destroyed due to human activity.[88] As in the United States, drainage of wetlands for agriculture is the primary reason for global losses.

The Convention on Wetlands of International Importance Especially as Waterfowl Habitat (known as the Ramsar Convention for the town in Iran where it was concluded) is an international treaty with the goal of stopping the global loss of wetlands. The convention came into force in 1975 and has 163 parties, including the United States. A conference of the parties meets every three years to adopt recommendations or resolutions.[89]

The Ramsar Convention was the first environmental treaty with the purpose of protecting specific habitats, not individual species. The convention defines wetlands as "areas of marsh, fen, peatland or water, whether natural or artificial, permanent or temporary, with water that is static or flowing, fresh, brackish or salt, including areas of marine water the depth of which at low tide does not exceed six metres."[90] This is a very broad definition of wetlands, and includes many different types of wetland habitats. The primary way in which the Ramsar Convention protects wetlands is by creating the List of Wetlands of International Importance.[91] Each party to the convention may unilaterally designate wetlands within its territory for inclusion in the list.[92] When a country agrees to join the convention, it must designate at least one wetland. The country, however, retains exclusive sovereignty over the wetland.[93] The parties to Ramsar have designated 2,065 sites for the List of Wetlands of International Importance. These 2,065 sites cover nearly 200 million hectares (494 million acres). The United States has designated 34 sites covering roughly 1.7 million hectares (4.2 million acres).

The Ramsar Convention sets the criteria for determining when a wetland may be designated for the List of Wetlands of International Importance. The convention states wetlands selected for the list should have international significance in terms of "ecology, botany, zoology, limnology or hydrology."[94] The convention also states that wetlands of international importance to waterfowl should be included on the list. Through several resolutions, the parties to the convention have adopted more specific criteria for what constitutes a wetland of international significance.[95] The criteria include wetland types that are rare or unique, support endangered species or communities, are important sources of food or spawning grounds for fish stocks, or support populations that contribute to global biological diversity. There are also more quantitative criteria. The criteria state wetlands should be included in the list if they regularly support 20,000 or more waterbirds, regularly support 1% of the individuals in a population of one species or subspecies of waterbirds, or regularly support 1% of the individuals in a population of one species or subspecies of wetland-dependent nonbird animal species.

Unlike the World Heritage Convention (discussed in chapter 4), the Ramsar Convention does not require that a wetland be a nationally protected area before a party to the convention may designate it for the list. So what obligations does the Ramsar Convention place on parties? The convention requires that parties formulate and implement planning to "promote the conservation" of the wetlands on the list.[96] The convention also requires that parties undertake, as far as possible, the "wise use" of the wetlands in their territory. Wise use is the central philosophy of the Ramsar Convention. A subsequent resolution defined "wise use" of wetlands as the maintenance of their "ecological character, achieved through the implementation of ecosystem approaches," but within the "context of sustainable development."[97] Wise use therefore allows for development of wetlands, if done in a way that maintains the ecological characteristics of the wetland. The parties to the convention treat the wise-use requirement as applying to all the wetlands within their jurisdictions, whether designated under Ramsar or not.[98]

Parties also have an obligation under the convention to report when a designated wetland undergoes a change to its ecological character as a result of human interference.[99] Parties with wetlands that are undergoing adverse ecological changes are expected to take action to remedy those changes. The Scientific and Technical Review Panel created under Ramsar may provide guidance and advice to parties with wetlands undergoing such changes. There is no specific obligation in the convention that requires parties to remedy adverse changes, though. The requirement to report on designated wetlands that are undergoing adverse changes acts primarily as a way to shame nations into protecting wetlands that are being degraded by human activities.

Finally, parties are instructed under the Ramsar Convention to promote conservation of wetlands by establishing nature reserves on their wetlands, whether a wetland is designated in the list or not.[100] The parties must also provide wardens for the reserves they create. There is again no provision in the convention to force parties to create such reserves.

As this chapter has discussed, the CWA is the primary federal statute for protecting wetlands in the United States. Does designating a wetland in the United States under the Ramsar Convention add to its protection? A study by Gardner and Connolly suggests that Ramsar designation does indeed add to wetland protection.[101] One of the most important benefits of designation in the United States is that it makes it easier to obtain management grants or funding once a wetland has been listed under Ramsar. Being able to claim international significance for a wetland is helpful in obtaining money for that wetland. Ramsar designation can also help build popular and political opposition to proposed development projects that would adversely affect a listed wetland. For example, DuPont planned to begin strip-mining for titanium at a site near the Okefenokee Swamp. The Okefenokee Swamp is a Ramsar-designated wetland, and popular opposition to a project that would potentially harm a listed wetland led DuPont to abandon its mining project. Finally, scientific study of a wetland often increases after it is listed under Ramsar, which may in turn lead to better management of the wetland.

Increasing the visibility and perceived importance of wetlands is the primary way the Ramsar Convention protects such habitats. As the study by Gardner and Connolly suggests, these benefits extend to wetlands in the United States. Ramsar designation is a very powerful tool for promoting the importance of wetlands to local politicians and the public.

CHAPTER NINE

Oceans and Coasts

Sea lions appreciate an easy meal just as much as the next species—and meals rarely come easier than thousands of fish congregating at the bottom of a dam. Congregating at the bottom of the Bonneville Dam is exactly what several populations of salmonids do as they migrate up the Columbia River. The fish are slowed by the fish ladders they must jump up to get past the dam on their way to spawn. The promise of an easy meal draws around 100 California sea lions (*Zalophus californianus*) to the Bonneville Dam every year (figure 9.1). The sea lions prey on the fish at the bottom of the dam, taking between 0.4% and 4.2% of the salmonids migrating past each year.

This predation is noteworthy because the salmonid populations being preyed upon are listed as threatened or endangered under the Endangered Species Act (ESA), and are also important to local fisheries. The predation included threatened and endangered populations of Chinook salmon (*Oncorhynchus tshawytscha*) and threatened populations of steelhead (*Oncorhynchus mykiss*). Nonlethal deterrence of sea lions is usually unsuccessful, as the sea lions quickly adapt. As a result, the states of Washington, Oregon, and Idaho applied to the Fisheries office of the National Oceanic and Atmospheric Administration (NOAA Fisheries) in 2006 for authorization to kill several California sea lions.

The reason the states had to apply for permission is that the Marine Mammal Protection Act (MMPA) prohibits the killing of marine mammals in almost all cases. There is an exception to this rule, though, for pinnipeds (such as sea lions) that are killing salmonids. Originally, the MMPA did not contain this exception, but in the 1980s sea lions began preying on steelhead

FIGURE 9.1. California sea lions on a buoy. Photograph from NOAA/Claire Fackler.

at Seattle's Ballard Locks. The sea lions were threatening the existence of the local steelhead population, but wildlife managers could not kill the sea lions because it was illegal under the MMPA. In response, Congress added an exception to the MMPA that allows intentional lethal taking of individual pinnipeds that are having a "significant negative impact" on the decline or recovery of salmonid fish stocks that are threatened or endangered or approaching that status.[1]

In response to the application by the states, NOAA Fisheries authorized them to reduce sea lion predation of salmonids by killing either the lesser of 85 sea lions per year or the number that would reduce predation to 1% of the salmonid run at the Bonneville Dam each year. For comparison, the total California sea lion population at the time was approximately 237,000 individuals.

There has been vociferous criticism of the exception added to the MMPA for killing pinnipeds. Critics point out that it is the building of dams and overfishing by humans that have threatened salmonid populations. Additionally, it is the dam slowing down the migration of the fish that makes the salmonids such easy prey for the sea lions in the first place. Killing pinnipeds for preying on salmonids that are threatened or endangered because of human activities seems somehow unfair.[2]

After approval of the application, several sea lions at the Bonneville Dam were euthanized, and many more would likely have been killed if NOAA Fisheries hadn't been caught by the inconsistency of its own numbers. NOAA Fisheries already permitted commercial, recreational, and tribal fisheries to take between 5.5% and 17% of the salmonid run in the Columbia River. NOAA Fisheries had even concluded in a 2005 report that the mortality from fishing had minimal adverse impacts on the viability of the salmonid populations. Additionally, the dams on the Columbia River kill a similar percentage of salmonids each year, and this was found by the US Army Corps of Engineers in 2007 not to harm salmonid populations. The sea lions were taking a much smaller percentage of the salmonid run each year than were being killed by fishing and dams, and fishing and dams had already been found not to harm salmonid populations.

The Ninth Circuit Court of Appeals in 2010 held that the NOAA Fisheries decision to authorize the killing of sea lions was arbitrary, capricious, and an abuse of discretion.[3] The court stated that NOAA Fisheries had not adequately explained why sea lions are supposedly having a significant negative impact on the salmonid populations in the Columbia River when fisheries that cause greater salmonid mortality are supposedly not having significant negative impacts. The court went on to say that NOAA Fisheries had also not adequately explained why 1% salmonid mortality was a magic number above which sea lions were having a significant negative impact on salmonids, but below which the impact was insignificant. As a consequence, the court stopped the killing of sea lions at the Bonneville Dam.

As this example suggests, management of the marine environment often pits species that are not commercially important against fish species that are commercially important. Commercial fishing operations negatively affect marine ecosystems in three primary ways: (1) overfishing; (2) physically damaging habitats, often with fishing gear such as trawls; and (3) incidentally killing nontarget species during fishing (bycatch).[4] Management of the marine environment must also deal with the stresses to marine ecosystems that come with the dumping of pollutants into the ocean. Pollutants enter the ocean when they are dumped or spilled from vessels and ocean drilling operations, but most pollutants make their way to the ocean when they are washed there from land. To be successful, laws that regulate the ocean must deal with all of these forms of marine ecosystem degradation. This chapter focuses on the effects that commercial fishing operations and pollution have on the marine environment.

The chapter first describes how international law carves up the ocean into different zones. Beginning with international law makes it easier to understand why US law asserts different levels of control in different parts of the ocean. The chapter then describes the MMPA. The law manages marine mammal populations, but also helps protect them from commercial fishing operations. Next is a look at the laws directly regulating commercial fisheries, followed by a discussion of marine protected areas. The last part of the chapter examines statutes that regulate the pollutants that enter the ocean, including the Clean Water Act and the Coastal Zone Management Act.

UNITED NATIONS CONVENTION ON THE LAW OF THE SEA

The ocean covers 71% of the earth's surface, and historically its sheer magnitude made the resources within seem unlimited. Most of the ocean was viewed as a global commons, to be freely used by all nations. We now understand, of course, that ocean resources are not unlimited, and that resource extraction by one country limits the resources available to other countries. As a consequence, countries have begun to negotiate treaties to address how and when the ocean may be used. The most important of these treaties is the United Nations Convention on the Law of the Sea (UNCLOS). UNCLOS addresses a wide range of issues regarding the world's oceans, including economic issues, passage of ships, and navigational rights. The convention also addresses the management and protection of marine species and ecosystems. UNCLOS is often described as the "constitution of the oceans."

Entering into force in 1994, the convention has 162 countries as parties. The United States is not one of those parties, though. The United States has not ratified UNCLOS because of Senate concerns that ratifying the convention will infringe on American sovereignty, specifically the ability of the United States to perform deep-sea mining. The United States, however, treats most of the other provisions of UNCLOS as customary international law. Recall from chapter 4 that customary international law is practiced by countries out of a sense of legal obligation. This means that the United States views itself as legally bound by most provisions in UNCLOS. Consequently, it is worthwhile to begin this chapter by understanding how UNCLOS carves up the ocean.

UNCLOS divides the ocean into four major zones: (1) the territorial sea,

(2) the exclusive economic zone, (3) the continental shelf, and (4) the high seas.

The territorial sea is the area of the ocean that extends 22 kilometers (12 nautical miles) from a country's coastline.[5] The territorial sea is controlled by the laws and regulations of the country that controls the coastline. That country also controls all the marine resources in the territorial sea; most major fish stocks, however, exist beyond the 22-kilometer territorial sea limit.

More important for fisheries is the exclusive economic zone, which extends 370 kilometers (200 nautical miles) from a country's coastline.[6] In the exclusive economic zone, the coastal country again controls the marine resources, but movement of foreign vessels or aircraft through the zone is not restricted.[7] In return for the coastal country's having control over resources in the exclusive economic zone, UNCLOS requires the country to manage and conserve the marine species in that zone.[8] The US exclusive economic zone is the largest in the world, making up 11.7 million square kilometers (3.4 million square nautical miles) of ocean, which is larger than the total land area of the 50 states.[9]

Extending the exclusive economic zone out to 370 kilometers is an attempt to reduce the "tragedy of the commons" for commercially exploited marine species. Over 90% of the world's fisheries fall within the exclusive economic zones of the world's coastal countries. If these fisheries were open for any country to fish, there would be a race to catch as many fish as possible each year because any reduction in fishing by one country would simply allow other countries to catch more fish. By allowing the coastal country exclusive access to the marine species in its exclusive economic zone, the country has an incentive to manage those species in ways that promote sustainable harvesting.

UNCLOS lists several requirements that constitute the management and conservation duty of coastal countries. Coastal countries must determine the total allowable catch of species in their exclusive economic zones that ensures the species are not endangered by overexploitation.[10] They must also consider how the catch of harvested species will affect other species that interact with harvested species. Finally, they must attempt to restore populations of harvested species to levels that allow a "maximum sustainable yield," taking into account environmental and economic factors.[11]

Although UNCLOS has provisions for the conservation of species, it also places great emphasis on using ocean resources. UNCLOS requires that parties promote "optimum utilization" of living resources.[12] This suggests that the convention expects countries to use an optimum yield approach to setting the total allowable catch in their exclusive economic zone. Box 9.1 describes maximum and optimum yields. After a coastal country has set the total allowable catch for commercial species, the country may take the entire catch. If it does not take the entire catch, though, UNCLOS requires that it allow other countries to fish in the exclusive economic zone until the allowable catch is reached.[13] One result of UNCLOS's emphasis on using marine resources is the finding that 91% of European fish stocks are below the population levels that would allow a maximum sustainable yield, and will likely remain that way for at least the next 30 years.[14]

BOX 9.1. MAXIMUM AND OPTIMUM YIELDS

Many laws that regulate the management of living marine resources base that regulation on the concepts of maximum sustainable yield and optimum yield. *Maximum sustainable yield* is the largest harvest that can be taken from a species' stock over an indefinite period of time. If one makes the simplifying assumption that a population has logistic (S-shaped) growth, then the maximum sustainable yield will occur when the population of the stock is held at one-half its carrying capacity in the environment. Almost all populations display growth that is more complex than logistic growth, though, making it difficult to determine the maximum sustainable yield of most marine populations.

The *optimum yield* is the harvest level that best benefits a country when taking into account such factors as the economics of harvesting, protection of the marine environment, and social values. While the maximum sustainable yield considers only the biology of the species being harvested, the optimum yield considers all other factors that are influenced by the harvest. The optimum yield is often found by determining the maximum sustainable yield, and then lowering it to account for economic, ecological, and social factors. Because a country can consider so many factors, almost any harvest level may be justified as being the optimum yield.

The next ocean zone described in UNCLOS is the continental shelf, the seabed and subsoil extending to the outer edge of the continental margin.[15] Coastal countries have sovereignty over the continental shelf and the natural resources of the shelf, and they have exclusive control of the sedentary species on the shelf. Frequently the continental shelf is within the exclusive economic zone of a coastal country, so the country has exclusive control of both the continental shelf and the water above it. Occasionally the continental shelf extends beyond 370 kilometers. In those instances, the coastal country controls the shelf, while the water above is defined as part of the high seas.

This leads to the last major ocean zone in UNCLOS, the high seas. The high seas are the areas of the ocean beyond the exclusive economic zones of coastal countries.[16] All countries are considered to have a right to fish in any area of the high seas.[17] UNCLOS does state, however, that parties to the convention have a duty to manage and conserve the species of the high seas.[18] UNCLOS requires parties to maintain or restore harvested species in the high seas to population levels that can produce their maximum sustainable yields.[19] Parties are also required to consider the effects of harvesting on other species, "with a view to" maintaining or restoring such populations to levels above which reproduction is no longer seriously threatened.[20] Overexploitation of fisheries in the high seas may be less of an overall threat to marine ecosystems than in the exclusive economic zones, where most fisheries are located. For individual species that spend much of their life in the high seas, though, the potential for overfishing by every country with a fishing fleet is significant. To help counter overfishing, many countries have signed a treaty to help conserve fish stocks that migrate through the high seas. This Fish Stocks Agreement is discussed later in the chapter.

Contrary to popular belief, once a vessel is on the high seas it is not beyond the reach of law. Many US statutes apply to Americans and American vessels on the high seas. For example, the MMPA states that it is illegal for any person under the jurisdiction of the United States to kill any marine mammal on the high seas.[21]

MARINE SPECIES AND ECOSYSTEMS

The chapter now turns to US law and its effect on marine ecosystems.[22] Statutes that regulate and manage individual species are discussed first.

Marine Mammal Protection Act

The Marine Mammal Protection Act of 1972 (MMPA) states as its primary goal the protection of marine mammals by maintaining the health and stability of the marine ecosystem.[23] The primary means by which the act accomplishes this is by establishing a moratorium on the taking and importing of marine mammals and marine mammal products.[24] This means that there may be no killing of marine mammals within the areas subject to US jurisdiction, or by any person subject to US jurisdiction on the high seas.

According to the MMPA, "take" means to "harass, hunt, capture, or kill, or attempt to harass, hunt, capture, or kill" any marine mammal.[25] The act defines "harassment" as any act of "pursuit, torment, or annoyance" that has the potential to injure a marine mammal in the wild, or has the potential to disturb a marine mammal in the wild by disrupting behavioral patterns such as "migration, breathing, nursing, breeding, feeding, or sheltering."[26] A US Fish and Wildlife Service (FWS) regulation expands "take" to include collecting dead marine mammals or parts thereof, restraining or detaining a marine mammal, tagging a marine mammal, and the operation of an aircraft or vessel in such a way as to disturb or molest a marine mammal.[27] A National Oceanic and Atmospheric Administration (NOAA) regulation adds to the definition of "take" the feeding or attempted feeding of a marine mammal in the wild.[28] This was added to prevent operators of commercial tour boats from feeding wild marine mammals in an attempt to lure them closer to the vessel for viewing by tourists. The MMPA is thus similar to the ESA in prohibiting the take of organisms protected by the acts. The difference is that the ESA applies only to threatened or endangered species listed under the act, while the MMPA applies to all marine mammals whether they are confronting extinction or not.

Many mammals that spend time in the ocean also spend significant time on land—so which mammals are covered by the MMPA? The act defines "marine mammal" as any mammal that is morphologically adapted to the marine environment, such as sea otters and whales, or any mammal that primarily inhabits the marine environment, such as polar bears.[29] Included in the definition of "marine mammal" is any part of such an animal, including its fur or skin.[30] The secretary of commerce through NOAA Fisheries has responsibility for cetaceans (whales, dolphins, and porpoises) and pinnipeds except walruses (seals and sea lions), while the secretary of the interior mainly through FWS has responsibility for all other marine mammals.[31]

As might be expected, the MMPA allows for the taking of marine mammals in certain situations if a permit has been issued. Permits may be issued for purposes that include (1) scientific research, (2) public display, (3) photography for educational or commercial purposes, (4) to enhance the survival or recovery of a species, (5) importation of polar bears taken in sports hunts in Canada, or (6) incidental taking of small numbers of marine mammals.[32] Under the last category, a permit may be issued for the incidental take but not the intentional take of marine mammals, and only for periods of less than five consecutive years while engaged in an activity.[33] Additionally, the permit may be issued only if the total takings will have a negligible effect on the species or population of the species. Such permits are issued for activities such as oil and gas extraction where a marine mammal may be incidentally killed. In 1994 Congress created a streamlined process for permits when take of marine mammals would occur only through harassment.[34] Such authorization may be issued for up to one year, and only if the harassment will have a negligible impact on the marine mammal species or population. These permits are often issued when marine mammals may be harassed by noise.

The MMPA has other exceptions to the moratorium on taking marine mammals. There is an exception for any "Indian, Aleut, or Eskimo" living in Alaska on the coast of the North Pacific Ocean or Arctic Ocean if the taking is for subsistence or the creation of handicrafts.[35] The taking of individual pinnipeds that are killing salmonids is another exception, discussed at the beginning of this chapter. Perhaps most importantly for marine ecosystems, there is also an exception for the incidental take of marine mammals during commercial fishing operations. The requirements for commercial fisheries will be taken up below.

Besides a moratorium on taking marine mammals, the MMPA sets as one of its major objectives keeping each marine mammal species from falling below its optimum sustainable population size.[36] The act defines "optimum sustainable population" as the number of animals that result in the maximum productivity of the population, "keeping in mind" the carrying capacity of the habitat and the health of the ecosystem.[37] NOAA Fisheries and FWS regulations further define "optimum sustainable population" as a population size that falls within the range between the population level that is the largest supportable within the ecosystem (i.e., carrying capacity), and the level that results in maximum net productivity.[38] "Maximum net productivity" is then defined as the greatest net annual increase in population numbers or biomass. Because the population size that results in maximum net productivity is the lower bound of the range defining an optimum sustainable population, it has been the target for marine mammal population sizes. To make this more concrete, a species is usually considered to be below its optimum sustainable population size if it falls below 60% of its historical carrying capacity.

If a marine mammal species or population stock falls below its optimum sustainable population, the MMPA considers the species "depleted."[39] A species is also considered depleted if it is listed as threatened or endangered under the ESA. If a species or stock is depleted, the MMPA requires that a conservation plan be prepared for that species.[40] Additionally, authorizations for the incidental take of depleted species may be limited. As an example, in 1988 NOAA Fisheries declared the stock of northern fur seals (*Callorhinus ursinus*) on the Pribilof Islands in Alaska to be depleted because they had fallen to 50% of their carrying capacity.[41] The stock had fallen to such a low number primarily because of commercial harvesting of fur seals in the 1950s and 1960s. Since the end of commercial harvesting, reductions in fur seal stock have been driven by other anthropogenic causes, such as decreases in fish prey populations due to commercial fishing, marine debris from fishing operations, and harassment from vessels and construction. NOAA Fisheries published a conservation plan for the fur seal stock in 1993, and a revised plan in 2007. The conservation plan describes actions such as removing marine debris, making recommendations on development activities, and studying and reporting on the impact of commercial fisheries on fur seals. While all the actions listed in the conservation plan would likely help the fur seal stock, NOAA Fisheries must work with other local, state, and federal agencies to implement almost all of them. As of the 2007 revised plan, the northern fur seal stock in the Pribilof Islands was still declining.

An additional way the MMPA helps protect marine mammals is through greater oversight of federal activities affecting marine mammals. The act does this through the establishment of the Marine Mammal Commission.[42] The commission is charged with a continuing review of the conditions of marine mammal populations. It is composed of three members appointed by the president with the advice and consent of the Senate. The commission must establish a committee of nine scientists knowledgeable in marine ecology.[43]

The Marine Mammal Commission is unique in federal environmental law in that it has the power to oversee and make recommendations on all federal agency actions relating to marine mammals. The commission may recommend steps that it deems necessary for the protection and conservation of marine mammals.[44] If the commission makes a recommendation to a federal agency that is not followed, the agency must write a detailed explanation for not following it.[45] Similarly, if the commission does not accept a recommendation made by its scientific committee, the recommendation must still be sent to the relevant federal agency along with a detailed explanation for why the commission did not accept it.[46] Although the commission does not have the power to force an agency to accept any of its recommendations, having a body with oversight across the federal government for activities affecting marine mammals seems to have helped create a more uniform approach to marine mammal conservation.[47] It is unfortunate that Congress did not include the creation of similar commissions in the other environmental statutes it has passed.

The greatest threat to marine mammals across the world is accidental capture or entanglement in fishing gear.[48] Such bycatch kills hundreds of thousands of marine mammals every year. The MMPA regulates the incidental taking of marine mammals by commercial fishing operations, but the requirements are quite different than presented above for nonfishing activities. Congress passed amendments in 1994 stating that commercial fisheries do not have to obtain an incidental-take permit before engaging in a fishing operation that may take marine mammals.[49] Instead, commercial fishery operators must register their vessels and then abide by specific requirements, such as monitoring and reporting, limits on incidental mortality, and gear restrictions. Further, the amendments do not require commercial fisheries to stop the incidental take of a marine mammal species even if the species is found to be depleted. Instead, the amended MMPA states as its goal that incidental take of marine mammals during commercial fishery operations be less than the "potential biological removal level" established for that species.[50]

The MMPA defines "potential biological removal level" as the maximum number of organisms (excluding natural mortality) that can be removed from a population while still allowing the population to reach or maintain its optimum sustainable population.[51] The potential biological removal level is calculated as the product of (1) the minimum population estimate of the population, (2) one-half the estimated net productivity rate of the species at a small population size, and (3) a recovery factor of between 0.1 and 1.0.[52] The recovery factor is simply a multiplier that reduces the potential biological removal level depending on the condition of a species or stock. An endangered species or declining stock receives a recovery factor of 0.1 to reduce the potential biological removal level. A threatened or depleted stock receives a recovery factor of 0.5. A stock at optimum levels receives a recovery factor up to 1.0.

The 1994 amendments state an additional goal: that commercial fisheries reduce incidental mortality and serious injury of marine mammals to insignificant levels approaching a zero mortality and injury rate by 2001.[53] This goal has not been met. Remarkably, according to a regulation, a zero mortality and injury rate is not even defined as zero mortality; instead it is defined as mortality below 10% of the potential biological removal level.[54]

The MMPA does create a process for protecting marine mammal species that are in trouble. Under the MMPA, if a marine mammal stock is listed as endangered or threatened under the ESA, designated as depleted, or has human-caused mortality above its potential biological removal level, then the stock is designated as a "strategic stock."[55] The MMPA requires stock assessments of strategic stocks every year, and stock assessments every three years for all other stocks.[56] The MMPA then requires NOAA Fisheries to create "take reduction plans" for each strategic stock that interacts with a commercial fishery.[57] The goal of a take reduction plan is to reduce incidental takings from commercial fisheries to levels less than the potential biological removal level established for that species. To write a take reduction plan, the secretary of commerce may establish a take reduction team composed of individuals who have expertise with that species. The take reduction plan may include fishery-specific limits on incidental mortality and serious injury for the species, it may restrict commercial fisheries by time or area, and it may require use of alternative commercial fishing gear or techniques.[58] The MMPA states that the long-term goal of take reduction plans is also to reduce the incidental mortality and serious injury of marine mammals to insignificant levels approaching a zero mortality and serious injury rate, although in this case, taking into account the "economics of the fishery."[59]

Several scientists have been critical of the potential biological removal approach for setting the allowable level of human-caused mortality. For most marine mammal species, estimating population abundance, mortality, and productivity rate is very difficult. For many populations, the estimates are likely not very accurate.[60] Indeed, for several marine mammal stocks, there is no reliable abundance data at all. As for estimating mortality, part of the difficulty is that there are simply not enough observers on fishing vessels to accurately estimate bycatch of many marine mammals.[61]

A recent study examined whether the potential biological removal approach would be able to detect a precipitous decline in abundance of 50% in 15 years of several marine mammal stocks.[62] If the potential biological removal approach could not detect such a large decline, then the approach potentially allows the incidental take of many more individuals in a given stock than is sustainable, because the estimated size of the stock is much too high. The study found that the potential biological removal approach would not detect precipitous declines in 72% of large whale stocks, 90% of beaked whale stocks, 78% of dolphin and porpoise stocks, 55% of polar bear and sea otter stocks, and 100% of on-ice pinniped stocks. Only for pinnipeds on land, which are relatively easy to survey, would a precipitous decline be detected for 95% of stocks. The potential biological removal approach may not be able to prevent declines in marine mammal species if it cannot even detect such declines.

An additional criticism of the potential biological removal approach is that it assumes that direct human-caused mortality is the greatest threat to all marine mammal species. Several species, however, have faced steep declines in abundance not obviously caused by human activity, such as the western stock of the Steller sea lion, harbor seals in Alaska, sea otters in western Alaska, and Hawaiian monk seals.[63] These steep declines may ultimately be due to anthropogenic causes such as global climate change, but the potential biological removal approach considers only direct human-caused mortality. The focus on determining how many individuals of a stock may be killed each year comes at the expense of managing the stock's entire ecosystem.[64] Even though the MMPA specifically mentions ecosystem management as a goal of the act, NOAA Fisheries and FWS have placed most of their attention on managing individual species, and less on how entire ecosystems are functioning. The potential biological removal approach is part of the MMPA, and must be followed by these agencies. Nevertheless, take reduction plans should place greater emphasis on managing ecosystems as a means of increasing the populations of strategic stocks.

Fishery Management

As the last section suggests, commercial fisheries can have wide-ranging ecological impacts. These ecosystem effects are bound to be even greater if fisheries push fish populations below sustainable levels. We tend to think of overfishing as a recent phenomenon, but amazingly, humans have been overfishing, and thereby altering marine ecosystems, since at least the late aboriginal and early colonial stages of our history.[65] Overfishing has certainly increased dramatically as humans entered the modern era, and thus so has the need for laws that attempt to prevent overfishing and to manage fish stocks.

Magnuson-Stevens Fishery Conservation and Management Act

The Magnuson-Stevens Fishery Conservation and Management Act (Magnuson-Stevens Act) is the primary federal statute used for managing fishery stocks in US waters. In general, the coastal states regulate fishing activities from the coast out to 5.6 kilometers (3 nautical miles). The federal government regulates fishing activities from 5.6 kilometers out to the 370-kilometer limit of the exclusive economic zone.

The Magnuson-Stevens Act defines "fish" as all forms of marine animal and plant life, except marine mammals and birds.[66] A "fishery" is one or more stocks of fish that can be treated as a unit for conservation and management, and any fishing for such stocks.[67]

The principal way the Magnuson-Stevens Act regulates fisheries is through the creation of eight regional fishery management councils.[68] Each council manages the fishery stocks in a large section of the ocean under US jurisdiction. NOAA Fisheries has authority over highly migratory fish species that cross the geographical boundary of more than one council.[69] Highly migratory fish include tuna species, marlin (*Tetrapturus* spp. and *Makaira* spp.), oceanic sharks, sailfishes (*Istiophorus* spp.), and swordfish (*Xiphias gladius*).[70]

The Magnuson-Stevens Act requires that the councils prepare a "fishery management plan" for each fishery under its authority that requires "conservation and management."[71] As a result, if a council decides that a particular fishery does not require conservation and management, then that fishery will not be subject to federal regulation. In 2011 there were 45 fishery management plans covering 537 individual stocks and stock complexes.

The act states that each fishery management plan be consistent with ten "national standards" for fishery conservation and management.[72] The first standard requires conservation and management measures to prevent overfishing, while achieving the optimum yield from each fishery in the US fishing industry. The first national standard clearly lays out the balance that must be achieved by fishery management plans between conserving fish species and allowing the largest yield of fish. The second standard requires that conservation and management measures be based on the "best scientific information available." The fifth standard states that conservation and management efforts shall consider efficiency in the utilization of fishery resources, but shall not have economic allocation as its sole purpose. Standard 8 requires that conservation and management measures, to the extent practicable, minimize adverse economic impacts on fishing communities. Finally, standard 9 requires that conservation and management measures shall, to the extent practicable, minimize bycatch, and when bycatch cannot be avoided, minimize mortality from bycatch.

Fishery management plans must specify objective criteria for when a fishery is overfished. Then the plan must contain conservation and management measures to end overfishing and rebuild the fishery. The plan must also specify the maximum sustainable yield and the optimum yield from the fisheries.[73] The act defines "optimum yield" as the amount of fish that will provide the greatest overall benefit to the nation, particularly with respect to food production and recreation, but taking into account protection of marine ecosystems.[74] The optimum yield is found by taking the maximum sustainable yield and reducing it by "any relevant social, economic, or ecological factor." For an overfished fishery, the optimum yield should provide for rebuilding the fishery to a level that produces the maximum sustainable yield for that fishery. The Magnuson-Stevens Act then requires fishery management plans to specify the extent to which fishing vessels will harvest the optimum yield.

Lastly, the plans must identify essential fish habitat for each managed fish species, and minimize, to the extent practicable, adverse effects on that habitat. The act defines "essential fish habitat" as the waters and substrate necessary for fish to spawn, breed, feed, or grow to maturity.[75] Options for reducing the adverse effects of fishing on such habitat include equipment restrictions, closing areas to all fishing during specific times, and limits on the take of prey species or species that provide structural habitat.[76] If another federal agency is planning to undertake an action that may adversely affect any essential fish habitat, the agency must consult NOAA Fisheries.[77] Federal agencies do not have to follow the recommendations of NOAA Fisheries, but they must provide an explanation for why they are not. The councils may also identify "habitat areas of particular concern" that are particularly important ecologically or particularly sensitive to human-induced degradation.[78] While these areas receive higher priority for conservation and management, only about 100 such areas have been identified.

The Magnuson-Stevens Act then describes several discretionary measures that may be included in a fishery management plan.[79] They include requiring a permit to fish, setting a cap on the number of vessels that may fish a given fishery, designating zones or periods when fishing will not be permitted, requiring certain types of fishing gear or fishing vessels, and requiring observers be carried on board vessels.

Voting members of each regional fishery management council include federal officials and officials from the coastal states adjacent to the ocean area the council manages. Voting members also include individuals who have been nominated by coastal state governors and then appointed by the secretary of commerce. These appointed members are almost exclusively representatives from fishing industries. Under the act, each council must establish a scientific and statistical committee to help develop and evaluate scientific information relevant to creating or amending fishery management plans.[80] Committee members may include state or federal employees, academicians, and independent experts. The committee is required to provide ongoing scientific advice for fishery management decisions.

The reauthorization of the Magnuson-Stevens Act by Congress in 2006 gave considerable new power to the scientific and statistical committees. The reauthorization states that each regional fishery management council must develop annual catch limits for its managed fisheries. These annual catch limits may not exceed the fishing-level recommendations of the scientific and statistical committees, or the "peer review process."[81] The requirement for annual catch limits was an attempt by Congress to make the Magnuson-Stevens Act more focused on conservation, with a greater basis in science. Most fisheries are only beginning to incorporate the annual catch limits required by the reauthorization, so it may be several years before their impacts on fisheries are known.

Once a regional fishery management council has written a fishery management plan, NOAA Fisheries must review and approve the plan.[82] NOAA Fisheries determines whether the plan is consistent with all the requirements of the Magnuson-Stevens Act. In practice, NOAA Fisheries is very deferential toward the councils. Between 1980 and 2000, NOAA Fisheries partially or wholly rejected only 7% of all fishery management plans.[83]

NOAA Fisheries does not necessarily have the last word on fishery management plans, though—the Magnuson-Stevens Act allows citizens to challenge the legality of a plan in federal court.[84] Winning a challenge to a fishery management plan is difficult. The court will overturn a plan only if the plaintiff can prove that a NOAA Fisheries action was arbitrary and capricious, an abuse of discretion, unconstitutional, in excess of statutory authority, or did not follow all procedures required by law.[85] Nevertheless, several lawsuits have alleged that NOAA Fisheries actions were arbitrary and capricious because they were not based upon the best scientific information available, thereby violating the second national standard required by the act. As discussed in the first chapter, courts give great deference to agencies to determine what is the best available science. Courts have also held that the second national standard does not require NOAA Fisheries to conduct new scientific studies before taking action.[86] In essence, courts require plaintiffs to show that there was no scientific basis whatsoever for a NOAA Fisheries action. This is a very high burden for plaintiffs to meet, although plaintiffs have won on occasion.[87]

If overfishing is severe enough, it may result in the local extinction of species, losses of entire trophic levels, and a considerably less productive ecosystem. Even if overfishing is not so severe, it still tends to be particularly damaging to marine ecosystems because it amplifies other stressors. Overfishing increases the damage done to marine ecosystems by pollution, disease, physical destruction of habitat, storms, introduction of invasive species, and climate change.[88] As an example, when European colonists began widespread land clearing for agriculture in the mid-eighteenth century, Chesapeake Bay started receiving large amounts of nutrient-rich sediment.[89] The Chesapeake Bay ecosystem changed slowly, however, and there was little evidence of eutrophication (see box 7.1 for a description of eutrophication). The large oyster populations in the bay were able to quickly filter the water column, limiting blooms of phytoplankton that would otherwise have been frequent occurrences. The oysters were thereby preventing eutrophication. In the 1870s, though, mechanical harvesting of oysters in Chesapeake Bay began. Because the mechanical harvesting was so efficient, oyster populations were quickly and dramatically reduced. With oyster populations decimated, phytoplankton blooms became common, and Chesapeake Bay faced widespread eutrophication by the 1930s. By the 1950s there were regular outbreaks of oyster parasites. Gray whales, dolphins, manatees, river otters, sea turtles, and several other species that used to be abundant in the Chesapeake Bay are now all but extirpated.[90] Overfishing of oysters amplified the effects of pollution and disease to greatly alter the Chesapeake Bay ecosystem.

A second example is Georges Bank off the New England coast. The region is a highly productive submarine plateau that has supported commercial fisheries for over 400 years.[91] In the 1960s, however, distant-water fleets began fishing in Georges Bank, greatly increasing the catch of fish. By the late 1980s and early 1990s, the populations of commercial fish species in Georges Bank plummeted. Total fish biomass declined by more than 50%.[92] The decrease in fish biomass greatly altered the community dynamics of Georges Bank. Cod and flounder populations declined, easing competitive pressures on dogfish shark and skate populations. Consequently, dogfish shark and skate populations increased dramatically. Dogfish sharks prey on commercially important species such as mackerel and herring. While mackerel and herring faced increased predation, they continued to be caught by fishing vessels, greatly reducing the populations of those species. The fishery management plan covering Georges Bank was finally amended in the early 1990s to limit the catch of depleted fish stocks.[93] The amended management plan reduced the days vessels could fish, and closed areas in Georges Bank to fishing altogether. The management plan helped slowly increase the populations of the commercially exploited species.[94]

How have the regional fishery management councils and NOAA Fisheries done overall in preventing overfishing? In general, fish stocks in US waters are among the best managed in the world. In a 2011 report to Congress, NOAA Fisheries reported on the 537 fish stocks managed that year.[95] The heart of the report consists of detailing which fish stocks were "overfished" and which stocks were "subject to overfishing." Regulations state that a stock is overfished when it has a biomass level below the level necessary to produce the maximum sustainable yield on a continuing basis.[96] A stock is subject to overfishing when it is being harvested at a rate above the level necessary for the stock to produce its maximum sustainable yield on a continuing basis.[97] Of the 537 managed stocks, there was not enough information on 318 stocks to determine if they were overfished or not, and not enough information on 279 stocks to determine if they were subject to overfishing or not. Of the 219 stocks where a determination of being overfished could be made, 45 (21%) were overfished. Of the 258 stocks where a determination of being subject to overfishing could be made, 36 (14%) were subject to overfishing. The report also declares that 27 fish stocks that were overfished in the past have been rebuilt, with 51 stocks in the rebuilding phase.

The determination by NOAA Fisheries of whether a fish stock is overfished does have some problems. The regional fishery management councils get to decide when a stock is overfished. Following the regulations, most fishery management plans define a stock as overfished when it is below half the biomass that would produce the stock's maximum sustainable yield.[98] Many ecologists, however, consider a stock overfished when its biomass falls below the amount necessary to produce the stock's maximum sustainable yield.[99] Consequently, fishery management plans do not define a stock as overfished until it is well below the biomass that ecologists would consider overfished. If this more stringent ecological standard were used, many more fish stocks would be described by NOAA Fisheries as overfished. Perhaps more importantly, not enough is known about more than half the fish stocks in US waters to determine whether they are overfished. Managing and conserving a fish stock is immeasurably more difficult if not enough is known about the stock to determine whether it is overfished.

Additionally, stating whether individual fisheries are overfished does little to indicate how marine ecosystems are responding to fishing activity.[100] Commercial fishing activity, even for stocks that are not overfished, can impact marine ecosystems in diverse ways. For instance, if a fishery preferentially catches large individuals, there will be selective pressure for fish to evolve smaller sizes and earlier ages at maturation.[101] Atlantic cod populations show genetic changes in growth as the result of fishing, producing significantly smaller individuals at maturity than 30 years ago.[102] Such evolution may ultimately reduce the productivity and stability of the fish stock. The evolutionary reduction in size can also affect community and ecosystem function. Reduction in body size may result in a trophic cascade: as the predatory fish that is being harvested evolves a smaller body, this frees its prey species from predation pressure, allowing the prey to overexploit resources at the bottom of the food web. Fishery management plans should do a better job of predicting how fishing will affect species beyond commercially important fish stocks. Marine protected areas, discussed below, may be an important tool in counteracting many of the ecosystem-level effects of commercial fishing operations.

Straddling Fish Stocks and Highly Migratory Fish Stocks

The Magnuson-Stevens Act regulates fish stocks in the US exclusive economic zone, but has no control over foreign vessels fishing on the high seas. To help prevent overfishing on the high seas, 78 countries, including the United States, are parties to a treaty called the Agreement for the Implementation of the Provisions of the United Nations Convention on the Law of the Sea relating to the Conservation and Management of Straddling Fish Stocks and Highly Migratory Fish Stocks (Fish Stocks Agreement). Straddling stocks are species that occur both in a country's exclusive economic zone and in the adjacent high seas. Highly migratory stocks are species that migrate great distances, moving through both exclusive economic zones and the high seas. Bluefin tuna (*Thunnus thynnus*) is an example of a highly migratory fish species. The Fish Stocks Agreement entered into force in 2001.

The Fish Stocks Agreement requires that countries cooperate to ensure the long-term sustainability of straddling stocks and highly migratory fish stocks.[103] The agreement states that stocks are to be restored to levels that can produce the maximum sustainable yield, taking into account environmental and economic considerations.[104] Ensuring sustainability of straddling stocks and highly migratory stocks often entails negotiations between coastal countries and other countries that want to fish on the high seas in that region.[105] To support such negotiations, the Fish Stocks Agreement encourages countries to form regional or subregional fisheries management organizations to manage such fish stocks.

The convention includes a mechanism to enforce agreements made between countries fishing on the high seas. Inspectors of a country that is a member of a regional or subregional fishery organization may board a ship flying the flag of a different country that is a party to the Fish Stocks

Agreement.[106] The inspectors may then inspect the ship to ensure that it is complying with conservation and management measures while fishing on the high seas.

Marine Protected Areas

As the above sections suggest, commercial fisheries have enormous impacts on the marine environment. One of the best ways to protect marine ecosystems from commercial fishing operations and other stressors is through the creation of marine protected areas. These geographical areas have been set aside by law to provide protection to all or part of the natural or cultural resources that they contain. Marine protected areas often lead to rapid increases in abundance, diversity, and productivity of marine species. The reason is that the protected areas decrease the mortality of organisms, and lessen habitat destruction. Marine protected areas lower the probability that species that use the area will go extinct.

Several federal and state laws are used to create marine protected areas. Under the Magnuson-Stevens Act, an area of the ocean may be closed to fishing for a certain species of fish, thereby creating a marine protected area for that species. Other federal and state laws place various restrictions on the protected areas they create, such as allowing or prohibiting oil and gas drilling, or allowing or prohibiting commercial fishing. A marine protected area that prohibits all resource extraction, including fishing, is called a marine reserve. This section focuses on the two most important federal statutes that are used to create marine protected areas, the National Marine Sanctuaries Act and the Antiquities Act.

National Marine Sanctuaries Act

The main federal statute for creating marine protected areas is the Marine Protection, Research, and Sanctuaries Act of 1972. The act has two primary purposes: to create marine sanctuaries and to regulate ocean dumping. In 1992 Congress renamed the section of the act that regulates the creation of sanctuaries as the National Marine Sanctuaries Act. The section of the act that regulates dumping is frequently called the Ocean Dumping Act (discussed later in the chapter).

Under the National Marine Sanctuaries Act, the secretary of commerce may designate an area of the marine environment as a national marine sanctuary.[107] The area designated must be of "national significance" because of its conservation value, ecological or scientific qualities, the communities of marine species it harbors, or its resource or human-use values. In deciding on areas to designate, the secretary must consider factors such as an area's contribution to biological productivity, maintenance of ecosystem structure, maintenance of commercially important or threatened species, and maintenance of critical habitat for endangered species.[108] The secretary must also consider the present and potential activities that occur in an area that may adversely affect the marine environment. Finally, the secretary must consider the public benefits to designating an area as a sanctuary, along with the socioeconomic effects of restricting income-generating activities in that area. There are 13 national marine sanctuaries that cover approximately 47,000 square kilometers (18,000 square miles).

Before the secretary of commerce may designate an area as a sanctuary, the National Marine Sanctuaries Act requires a time-consuming process of consultation with other federal and state agencies and with the relevant regional fishery management council, and opportunities for public comment.[109] The act also requires the secretary to prepare a draft environmental impact statement.[110] After this lengthy process, Congress may then reject a designation during a 45-day review period, and any proposed sanctuary in state waters may be rejected by the governor of that state.[111]

The National Marine Sanctuaries Act makes it unlawful for any person to injure or cause the loss of any sanctuary resource.[112] The act also makes it unlawful to possess, sell, import, export, or transport any sanctuary resource. NOAA has authority to create general regulations that apply to all the sanctuaries, and specific regulations that apply to individual sanctuaries. The general regulations NOAA has promulgated allow activities such as fishing, boating, diving, and research in the sanctuaries, unless prohibited by a sanctuary-specific regulation.[113] The general regulations also require development and implementation of a management plan for each sanctuary.[114] Several of the sanctuary-specific regulations prohibit activities such as drilling in the seabed for oil or gas, anchoring a vessel, releasing a nonnative species, or taking sea turtles or seabirds.[115] Other regulations are tailored to a given sanctuary, such as the regulation in the Hawaiian Islands Humpback Whale National Marine Sanctuary that vessels are not allowed to approach within 91 meters (100 yards) of a humpback whale.

Under the National Marine Sanctuaries Act, NOAA may issue permits to allow persons to conduct activities normally prohibited by sanctuary-specific regulations.[116] The permitted activity must be "compatible with the purposes for which the sanctuary is designated," and designed to protect sanctuary resources.[117] Permits last for five years, but may be renewed by NOAA.

While the National Marine Sanctuaries Act is the primary federal statute for designating marine protected areas, the act has been used to create only a handful of sanctuaries, and the sanctuaries do not restrict many potentially damaging activities, such as fishing. The next statute discussed, though not passed by Congress with the intention of being used to create marine protected areas, has had a much larger role in creating such protected areas than the National Marine Sanctuaries Act.

Antiquities Act

The Antiquities Act of 1906 authorizes the president to declare objects of "historic or scientific interest" situated on land owned or controlled by the federal government to be national monuments.[118] The act requires that land included as part of the monument take up the smallest amount of area compatible with proper care and management of the monument. Congress originally passed the Antiquities Act to protect archeological sites and artifacts on federal lands. Successive presidents have greatly broadened the use of the act in creating national monuments.

In 2006 President George W. Bush used the Antiquities Act to create the Papahānaumokuākea Marine National Monument in Hawaii. The monument covers 362,061 square kilometers (139,793 square miles) of ocean, including several islands and atolls. The monument is 7 times larger than all the national marine sanctuaries combined, and 100 times larger than Yosemite National Park. While not technically a marine protected area, the monument is functionally equivalent to one, and is one of the largest marine conservation areas in the world.

Exploring for or producing oil, gas, or minerals within the monument is prohibited, as is releasing nonnative species and anchoring a vessel to a living or dead coral.[119] Almost all other activities within the monument require a permit. Unless individuals receive a permit, they may not take, possess, injure, or disturb any living or nonliving resource in the monument. Individuals also may not drill into the seabed, anchor a vessel, touch living or dead coral, attract any organisms, or possess unstowed fishing gear. Only responses to emergencies, law enforcement, military activities, and uninterrupted passage do not require a permit.

Working together, NOAA and FWS created a 15-year management plan for the monument. The three primary goals of the management plan are to protect threatened and endangered species in the monument, conserve migratory bird habitats, and restore native ecosystems. To achieve its goals, the management plan focuses on monitoring populations, removing or mitigating sources of anthropogenic harm, and restoring native species while removing introduced species.

There has been fierce criticism of using the Antiquities Act to create marine national monuments.[120] Part of the reason is because the monuments tend to be huge. In one of his last acts as president, President Bush in 2009 used the Antiquities Act to create the Marianas Trench Marine National Monument, the Rose Atoll Marine National Monument, and the Pacific Remote Islands Marine National Monument. The monuments cover approximately 505,770 square kilometers (195,280 square miles), and all of them prohibit commercial fishing. For a size comparison, the country of Spain has an area of 504,782 square kilometers (194,897 square miles). Critics point out that the creation of these huge marine national monuments, along with Papahānaumokuākea Marine National Monument, did not require any consultation or public feedback. Recall that the National Marine Sanctuaries Act requires consultation with other agencies, an opportunity for public comment, a draft environmental impact statement, and time for Congress and the governors of coastal states to reject a proposed sanctuary. The Antiquities Act requires none of this. Critics argue that if coastal communities feel they did not have a say in the creation of these large national monuments, they will be less likely to abide by restrictions placed on the monuments.

Supporters of using the Antiquities Act to create marine national monuments respond that under the act, a national monument can be created much more quickly than through the National Marine Sanctuaries Act. For ecosystems facing daily threats, the quickness of the Antiquities Act may be crucial in helping protect a threatened ecosystem. The Antiquities Act also sidesteps any potential political gridlock that may occur through consultation with other agencies, Congress, or coastal governors.[121] Furthermore, Congress has the power to reverse or expand the designation of a national monument through the normal legislative process if it does not approve of a monument (although a president would likely veto any attempt by Congress to reverse the designation of a monument he or she personally created, thereby requiring a supermajority of Congress, or requiring Congress to wait until the next presidency). Finally, supporters of the Antiquities Act point out that, over time, national monuments almost always become very popular with the public.

Ecology of Marine Protected Areas

Empirical studies show that marine protected areas increase the abundance and reproductive rate of species in the protected area. This is true, though, mainly for species that have low or moderate movement.[122] Species that move relatively little remain within the protected area and confront less anthropogenic stress.

Marine protected areas are less likely to directly benefit highly mobile species, as they spend more time outside of the protected areas and therefore confront greater anthropogenic stress. Reducing the catch of mobile species, such as through the Fish Stocks Agreement discussed above, is an important tool for protecting highly mobile species. Marine protected areas, though, may also provide some protection to highly mobile species. Restricting fishing in the spawning grounds, or the areas where juveniles tend to feed, may provide the best protection for such species. Fishing, however, is not the only stressor for mobile species. Other anthropogenic stresses, such as pollutants or the drilling for oil or gas, may also negatively affect highly mobile species. To help highly mobile species, future marine protected areas should include areas where such species are most vulnerable to anthropogenic stress other than fishing. There is a need, though, for more empirical research to suggest the best way to design marine protected areas for highly mobile species.[123]

For the species that are helped by marine protected areas, their increase in abundance often leads to spillover of individuals to regions outside of the protected area. If a particular species is used by a commercial fishery, the spillover effect can greatly increase catches by that fishery. Occasionally, though, to take advantage of the spillover from a marine protected area where fishing is prohibited, fishermen may fish along the protected area boundary line.[124] Modeling suggests that intensive "fishing the line" acts as an edge effect that may significantly reduce the abundance of populations in the protected area. Most marine protected areas that prohibit fishing do not also prohibit fishing the line.

Marine protected areas may help commercial fisheries in a different way: they help counteract the evolutionary pressure that is frequently created by those fisheries. Recall that commercial fishing operations often create evolutionary pressure on fish species to mature at younger ages and sizes. Marine protected areas help protect larger fish, or those fish that mature at a later age, from being caught, thereby reducing the selection pressure from commercial fisheries. Again, though, the greatest benefit may be achieved for less mobile species, with highly mobile species such as Atlantic cod still facing significant selection pressure from fishing.[125]

Similar to criticism for land-based protected areas, marine protected areas are criticized for being static protectors of the marine ecosystems. Marine protected areas are designated to protect marine ecosystems, but the boundaries of protected area are defined on a map, not by ecosystem. If the ecosystem the sanctuary is meant to protect moves because of climate change, the boundaries of the protected area do not move with the ecosystem; they remain frozen in place.

Marine protected areas do serve an important role in conserving ecosystems in the age of global climate change, though—and this may be especially true for coral reefs. Many coral species are symbiotic with zooxanthellae, a type of algae; when ocean temperatures rise, the corals become stressed and expel the zooxanthellae, leaving the coral white and potentially killing it. This process is known as coral bleaching. As if coral bleaching were not bad enough, corals also face ocean acidification. The ocean absorbs one-quarter of the carbon dioxide released by humans into the atmosphere each year. The increase in carbon dioxide in the ocean lowers its pH, making it more acidic. This lower pH reduces the ability of corals to grow the calcium carbonate shells they need to live. Ocean acidification also negatively affects other organisms that grow shells, such as oysters, clams, sea snails, and many species of plankton.

The threats to corals are important because coral reefs are extremely important in many marine ecosystems. Coral reefs provide resources and habitats for thousands of marine species. Indeed, roughly half of US commercially fished species rely on coral reefs for part of their life cycle.[126] Coral reefs also directly benefit humans by acting as a buffer for shorelines during storms.

Although marine protected areas will do nothing to stop warming oceans or acidification, they may still protect coral species by reducing other stressors. By reducing stressors such as fishing and pollution, protected areas may provide an opportunity for coral species to adapt to the stressors that the protected areas cannot control.[127] Rapid evolution has been found to be a widespread phenomenon, and may allow many species to escape extinction caused by climate change. Species and ecosystems may be weakened by warming ocean temperatures and acidification, but they may be able to adapt if protected from other human impacts.[128] Most marine protected areas that now exist protect the corals within their boundaries, but do not limit commercial fishing. More marine protected areas should be created as marine reserves so that all other stresses on the corals are eliminated, giving them a greater opportunity for adaptation to warming temperatures and ocean acidification.

Ultimately, species and ecosystems will move as global climate change progresses. To protect moving ecosystems, marine protected areas may also need to move.[129] Creating dynamic marine protected areas under the

National Marine Sanctuaries Act or the Antiquities Act would require Congress to amend those acts, but would likely provide the best chance for protecting moving ecosystems.

Roughly 1,700 marine protected areas in the United States have been created by federal or state governments. While this seems impressive, many of these protected areas are quite small, and consequently marine protected areas cover less than 0.1% of US waters.[130] For comparison, 4.6% of US land is designated as wilderness area. The best way to make marine protected areas more ecologically beneficial is to create more of them.

THE OCEAN AND POLLUTION

The chapter now turns to the other enormous source of stress on marine ecosystems, pollution. A significant amount of pollutants that enter the ocean come from vessels or drilling operations that are physically located on the ocean, but the vast majority of marine pollutants come from land. Pollutants from land may be pumped into the ocean through a pipe, they may be washed off the coasts directly into the ocean, or they may be washed off land into streams and rivers that eventually empty into the ocean. Much of the land-based pollutants that enter the ocean come from smokestacks or cars that release pollutants into the air, which settle out on the surface of the ocean. Air pollution is examined in the next chapter, leaving the other sources of pollution to be discussed here.

Clean Water Act

As examined in detail in chapters 7 and 8, the CWA prohibits the unpermitted discharge of pollutants into US waters by any person. The act defines "discharge of pollutants" as any addition of any pollutant to "navigable waters" from any point source. The act also includes addition of any pollutant to "waters of the contiguous zone or the ocean" from any point source other than a "vessel or other floating craft."[131] The CWA defines "navigable waters" as "waters of the United States, including the territorial seas."[132] The term "territorial seas" is defined as the ocean "extending seaward a distance of three miles."[133] The CWA defines "contiguous zone" as "the entire zone established or to be established by the United States under article 24 of the Convention of the Territorial Sea and the Contiguous Zone."[134] That convention created a contiguous zone out to 22 kilometers (12 nautical miles). The contiguous zone as defined in the CWA therefore extends out to sea from 5.6

to 22 kilometers.[135] Finally, the CWA defines "ocean" as any portion of the high seas beyond the contiguous zone.[136] What all this means is that under the CWA, navigable waters include the sea out to 5.6 kilometers, the contiguous zone is the sea out to 22 kilometers, and the ocean is the 370-kilometer exclusive economic zone.

These definitions are important in applying the CWA. The Section 404 permit program for dredge and fill, discussed in chapter 8, applies only to discharges into navigable waters.[137] As "navigable waters" includes only waters of the United States and the territorial seas, the Section 404 permit program does not apply beyond 5.6 kilometers.

The National Pollutant Discharge Elimination System (NPDES), on the other hand, applies to the "discharge of any pollutant."[138] As the discharge of pollutants is defined by the CWA to include navigable waters, the contiguous zone, and the ocean, the NPDES applies all the way out to the edge of the 370-kilometer exclusive economic zone. Recall that the CWA also requires states to set water-quality standards to maintain ambient water quality. Point sources may be required to reduce their discharge of pollutants below the technology-based effluent limitations if it is necessary to maintain the water-quality standard of a particular water body. The CWA requires the states to create water-quality standards for their navigable waters.[139] Given the definition in the CWA, this means that water-quality standards must be set only for the 5.6 kilometers of the territorial sea.

In a different section of the act, the CWA requires the EPA to set specific discharge criteria for point sources discharging into the ocean under the NPDES.[140] These criteria apply out to the edge of the exclusive economic zone. The ocean discharge criteria are meant to recognize that marine ecosystems are very different from inland water bodies, and therefore to go beyond technology-based and water-quality effluent limitations. The CWA requires that the criteria consider the effect of the disposal of pollutants on factors including "plankton, fish, shellfish, wildlife, shorelines, and beaches"; on "changes in marine ecosystem diversity, productivity, and stability"; and on "species and community population changes."[141]

As required, the EPA has promulgated regulations to set discharge criteria for the ocean.[142] The regulations command that a person applying for an NPDES permit must provide the EPA with an analysis of the chemical constituents of the proposed discharge, and an analysis of the biological community at the proposed discharge site. The EPA then determines whether the discharge will cause "unreasonable degradation of the marine environment."[143] "Unreasonable degradation" is defined as including significant adverse changes in ecosystem "diversity, productivity and stability," or loss of scientific or economic values that is "unreasonable in relation to the benefit derived from the discharge."[144] The EPA determines whether unreasonable degradation will occur by considering such factors as the potential for bio-accumulation of pollutants, the vulnerability of biological communities to be exposed to pollutants, the presence of unique species or those critical to the structure or function of the ecosystem, the presence of spawning sites or migratory pathways, and the presence of marine sanctuaries or coral reefs.[145] However, the regulations then state that if a discharge would comply with state water-quality standards, the discharge will be presumed not to cause unreasonable degradation of the marine environment.

Under the CWA, the rules for where vessels may discharge pollutants into the ocean get special consideration. As noted above, the CWA defines the discharge of pollutants, in part, as the addition of any pollutant to the contiguous zone or ocean from point sources, except for a "vessel or other floating craft."[146] Consequently, the NPDES applies to vessels discharging pollutants into the territorial sea, but does not apply to vessels beyond 5.6 kilometers of sea. An EPA regulation exempted all discharges from vessels from the NPDES, including in the territorial sea.[147] A federal appeals court in 2008, however, held that exempting discharges from vessels from the NPDES was contrary to the requirements of the CWA.[148] In response to the court's decision, the EPA now requires NPDES permits for discharges from vessels, including discharges of ballast water (see chapter 2 for a discussion of ballast water).[149] Vessels may not simply go out beyond 5.6 kilometers and then dump whatever wastes they want, however. As examined in the next section, the Ocean Dumping Act regulates the dumping of pollutants by vessels beyond the 5.6-kilometer territorial sea.

The CWA also states that the definition of "pollutant" does not include "sewage from vessels."[150] "Sewage" is defined in the act as human body waste.[151] Consequently, the NPDES does not apply to the discharge of sewage by vessels. In a different section of the act, though, the CWA does require the EPA, in consultation with the Coast Guard, to create standards for marine sanitation devices for treating sewage on vessels.[152] Vessels must treat their sewage with an appropriate marine sanitation device before discharging it into the water. The requirement for treatment of sewage applies only to the navigable waters, and thus only out to 5.6 kilometers of sea.[153] States may request that the EPA create no-discharge zones for the waters within their borders, including the territorial sea, if it is necessary to enhance the state's water quality.[154]

As discussed in detail in chapter 7, the CWA does a poor job of regulating nonpoint sources of pollution. This is especially true for agricultural pollutants such as fertilizer, which can be carried by storm-water runoff into nearby bodies of water. The ecological consequence of this lack of regulation can be seen at its biggest scale in the "dead zone" in the Gulf of Mexico.

Every summer, storm-water runoff carries enormous amounts of nutrients from agricultural activities into the Mississippi River, which then carries them into the Gulf of Mexico.[155] The nutrients result in eutrophication in the gulf, which leads to a large drop in dissolved oxygen in the waters of the northern Gulf of Mexico. Because of the large drop in oxygen, the area is called a dead zone. The size of the Gulf of Mexico dead zone changes with changes in the flow of the Mississippi River, but at its largest the dead zone is greater than 17,000 square kilometers (6,700 square miles), which is larger than the state of Connecticut.

Organisms move out of the dead zone if they can, but the enormous size of the dead zone leaves them with less habitat to occupy. Many organisms that cannot move out of the dead zone die. As would be expected, the dead zone wreaks havoc on the ecological communities within the zone. Higher trophic levels are decimated, leading to a flow of energy to microbial pathways.[156]

There are several other dead zones in US waters, such as in Chesapeake Bay. It should be remembered, though, that while pollutants from nonpoint sources are what create these dead zones, years of overfishing have almost certainly amplified the effects of the pollutants.

Ocean Dumping Act

Recall that in 1972 Congress passed the Marine Protection, Research, and Sanctuaries Act. One section of that act is frequently called the Ocean Dumping Act.[157] The act prohibits any person from transporting from the United States any material for the purpose of dumping it into ocean waters.[158] The act also prohibits any vessel registered in the United States or flying the US flag from transporting material from any location for the purpose of dumping it into ocean waters. Lastly, the act prohibits anybody from transporting material from outside the United States and dumping it into the territorial sea of the United States, or in the 22-kilometer contiguous zone of the United States if it may affect the territorial sea.[159]

The Ocean Dumping Act and associated regulations define "ocean waters" as the territorial sea, contiguous zone, and exclusive economic zone.[160] The act therefore applies for 370 kilometers out to sea. The Ocean Dumping Act then defines "material" as matter of any kind, including garbage, sewage, dredged material, biological and laboratory waste, and municipal waste.[161] The definition specifically excludes sewage from vessels, though.

Similar to the CWA, the EPA and the Army Corps of Engineers may issue permits under the Ocean Dumping Act to allow dumping that would otherwise be illegal. The EPA issues most permits, while the Corps issues permits for dredged material.[162] The EPA may issue a permit if the dumping will not "unreasonably degrade or endanger" human health, the marine environment, or ecological systems. The act instructs the EPA when evaluating permit applications to consider such factors as the effect of dumping on fisheries resources, plankton, fish, shellfish, and wildlife. The EPA must also consider the effect of dumping on marine ecosystems, particularly with respect to the transfer and concentration of material through biological processes. Potential effects on ecosystems also include changes in marine ecosystem diversity, productivity, stability, and species and community population dynamics. EPA regulations state that the EPA will not issue a dumping permit if there is an alternative means of disposal for the material. The EPA will issue a permit only if the dumping will not "unduly degrade or endanger the marine environment" and will not have unacceptable adverse effects on human health or the marine ecosystem.[163]

In granting a permit, the EPA designates the geographical site in the ocean where the dumping may occur.[164] The dumping site generally has to avoid areas of existing fisheries or shellfisheries and, where feasible, be beyond the edge of the continental shelf.[165] The EPA may issue general permits (such as for burial at sea or disposal of vessels), or special permits on a case-by-case basis.

In 1988 Congress passed an amendment to the Ocean Dumping Act called the Ocean Dumping Ban Act, which phased out and then completely eliminated dumping permits for sewage sludge and industrial waste.[166] The Ocean Dumping Act does make one important exception for dumping that does not need a permit. The act specifically states that a permit is not required to dump fish wastes into the ocean.[167] The only time a permit may be needed to dump fish wastes is if the dumping will take place in a harbor, protected coastal area, or location in the ocean where the wastes could endanger the ecological system.

Virtually all dumping that occurs in the ocean today is dredged material. Sediment dredged from waterways to maintain navigational channels is the largest mass of waste discharged into the ocean.[168] The Ocean Dumping Act regulates the dumping of dredged material into the ocean, while the CWA regulates the dumping of dredged material into inland and coastal waters (see chapter 8). As under the CWA, the Corps must use the criteria set by the EPA in determining whether to issue a permit for dumping dredged material.[169] The EPA is also responsible for recommending the ocean sites where the dumping of dredged material may occur. If the EPA does not concur with the Corps in approving a permit application, the permit is not issued.[170]

As might be expected, dumping dredged material into the ocean affects the ecosystem where the dumping occurs. Benthic communities at the disposal site are completely buried. Sediments also tend to migrate off the disposal site, carried by ocean currents. This sediment migration may partially bury habitats away from the disposal site, leading to reductions in diversity and abundance of fish in those nearby communities. Several studies, however, have found little evidence of long-term negative ecological effects of ocean dumping of dredged material.[171] Recolonization at the dumping site often occurs quite rapidly. Moreover, there appears to be little harm done by migration of sediments. For the sites that have been studied, ocean dumping of dredged material appears to produce relatively little lasting impact.

Coastal Zone Management Act

Coastal areas make up 17% of the conterminous US land area, but are home to more than 53% of the US population.[172] Not surprisingly, coastal areas are the most developed areas in the country. The density of people living in coastal areas puts enormous pressure on coastal waters.

Coastal waters include estuaries, coastal wetlands, sea-grass meadows, coral reefs, mangrove forests, and kelp forests, among other habitats.[173] Coastal habitats provide spawning grounds and shelter for many species of fish, and nesting, feeding, and breeding habitats for 75% of US waterfowl and migratory birds.

Congress passed the Coastal Zone Management Act (CZMA) in 1972 to manage coastal resources, including the Great Lakes. Prior to passage of the act, control over coastal zones tended to be highly fragmented within states. The CZMA promoted the development and implementation of state-level plans for coastal zones, to reduce agency fragmentation and encourage cooperation among those with control over coastal zones.[174]

The CZMA defines the "coastal zone" as the coastal waters, including the land underneath the water, and the adjacent shore lands.[175] The coastal zone extends offshore for 5.6 kilometers, and extends inland from the shoreline to the extent necessary to "control shorelands, the uses of which have a direct and significant impact on the coastal waters," and to control areas that are likely to be affected or vulnerable to sea-level rise. The definition specifically includes in the coastal zone areas such as islands, intertidal zones, salt marshes, wetlands, and beaches.

Unlike most federal environmental laws, state participation in the CZMA is voluntary. The act creates a set of guidelines for states to develop and implement coastal management programs. If a state creates a program that fits the guidelines, then the state is eligible for federal matching funds for the program.[176]

According to the CZMA, a state management program must meet several requirements.[177] The first requirement is that a coastal state must define what constitutes the landward boundary of its coastal zone. Because the states get to define their own landward boundaries, most states have different coastal zones. For example, California defines the landward portion of its coastal zone as a 1,000-yard strip extending in from the coastal waters; Louisiana's landward portion extends inland from 16 to 32 miles; and Hawaii's landward portion includes the entire state.[178]

Additional requirements for state management programs include defining what constitutes permissible land uses and water uses in the coastal zone, and identifying how the state plans to control those uses. This control over the coastal zone emerges from creating planning processes for the protection and access to beaches, for siting energy facilities in the coastal zone, and for lessening and restoring areas of shoreline erosion.

NOAA regulations specify that management plans must also include an inventory and designation of areas of particular concern in the coastal zone, and whether the areas require special management.[179] Areas of particular concern include areas of high natural productivity or essential habitat for living resources, including fish, wildlife, and endangered species, and the "trophic levels in the food web critical to their well-being." These areas are meant to receive special management attention, such as through more restrictive permit or regulatory requirements.[180]

After a coastal state has submitted a proposed management program, NOAA determines whether the program meets all the requirements of the CZMA and the regulations. NOAA also conducts continuing reviews to determine how well the states are implementing their management programs.[181]

For many coastal states, management programs contain measures such as restricting construction on shorelines and removing harmful erosion-control devices. Programs may also include plans for purchasing land in the coastal zone, removing invasive species, and restoring the habitat. States may add plans to establish marine protected areas in their coastal waters, and to establish public education and outreach activities. Management programs may also include measures, such as plans to create pedestrian trails, that allow the public greater access to beaches for recreational activities. Thirty-four states have approved coastal management programs.

Nonpoint-Source Pollution Control

As mentioned throughout this part of the book, the greatest source of pollutants entering aquatic environments is nonpoint. Nonpoint sources also contribute most pollutants in coastal waters, coming primarily from agricultural and urban runoff. Recognizing this fact, in 1990 Congress amended the CZMA by passing the Coastal Zone Act Reauthorization Amendments.[182] The amendments required the EPA and NOAA to create guidelines for nonpoint-source pollution control in the coastal zone. The coastal states were then directed to submit a control program for coastal nonpoint-source pollution for approval by the EPA and NOAA.

The management measures to control nonpoint-source pollution are defined in the amendments as "economically achievable" measures that reflect the best available control practices.[183] Management measures for agricultural land include such actions as improving the timing and efficiency of pesticide use, restricting livestock from stream banks, and preventing erosion through contour farming or terracing. For urban areas measures include siting roads and highways away from areas susceptible to erosion, reducing erosion and sediment from construction sites, and public education to promote proper disposal of hazardous chemicals.

Recall from chapter 7 that the CWA has a provision requiring all states to create management programs for nonpoint-source pollution. The CWA, however, has no way to make the states actually implement those programs. The amendments to the CZMA, on the other hand, go further in actually compelling coastal states to control nonpoint-source pollution. If a state does not submit an approvable program to control nonpoint-source pollution, NOAA may withhold 30% of the state's grants for its coastal management program, and the EPA may withhold 30% of water-pollution-control funds awarded under the CWA.[184] Once a coastal state has an approved management program for coastal nonpoint-source pollution, the CZMA declares that the state "shall implement" the program.[185] To help implement the programs, NOAA and the EPA may award grants to the states.[186]

The possibility of losing funds from the federal government is a powerful incentive in convincing coastal states to create management programs. Perhaps an even greater incentive, though, is the "federal consistency" requirement of the CZMA. If a state creates a coastal management program or a control program for nonpoint-source pollution, then when a federal agency undertakes an activity or issues a permit for an activity that affects the coastal zone, the activity must be consistent with the coastal management policies of the state.[187] The federal consistency requirement in essence gives states a veto over federal activities in the coastal zone. More precisely, the CZMA declares that federal agency activity within or outside the coastal zone that "affects any land or water use or natural resource of the coastal zone" must, to the "maximum extent practicable," comply with the state management program.[188] If a federal activity is not consistent with the policies of the management program, it must be halted, unless the secretary of commerce allows the activity to continue. An applicant for a federal license or permit to conduct an activity that may affect the coastal zone must provide a certification that the proposed activity complies with the management program.[189] The state where the proposed activity will occur may then object to the application, blocking the license or permit. As for federal activities, though, the secretary of commerce may override the state's objection.

Besides being an incentive for states to create coastal management programs, the federal consistency requirement also helps promote cooperation between the states and the federal government. Federal consistency means that federal agencies will be forced to consult with the states before undertaking activities that affect the coastal zone. As mentioned earlier, creating cooperation and reducing fragmentation of controlling agencies was one of the main reasons for the passage of the CZMA.

National Estuarine Research Reserve System

The passage of the CZMA created the National Estuarine Research Reserve System.[190] The act itself established a series of estuarine reserves, and allows the secretary of commerce to designate additional estuarine reserves if nominated by the governor of a coastal state. Estuarine reserves must be representative estuarine ecosystems suitable for long-term research, and contribute to the biogeographical balance of the reserve system. The estuarine reserve must also be protected under the laws of the coastal state. Once an area has been designated an estuarine reserve, NOAA may make grants to the coastal state to help manage the reserve, and to public and private persons to conduct research within the reserve. There are 28 reserves that cover 526,000 hectares (1.3 million acres).

As its name implies, the National Estuarine Research Reserve System was also designed to support high-quality research in estuarine function and ecology. The CZMA requires the secretary of commerce to write guidelines for how research may be conducted in the estuarine reserves.[191] The act requires that the guidelines for research incorporate several features, including identifying and prioritizing the coastal management issues research should address, and common research principles and objectives. The guidelines must also include measures to ensure uniform research methodologies, and performance standards to determine the effectiveness of the research in addressing coastal management issues. Regulations indicate that habitat manipulation as part of a research project is allowed as long as the manipulation is limited to the minimum amount necessary for the research objective.[192] The CZMA states that the secretary of commerce may withdraw designation for an estuarine reserve if a "substantial portion" of the research conducted in the reserve over a period of years has not been consistent with the research guidelines.[193] Finally, as one way to help fund research, NOAA supports graduate research fellowships for students interested in conducting research in an estuarine reserve.

PART IV

Air

CHAPTER TEN

Air Pollution

Happenstances of geology and wind are currently killing trees and fish in the northeastern United States. The reason is a third happenstance: the sulfur and nitrogen present in coal. When coal is burned, sulfur dioxide (SO_2) and nitrogen oxides (NO_x) are released into the air. SO_2 and NO_x react in the atmosphere to produce sulfuric acid and nitric acid, which then fall to the ground attached to dust or in rain. This is commonly known as acid rain.

On the ground, acid rain lowers the pH of soils and water, while promoting the release of aluminum from soils. The aluminum can remain in the soil, or it can be washed into nearby water bodies during rainstorms. The reduced pH and increased aluminum in soils and water bodies can harm and even kill trees and fish. Acid rain has the ability to alter the ecology of entire forests, and to leave lakes with no fish.

The happenstance of geology is important because some soils are less affected by acid rain than others. Certain soils have a high buffering capacity and can neutralize the acid rain falling on them. The buffering capacity of soil depends on the thickness and composition of the soil, as well as the underlying bedrock.[1] If a soil does not have a high buffering capacity, then acid rain will cause the soil to become more acidic, and it will release more aluminum. Additionally, water running off the soil into nearby water bodies will be more acidic and contain more aluminum. States in the Midwest tend to have soils with a high buffering capacity, while states in the Northeast tend to have soils with a low buffering capacity. This is the reason that in the New Jersey Pine Barrens, over 90% of streams are acidic.[2]

The happenstance of wind is equally important because in the United

States the wind tends to blow from the Midwest to the Northeast. This is in turn significant for a slightly nefarious reason—a reason that arose because of the Clean Air Act (CAA).

After passing the CAA in 1963, Congress greatly amended it in 1970, turning it into the primary statute regulating air pollution in the United States. The Environmental Protection Agency (EPA) implemented the CAA by setting limits on the concentrations of certain pollutants in the atmosphere. The limits were nationwide, meaning that each state had to meet the limits for those air pollutants. The CAA, though, allowed each state to devise its own plan detailing how it would meet the limits.

In creating their plans, each state had a decision to make: should the state require the facilities within its borders to reduce their emission of pollutants, or should it push the pollutants onto some other state? As you may have guessed, several states decided it would be easier to allow the industries within their states to keep polluting at the same rate, and instead send their pollutants to other states.[3] This is where the happenstance of wind pattern is important. To be in compliance with the CAA pollution limits, a state had only to make sure the air pollutant concentrations within that state were at or below the limits set by the EPA.[4] Pollutants that were blown out of the state by the wind did not count against the state. To get pollutants into the wind and out of the state only required a little engineering.

Pollutants that enter the atmosphere stay there for a while before falling back to the ground. Pollutants emitted from a tall smokestack are emitted higher in the atmosphere, allowing them to stay in the air for much longer before falling to the ground. The longer pollutants are in the air, the greater the potential for the wind to blow them very long distances.[5] Emissions from a very tall smokestack may be blown thousands of kilometers before falling back to the ground.[6]

To get air pollutants outside of state borders, several states decided to require facilities to build much taller smokestacks. In 1970 only 2 smokestacks in the United States were taller than 152 meters (500 feet).[7] By 1985, 180 smokestacks were taller than 152 meters, with 23 smokestacks taller than 305 meters (1,000 feet). For comparison, the Chrysler Building in New York City is 319 meters (1,046 feet) tall. Fifteen states created regulations that set the level of pollutants an industrial facility could release based on the height of its smokestack.[8] The taller the smokestack, the more pollutants a facility could emit.

Building such tall smokestacks had a significant effect on acid rain. The increase in smokestack height came primarily from midwestern states, with northeastern states showing little increase.[9] These tall smokestacks were the primary source of SO_2 emitted in the United States. In 1977–78, 90% of SO_2 from stationary sources came from facilities with smokestacks taller than 100 meters, 63% came from smokestacks taller than 200 meters, and 38% came from smokestacks taller than 300 meters.[10] The wind took the SO_2 and NO_x emitted by the tall smokestacks in the Midwest and blew it to the northeastern states. As a consequence, most of the sulfur deposition in New England was from long-range transport from upwind states.[11]

The end result of these happenstances and slightly nefarious actions is that the early implementation of the CAA actually made the acid rain problem worse. The requirement in the CAA that states control the concentration of pollutants within their borders resulted in more SO_2 and NO_x being deposited in the northeastern United States, the region of the United States where acid rain has its most negative effects.

When it passed the amendments to the CAA in 1970, Congress realized that it had created an incentive for states to send their air pollutants downwind to other states. To counteract this, Congress included in the CAA a requirement that state plans contain provisions for limiting interstate pollution.[12] The EPA, however, did not enforce this provision. The EPA required only that states share information with neighboring states when new industrial facilities that could contribute to interstate pollution were being built.[13] The EPA did not require the states to actually do anything to prevent interstate air pollution.

In 1977 Congress passed new amendments to the CAA that declared that the states may not emit any air pollutants in quantities that would "prevent attainment" of air pollution limits in another state.[14] Congress further included a provision allowing a state to petition the EPA for a finding that another state was emitting air pollutants in quantities that were preventing downwind states from being in compliance.[15] Congress also added an amendment to the CAA declaring that air-quality standards could not be met by building smokestacks that exceeded "good engineering practice."[16] The amendment stated, however, that the EPA could in no way prohibit or restrict the height of a facility's smokestack.[17]

Despite the new amendments by Congress, the EPA still did relatively little to prevent the interstate transport of air pollutants. It was not until Congress again passed amendments to the CAA in 1990 that interstate transport of pollutants began to face greater regulation. The 1990 amendments also created a cap-and-trade program meant to specifically reduce acid rain. The cap-and-trade program will be discussed in greater detail below.

Although the interstate transport of air pollutants is less of a problem today than it was in the past, the legacy of states simply sending their pollutants downwind remains. Many ecosystems that underwent acidification of soil and water are just beginning to recover. In some areas of the country, most streams still show signs of acidification, while some forests still show signs of reduced growth. Ironically, the prohibition on interstate pollution in the CAA was used by a federal appellate court to halt an EPA program meant to further reduce SO_2 emissions.[18] This turn of events will be discussed later in the chapter.

The chapter first gives an overview of the CAA. Then it addresses the ecological effects of the six "criteria pollutants" that receive the greatest attention under the CAA. Certainly the single most important air pollutant being emitted in the United States today is carbon dioxide. The ways in which the CAA is being used to regulate the emission of carbon dioxide is left for chapter 11.

CLEAN AIR ACT

The core of the CAA is the National Ambient Air Quality Standards (NAAQS). The CAA requires the EPA to create NAAQS for air pollutants "which may reasonably be anticipated to endanger public health or welfare," and that occur in the air due to "numerous or diverse mobile or stationary sources."[19] The EPA has created NAAQS for six air pollutants: ozone, particulate matter, carbon monoxide, nitrogen dioxide, sulfur dioxide, and lead. The EPA focuses on these six pollutants because they are widespread in the United States and can do considerable harm to human health and the environment. These six pollutants are termed "criteria pollutants."[20]

Each NAAQS is split into a primary standard and a secondary standard.[21] The primary standard sets the level of a criteria pollutant in the ambient air low enough to "protect the public health" while "allowing an adequate margin of safety."[22] The secondary standard then sets the level low enough to "protect the public welfare" from any "adverse effects associated with the presence of such air pollution in the ambient air."[23] The CAA defines "welfare" as including effects on soil, vegetation, animals, wildlife, weather, visibility, and climate.[24] Importantly, the EPA may not consider cost when setting the NAAQS levels, a requirement reaffirmed by the Supreme Court in *Whitman v. American Trucking Associations, Inc.*[25]

The CAA requires the EPA to review and potentially revise the NAAQS every five years.[26] As part of this review, the EPA writes an integrated science assessment for each criteria pollutant. The integrated science assessments compile and evaluate the latest scientific literature relevant to each pollutant. The EPA then writes a risk and exposure assessment and a policy assessment. The policy assessment explains the EPA's reasons for revising or not revising the NAAQS. The CAA further requires that an independent scientific committee, named the Clean Air Scientific Advisory Committee, also review the NAAQS every five years and make recommendations to the EPA on revisions.[27] Both the EPA documents and scientific committee documents are helpful in explicitly pointing out where additional research would help fill gaps in the scientific understanding of the effects of the criteria pollutants on humans and ecosystems.

Once the EPA has set the NAAQS, the individual states must meet those standards. Each state creates its own state implementation plan (SIP) that sets emission standards within that state to meet the NAAQS.[28] As discussed in the introduction to this chapter, SIPs must also prevent emissions that "contribute significantly to nonattainment in, or interfere with maintenance" of, the primary or secondary NAAQS in any other state.[29] This is frequently called the good neighbor provision. If a state does not submit a SIP, or submits one that the EPA determines is inadequate, the EPA may issue a federal implementation plan (FIP) for that state.[30]

As part of a SIP, any stationary source that is a major emitter of air pollutants must receive an operating permit from the state.[31] A major source is one that emits 10 tons per year of a hazardous air pollutant (discussed later), 25 tons a year of any combination of hazardous pollutants, or 100 tons per year of any other pollutant, such as a criteria pollutant.[32] The requirements for the permits are described in Title V of the CAA, and so are often called Title V operating permits. The permits set limits on the amount of pollutants sources may emit, or prescribe the pollution-control devices they must use.

Even after creating a SIP and requiring major sources to obtain Title V permits, the air in certain areas of a state may still not achieve the NAAQS. If an area of a state does not achieve the NAAQS for a particular criteria pollutant, then the area is in nonattainment for that pollutant. (It can nevertheless be in attainment for other pollutants). To deal with nonattainment areas, the CAA requires that states create plans to bring their nonattainment areas into attainment within a certain number of years, with more time given for more polluted areas. As an example, the 1990 amendments to the CAA gave states with areas of extremely high ozone levels 20 years to achieve the NAAQS for ozone.[33]

The CAA also creates other requirements for nonattainment areas. Existing stationary sources, such as industrial facilities, must use "reasonably available control technology" in nonattainment areas.[34] There is also a pre-construction permit process for new sources in nonattainment areas.[35] Under the permit process, facilities that are being built must install pollution-control technology that is at least as effective as the best technology used by an existing facility of the same kind. This is called the "lowest achievable emission rate."[36] Additionally, the permit process requires that a facility that wants to start emitting air pollutants in a nonattainment area must offset more pollutants in that area than it emits.[37] For instance, a new facility may have to offset 1.5 tons of pollutants for every ton it emits. Offsets usually occur when the owner of the new facility reduces emissions at another facility that it owns in the nonattainment area. An owner may also create offsets by paying another facility in the nonattainment area to reduce its emissions, or by buying and shutting down a facility in the area. As a consequence of the offsets, a new polluter entering a nonattainment area should result in a net decrease in pollution. Although several areas of the country remain in nonattainment for particular criteria pollutants, overall the air quality in the United States has greatly improved since passage of the CAA.

The CAA regulates more than just the criteria pollutants. Potentially any air pollutant may be regulated through the new source performance standards (NSPS).[38] All new stationary sources that are a "major" source of an air pollutant must meet the NSPS limits. Regulated pollutants include the criteria pollutants and other air pollutants (such as greenhouse gases, which will be discussed in the next chapter). Instead of specifying a particular control technology, NSPS set emission limits for newly constructed stationary sources, with limits based on the best pollution-control technology available for a given industry. There are 75 categories of industries, and the emission limits for a new facility depend on which category it falls into.

The NSPS program applies not only to new sources but also to existing sources that undergo modifications.[39] The CAA defines a modification as a physical change or change in operation of a stationary source that "increases the amount of any air pollutant emitted by such source or which results in the emission of any air pollutant not previously emitted."[40] Sources that already exist but that do not undergo a modification do not have to meet the NSPS emission limits—they have been grandfathered in. When Congress passed the CAA, it was thought that existing sources would eventually shut down or have to undergo modifications to keep operating, and at that point would fall under NSPS.[41] Owners of existing facilities quickly understood the loophole, though. Seeing that making modifications would require installing expensive new pollution-control technology, many owners kept old facilities up and running by continually making minor repairs and never undertaking modifications as defined by the CAA. The result is that these old grandfathered facilities continue to release considerably more pollutants than newer facilities.

Although the NAAQS and NSPS work together to set a limit on how dirty the air can get, they also create an incentive to pollute areas with air cleaner than the limit. Suppose a business owner is deciding where to build a new industrial facility and has a choice between a nonattainment area where she would be required to buy offsets, and an area in attainment where she would not need to buy offsets. The rational choice would be to build the facility in the attainment area and avoid the cost of buying offsets. Congress realized it was creating an incentive in the CAA to preferentially pollute areas with clean air, and so added to the CAA a section called the Prevention of Significant Deterioration (PSD).[42]

The PSD program is meant to protect the air quality of areas that are already in compliance with the NAAQS. To do so, it creates Class I, Class II, and Class III areas.[43] Class I includes special areas such as national wilderness areas and national parks, and Class II includes attainment areas. No Class III areas have been designated. The PSD program then limits how much dirtier the air in Class I and Class II areas can get. The program sets maximum incremental increases in the concentration of air pollutants, with Class I areas having much lower incremental increases than Class II areas.[44] For example, in a Class I area, the mean annual emission of SO_2 may not increase more than 2 micrograms per cubic meter above the baseline concentration in the area.[45] The increase in a Class II area, however, is set at 20 micrograms per cubic meter above the baseline concentration.[46]

Like facilities being built in nonattainment areas, new or modified stationary sources in the PSD program must obtain a preconstruction permit. This permitting is known generally as new source review. New major emitting facilities under the PSD program must use the "best available control technology" for each pollutant.[47] The standards for new or modified sources under the PSD program are more stringent than those under NSPS.

The CAA next requires the EPA to regulate compounds deemed to be hazardous air pollutants.[48] These pollutants include compounds that are carcinogenic, mutagenic, or neurotoxic; cause reproductive problems; or can kill a human acutely or chronically.[49] The CAA requires new and existing facilities to use the "maximum achievable control technology." The maximum achievable level, though, is set higher for new sources than for existing sources. These standards are called national emission standards for hazardous air pollutants (NESHAP).

The list of hazardous air pollutants used to be quite short—between 1970 and 1990 the EPA listed only seven pollutants as hazardous. Growing frustrated by the lack of EPA action, Congress in the 1990 amendments to the CAA listed 187 pollutants as hazardous.[50] As a result of the prompting by Congress, the EPA has set NESHAP for several different industrial categories. Incredibly, the EPA did not set NESHAP for coal and oil-fired power plants until 2011. There were no national limits on the amount of hazardous air pollutants that coal and oil-fired power plants could emit into the air. The 2011 NESHAP are known as the mercury and air toxics standards, and will be discussed near the end of the chapter.

Lastly, the CAA requires the EPA to set emission standards for new motor vehicles or new motor-vehicle engines.[51] These so-called mobile sources include cars, trucks, buses, tractors, lawn mowers, trains, boats, and airplanes. The CAA states that the EPA "shall" prescribe standards for motor vehicles, and a class of motor vehicles is to be included under the standards if it emits pollutants that "may reasonably be anticipated to endanger public health or welfare."[52] As will be discussed in the next chapter, this language is important in regulating greenhouse gases. The CAA gives the EPA authority to set emission limits for mobile sources, but specifies that the EPA must consider cost in doing so.[53] The CAA also allows California to request an exemption from the EPA standards to increase the stringency of the requirements in California. Other states may then choose to implement California's mobile source standards within their own state.

Motor-vehicle emissions are regulated in another way that is separate from the CAA. The 1975 Energy Policy and Conservation Act allows the National Highway Traffic Safety Administration to set a corporate average fuel economy (CAFE). CAFE is the average miles per gallon cars and light trucks must obtain. Increasing the CAFE standards reduces the average emission of pollutants by cars and trucks.

CRITERIA POLLUTANTS

The six criteria pollutants are all widespread in the United States and have the potential to cause considerable harm both to human health and to ecosystems. For most of the criteria pollutants, it is not the concentration of pollutants in the air that causes negative effects to ecosystems, but instead the rate at which pollutants fall from the air onto plants, soil, or water. The

SO_2 and NO_x discussed in the introduction have their greatest effect on ecosystems not when they are in the air, but when they are deposited on the ground. Air pollutants are deposited in two ways. Wet deposition occurs when pollutants fall to the ground as part of rain, snow, or fog. Dry deposition occurs when pollutants become incorporated into dust that then falls to the ground.

The amount of a pollutant that gets deposited on the ground in a particular area will determine the effects on the ecosystems in that area. A small amount may have no visible effect, whereas a greater amount of pollutant may have significant effects. The concept of a critical load was created to delineate the threshold for harm. "Critical load" is defined as the amount of a pollutant that may be deposited on an ecosystem before harmful effects occur.[54] Determining the critical load for an ecosystem would ideally allow regulators to set the maximum amount of an air pollutant that could be emitted without harming the ecosystem.

There is considerable difficulty, though, in using the NAAQS to implement the critical-load concept. Different ecosystems have different critical loads for the same pollutant.[55] The NAAQS, however, apply throughout the United States. Additionally, critical loads must take into account the cumulative effect of pollutants.[56] An ecosystem may show little effect from relatively low levels of deposition at first, but the cumulative effect of even small amounts of deposition may began to negatively affect an ecosystem over time. For instance, low levels of nitrogen inputs in experimental plots do not initially alter the biodiversity of those plots, but as nitrogen begins to accumulate in the plots, there is a large loss of biodiversity.[57] The heterogeneity of ecosystems in the United States, the requirement for national standards, and the need to consider cumulative effects all make the concept of critical loads difficult to apply in practice. The NAAQS are frequently too broad a tool to keep the emission of pollutants below the critical loads for all ecosystems in the United States. It is worth thinking about the critical-load concept and its relation to the NAAQS as each criteria pollutant is discussed.

Particulate Matter

Particulate matter (PM) is very small solid particles or liquid droplets suspended in the air. PM can be emitted directly by a source, or can form when gaseous pollutants are emitted and undergo chemical reactions in the atmosphere. For instance, sulfates (SO_4) and nitrates (NO_3) are PM that forms in the atmosphere after the emission of SO_2 or NO_x from fossil-fuel combustion.

PM is often placed into different categories based on the size of the particles. PM with a diameter less than or equal to 2.5 micrometers is denoted $PM_{2.5}$ and is often referred to as fine particles. PM with a diameter less than or equal to 10 micrometers is denoted PM_{10}. PM with a diameter greater than 2.5 micrometers and less than or equal to 10 micrometers is denoted $PM_{10-2.5}$ and is referred to as the coarse fraction of PM_{10}. For comparison, a human hair has a diameter of approximately 70 micrometers. $PM_{2.5}$ is especially dangerous to human health as it can be inhaled deep into the lungs. The size classes are less important in determining how PM affects ecosystems.[58] Instead, it is the elements within PM that tend to determine how PM influences ecosystems.

$PM_{2.5}$ is mainly produced by the combustion of fossil fuels by vehicles and stationary sources.[59] The burning of organic matter, such as prescribed burns or wildfires, also contributes to $PM_{2.5}$. $PM_{2.5}$ can include SO_4 and NO_3, metals, organic carbon, and ammonium (NH_4). $PM_{10-2.5}$ is usually produced by abrasion and crushing processes. As a result, $PM_{10-2.5}$ is mainly composed of soil, road debris, construction and demolition debris, ash, plant and insect fragments, pollen, and microorganisms.[60]

There is considerable variation in PM concentrations throughout the United States. Because it is small, $PM_{2.5}$ tends to stay in the atmosphere for days to weeks, while $PM_{10-2.5}$ tends to stay in the atmosphere minutes to hours. As it stays in the atmosphere for a relatively short time, $PM_{10-2.5}$ tends to be highest near the sources producing the particles.[61] When PM does fall to the ground, it can do so through both wet and dry deposition.

While it is still in the atmosphere, PM scatters and absorbs light coming from the sun. This haze is the reason some city skylines seem to be covered in a blanket of smoke. Haze occurs not just in urban areas, though; it can also blow into more rural areas. In some parts of the world with considerable haze, the amount of solar radiation absorbed by the haze can result in reduced crop growth.[62] While haze can reduce the direct solar radiation reaching the plant leaves in an ecosystem, it can also scatter light, resulting in diffuse solar radiation. This diffuse radiation is more evenly distributed throughout a canopy, and thus better able to reach lower leaves in that canopy.[63] As a result, diffuse radiation may increase overall canopy photosynthetic productivity.

PM is more likely to impact ecosystems by the deposition of particles on soil, water, and plants. Sticking up high into the air, tree canopies tend to increase dry deposition from the atmosphere.[64] This is especially true of high-elevation canopies because the faster wind speeds at higher elevations increase the rate of PM impact with canopy leaves, which increases rates of deposition. In areas with particularly high concentrations of PM, the PM deposited onto leaf surfaces may reduce photosynthesis by blocking light from reaching the leaf. The PM may also block the stomata on leaves, and fine PM may even enter leaves through stomata and alter leaf chemistry.[65] Blocking of light and stomata is more likely to occur in areas that receive little rainfall to wash deposited PM off leaves. In areas with more typical PM concentrations, the deposit of PM on foliage appears to have a limited impact on plants.

The constituents of PM are the most important determinant in how PM affects ecosystems. The chemicals that tend to have the greatest effect on ecosystems are SO_4 and NO_3. Sulfur and nitrogen as pollutants will be examined in a later section. Metal in PM can also significantly impact ecosystems. PM containing metal is especially prevalent in road dust. In fact, tire wear is an important source of zinc in the environment. Metal in the soil may inhibit the growth of bacteria and mycorrhizal fungi (fungi that forms a symbiosis with plant roots), which may in turn inhibit soil nutrient cycling. When taken up by plants, metals can reduce photosynthesis and growth. Some metals bioaccumulate and biomagnify in the environment. Bioaccumulation occurs when an organism takes in a pollutant at a faster rate than it is lost from the organism, so that the pollutant accumulates within the organism. Biomagnification occurs when a pollutant becomes more concentrated in organisms as it moves up a food chain. For example, mercury tends to bioaccumulate in fish species, and then biomagnify as fish species at higher trophic levels eat species at lower levels. The bioaccumulation and biomagnification of metals deposited by PM can occur in both terrestrial and aquatic ecosystems.

The EPA first set the NAAQS for PM in 1971, and has since revised the standards several times. The latest revision came in 2012. The EPA lowered the primary standard for $PM_{2.5}$, but maintained the previous primary standard for PM_{10}.[66] The EPA did not change the secondary standards for $PM_{2.5}$ or PM_{10} from what they had been in the past. Recall that the primary standard is meant to protect public health, while the secondary standard is meant to protect public welfare, including vegetation and wildlife. The EPA concluded that there was insufficient evidence to lower the secondary standards for PM, arguing that there were large data gaps in understanding how current concentrations of PM in the atmosphere influence ecosystems. The EPA pointed out that most studies of the ecological effects of PM are done near industrial facilities that are releasing large amounts of PM. Conversely, there are relatively few studies that examine the effects of more typical concentrations of PM on ecosystems. The Clean Air Scientific Advisory Committee agreed with the EPA that there were not enough published studies to support lowering the secondary standards for PM. As a consequence, the EPA left the secondary standards for $PM_{2.5}$ and PM_{10} at the same level or higher than the primary standards.

Carbon Monoxide

Most carbon monoxide (CO) that is emitted into the atmosphere comes from the incomplete combustion of fuels. Mobile sources that burn fuel—including cars, ships, lawn mowers, and aircraft—make up approximately 80% of CO emissions in the United States.[67] Wildfires, burning biomass, and industrial processes also emit CO into the atmosphere. As most CO is released by mobile sources, areas with high concentrations of vehicles, such as cities, also tend to have the highest concentration of CO in the atmosphere.

The most important way in which CO affects ecosystems is by contributing to climate change. CO is not as efficient as other molecules in trapping energy from the sun, and is therefore not one of the main greenhouse gases. However, the primary way CO exits the atmosphere is by combining with hydroxyl radicals (OH) to produce carbon dioxide (CO_2), which is an important greenhouse gas. Additionally, because CO is the largest atmospheric sink for OH, CO indirectly increases the concentration of other greenhouse gases in the atmosphere, such as methane (CH_4) and ozone, that are also removed from the atmosphere through reactions involving OH. As a consequence, CO directly and indirectly contributes a small but meaningful amount to global climate change.

When the EPA established the first NAAQS for CO in 1971, the primary and secondary standards were identical. In the 1985 review of the NAAQS for CO, the EPA concluded that there was no evidence that CO negatively affected plants at ambient air concentrations. As a result, EPA revoked the secondary standard for CO in 1985.

In its latest policy assessment, published in 2010, the EPA concluded that the scientific evidence suggested retaining the primary standard for CO set in 1971. For the secondary standard, the policy assessment noted that there did not appear to be any scientific research indicating an ecological effect of CO at ambient air concentrations. The only effect on ecosystems came from the climate-related effects of CO.[68] The policy assessment concluded that CO does contribute to climate change, but that most of this effect is attributable to its increasing the concentration of ozone in the atmosphere. Ozone is itself a criteria pollutant (and discussed next). Consequently, EPA did not propose the creation of a new secondary standard for CO.

Ozone

Ozone (O_3) occurs naturally in the stratosphere, where it is important in absorbing ultraviolet radiation from the sun. Ozone also occurs in the troposphere, which is the layer of the atmosphere extending from the stratosphere down to the surface of the earth. In the troposphere, ozone acts as an oxidizing agent that can harm plants and animals.[69]

Ozone is not released from anthropogenic sources into the troposphere directly; instead it is created when precursor molecules such as volatile organic compounds, NO_x, and CO are released. These precursor molecules then react with other atmospheric molecules in the presence of sunlight to undergo a string of chemical reactions, eventually producing ozone.[70] The precursor molecules are emitted naturally from vegetation, fires, and lightning. They are also released from anthropogenic sources, with vehicles responsible for the greatest emission of NO_x and CO, and solvent usage responsible for the greatest emission of volatile organic compounds.[71] The highest levels of ozone are often found not in urban areas, where most precursor molecules are emitted. Instead, ozone levels are often highest in suburban and rural areas after being blown downwind of urban areas.[72]

Ozone has several ecological effects. It can enter through plant stomata, causing damage to plant leaves while decreasing photosynthetic rates and plant growth.[73] The stress from ozone damage may increase the susceptibility of forest vegetation to chewing insects, such as the southern pine beetle (*Dendroctonus frontalis*) and western pine beetle (*Dendroctonus brevicomis*).[74] Decreases in the growth of individual plants can lead to reduced productivity at the ecosystem level. Current ambient levels of ozone are estimated to decrease total annual biomass growth of forest tree species in the Northern Hemisphere by 7%.[75]

High levels of ozone can also lead to the loss of ozone-sensitive species from ecosystems. Ozone can thus reduce biodiversity and alter community composition. For instance, in the San Bernardino Mountains in California, high ozone levels caused a decline in ponderosa pine (*Pinus ponderosa*) and Jeffrey pine (*Pinus jeffreyi*). These ozone-sensitive pines were then largely replaced by white fir (*Abies concolor*), which is less sensitive to ozone.[76] Ozone can affect forests in another way. It tends to increase leaf turnover rates, which increases leaf litter on the forest floor. By increasing fuel on the forest floor, ozone can indirectly cause more frequent wildfires.[77]

The EPA last revised the NAAQS for ozone in 2008 when it slightly lowered the allowable ambient concentration. The primary and secondary standards for ozone have always been set at the same level, going back to 1971. In a draft policy assessment from 2014, the EPA wrote that the primary standard for ozone was not adequate to protect public health. In that same policy assessment, the EPA concluded that the current secondary standard was not adequate to protect vegetation, and that it should be revised.[78] The EPA wrote in its most recent science assessment for ozone that the current NAAQS allow for an ambient ozone concentration in many parts of the United States that "has been known to cause detrimental effects in plants."[79] As a result of the science and policy assessments, the EPA in 2014 proposed to lower both the primary and secondary NAAQS for ozone.

Lead

Petroleum refiners began adding lead to automobile gasoline in the 1920s as a way to improve engine performance.[80] Lead remained an additive in gasoline until well into the 1980s, meaning that tons of the metal were emitted into the air as automobiles became ubiquitous in the United States. Lead, however, can have very negative effects on human health. For example, in children lead exposure may result in lower IQs and behavioral problems, while in adults lead exposure may result in hypertension and decreased kidney function. Lead use in gasoline began to decline in 1974 when the EPA required the use of unleaded gasoline in cars with catalytic converters.[81] Eventually, lead as an additive in automobile gasoline was completely phased out. As a result of the phaseout, ambient lead levels fell 98% from 1970 to 1995.[82] Lead levels then fell an additional 77% from 1995 to 2008.

Currently, more than half of lead emissions come from piston-engine aircraft, which still use leaded aviation fuel. Other significant emissions of lead come from metals processing facilities and coal combustion.[83] There is considerable variation in lead emissions levels throughout the United States. Busy airports with many piston-engine aircraft flying in and out tend to have relatively high levels of lead emissions. Busy airports, as well as industrial facilities that emit lead, are often near metropolitan areas, meaning that cities are more likely to have higher lead emissions than rural areas.[84]

Large particles containing lead tend to deposit near the source of the lead emission, while smaller particles can be transported long distances in the atmosphere. Metropolitan areas are often the sites with the highest lead levels in soil and water. Lead also remains in soils and surface waters from deposition decades earlier. This means that much of the lead emitted by automobiles decades ago is still in the soil and water, so the soils near major roadways are often high in lead. The soil near former lead smelters and mine sites also tends to be particularly high in lead.

Lead in soil can diminish soil microbial activity, resulting in changes to the community composition of soil microbes. Additionally, plants can take up lead from the soil or directly from the atmosphere. Lead impairs photosynthesis in plants, and can decrease chlorophyll content in some species. Lead exposure thereby reduces growth in terrestrial plants. There tends to be considerable variability in how particular plant species respond to lead exposure, with some species being very sensitive and others seemingly unaffected. The difference in sensitivity can produce changes in plant community composition, with less sensitive species replacing more sensitive species in lead-contaminated ecosystems.[85] From plants, lead may move up the food chain. However, concentrations of lead tend to be lower as it moves up trophic levels, meaning that biomagnification of lead does not appear to occur.[86]

In animals, exposure to lead may produce decreased growth and reduced reproduction. Animals may also show behavioral abnormalities when exposed to lead. For instance, studies on laboratory animals show that lead exposure can produce changes in learning, memory, attention, and motor function.[87]

In aquatic ecosystems, lead can reduce plant photosynthesis and growth. Freshwater invertebrates seem to be particularly affected by lead, with exposure leading to reduced growth, reproduction, and survival.[88] Lead exposure in fish may produce behavioral abnormalities such as decreased prey-capture rates, slower swimming speed, and declines in startle response.[89]

The primary and secondary NAAQS for lead are identical, and were first set in 1978. The NAAQS for lead stayed the same for 30 years before being lowered tenfold in 2008. The new level was established mainly to protect highly exposed children from neurocognitive effects.

In its latest policy assessment, completed in 2014, the EPA concluded that the primary and secondary NAAQS set in 2008 were sufficient to protect public health and welfare.[90] The EPA noted, however, the lack of scientific studies examining the effects of lead on ecosystems that were conducted at lead concentrations reflecting current ambient atmospheric levels. Additionally, most studies on the effects of lead were conducted in laboratories, making them difficult to generalize to field conditions. Connecting current ambient lead concentrations to ecosystem effects is particularly difficult because lead concentrations were so much higher in the past. Recall that lead tends to persist in the environment for a long time. Current negative effects of lead in ecosystems may be caused primarily by lead released long ago, and not by current emissions of lead.[91] Consequently, the EPA concluded that there was not enough evidence to lower the secondary standard for lead.

Sulfur Dioxide and Nitrogen Dioxide

Sulfur dioxide (SO_2) and nitrogen dioxide (NO_2) are both criteria pollutants, but their effects on the environment are often linked. In its latest review of the NAAQS for SO_2 and NO_2, the EPA decided to review the pollutants together. This section follows the example of the EPA.

The EPA uses NO_2 as an indicator for the larger group of NO_x emitted by various anthropogenic sources. Similarly, SO_2 acts as an indicator for the larger group of sulfur oxides (SO_x). The primary anthropogenic source for NO_x emissions is fossil-fuel combustion, with vehicle emissions accounting for 56% and power plants accounting for 22%.[92] Combustion of fossil fuels is also the primary anthropogenic source for SO_2 emissions, with power plants contributing 66% and vehicles adding 5%.[93] Figure 10.1 shows the total emissions from various sources for both SO_2 and NO_x.

As mentioned in the introduction to the chapter, once SO_2 and NO_x are emitted into the air, they react with water in the atmosphere to produce sulfuric acid and nitric acid. These acids are then deposited on the ground through either wet deposition or dry deposition. Deposition of sulfur and nitrogen can alter ecosystems through three main processes: (1) acidification of soils and water bodies, (2) disruption of nutrient balance, and (3) increasing the concentration of methylmercury.

Acidification

As sulfuric acid and nitric acid fall on soils, they can cause soil acidification. There is a loss of base cations and an accumulation of acidic cations, including hydrogen ions (H^+) and aluminum ions (Al^{n+}). This change in cations results in a decrease in soil pH.[94] Acidification of soils often produces aluminum toxicity in plants, and a decreased ability for plants to take up base cations.[95] This is especially true of the base cation calcium, which is an important plant nutrient. Acidification can cause individual plants to die, or

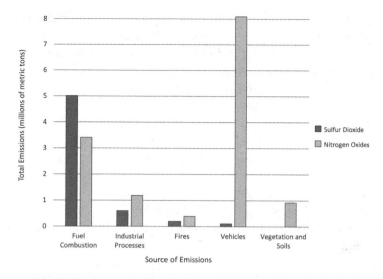

FIGURE 10.1. Major sources of sulfur dioxide and nitrogen oxides emissions in the United States in 2011. Data from Environmental Protection Agency, Air Emission Sources, http://www.epa.gov/air/emissions.

can make plants more sensitive to stress, such as cold temperatures, pests, and disease. The death of plants may eventually lead to reduced forest biodiversity and altered community dynamics, although there has been little research documenting such effects.[96]

Terrestrial areas that are most negatively affected by acid deposition are forests in the "Adirondack Mountains of New York State, Green Mountains of Vermont, White Mountains of New Hampshire, the Allegheny Plateau of Pennsylvania, and high-elevation forests in the southern Appalachians."[97] This is in part due to the soil-buffering capacity discussed at the beginning of the chapter. Red spruce (*Picea rubens*) and sugar maple (*Acer saccharum*) tend to be particularly sensitive to soil acidification. During the 1970s and 1980s more than 50% of red spruce canopy trees died at high elevations in the Adirondack and Green Mountains. The deaths were caused primarily by deficiencies in calcium in red spruce needles, which reduced cold tolerance by 3 to 10°C.[98]

The deposition of sulfuric and nitric acids also impacts aquatic ecosystems. Besides direct deposition into water bodies, water running off land and into adjacent water bodies may carry nitrogen and sulfur into aquatic ecosystems. Many lakes and streams in the United States are chronically acidic, meaning they have consistently low pH. Even more frequent, though, are cases of episodic acidification. Episodic acidification occurs when the pH of a lake or stream temporarily decreases due to the runoff from acidified soils that occurs during a heavy rainstorm or during snowmelt in the spring.

Aquatic areas that are most negatively affected by acid deposition are surface waters in the eastern United States, Florida, the upper Midwest, and mountainous regions in the West.[99] Acid deposition in water bodies decreases pH and increases aluminum concentration. Decreases in pH can cause the loss of acid-sensitive species, which are then replaced by more acid-tolerant species, altering the aquatic community.[100] Increases in aluminum may cause aluminum toxicity, reducing the reproductive success of fish species. Generally, as pH decreases and aluminum concentration increases in water bodies, the richness and abundance of zooplankton, macroinvertebrates, and fish species all decline.[101]

An example of the effects of acidification on an aquatic ecosystem is Honnedaga Lake. Honnedaga Lake is a 312-hectare (771-acre) lake located in the Adirondack Mountains of New York. As a result of heavy deposition of sulfur and nitrogen from the 1920s through the 1970s, the pH of the lake was historically below 5.[102] As will be discussed soon, emissions of SO_2 and NO_x have declined significantly since the 1970s. Consequently, the current summer surface pH of Honnedaga Lake is above 5. Aluminum concentrations have also declined significantly in the lake. Despite these improvements, Honnedaga Lake still shows evidence of being acid stressed. The pelagic zooplankton community continues to have low species richness.[103] For instance, *Daphnia* species tend to be fairly sensitive to pH, with decreasing abundance at a pH below 6.5, and absence at a pH below 5. In Honnedaga Lake, only one species of *Daphnia* is present (*D. pulicaria*), and it is a species that is relatively more acid tolerant than other *Daphnia* species.[104] All the other pelagic zooplankton currently in Honnedaga Lake are also acid-tolerant species. Brook trout (*Salvelinus fontinalis*) is native to the lake, and is currently the only fish species in the lake. Brook trout numbers have increased in Honnedaga Lake since the 1970s, but the increase has been smaller than expected, given the increase in pH in the lake. The reason appears to be that brook trout spawn in tributaries flowing into Honnedaga Lake. Many of these tributaries are still highly acidified and have high concentrations of aluminum. Given the current level of deposition of sulfur and nitrogen, these tributaries will remain acidified into the foreseeable future.[105] Even though the chemistry of Honnedaga Lake is recovering, the lake has still not recovered enough to allow the growth of acid-sensitive species, and the lack of recovery by other water bodies continues to negatively impact the Honnedaga Lake ecosystem.

Nutrient Balance

In many terrestrial ecosystems nitrogen is the nutrient that limits plant growth. As a result, the deposition of nitrogen from the atmosphere can increase primary productivity in terrestrial ecosystems. However, there are several ways in which the increased growth may be harmful. Nitrogen deposition usually increases aboveground growth of plants more than below-ground growth.[106] The greater aboveground growth can increase the vegetation fuel supply, increasing the susceptibility of an ecosystem to fire. As there are relatively fewer roots taking up water, drought may also be more likely to occur. Additionally, as there are relatively fewer roots anchoring plants to the ground, there is a greater chance that wind will damage plants. The increase in primary productivity can also alter community structure, as plant species that are better at using deposited nitrogen can outcompete other species, resulting in a decrease in biodiversity.

Alpine ecosystems are the most sensitive to nitrogen deposition. Plants in alpine ecosystems must often contend with a short growing season, steep terrain, and shallow soils. As a result, alpine herbaceous plants are generally very nitrogen limited. Nitrogen deposition in alpine ecosystems tends to cause changes in plant productivity, which in turn causes changes to the species composition of plants, lichens, and mycorrhizae.[107] Epiphytic lichens, which grow on trees and shrubs, are especially sensitive.[108] Epiphytic lichens survive almost entirely on the nutrients and moisture in the atmosphere. Nitrogen deposition can result in the loss of nitrogen-sensitive species, reducing lichen diversity in forests.[109]

As in terrestrial ecosystems, nitrogen deposition in aquatic ecosystems can alter community dynamics and lead to a loss of biodiversity. In aquatic ecosystems nitrogen deposition can promote eutrophication (described in box 7.1). Recall from chapter 9 that Chesapeake Bay has experienced considerable eutrophication. Approximately 30% of Chesapeake Bay's inorganic nitrogen comes from atmospheric deposition.[110]

Mercury

Mercury is frequently emitted into the atmosphere through human activity, and then deposits on aquatic ecosystems. Power plants, especially those that burn coal, are the largest source of mercury emissions in the United States.[111] Mercury from stationary sources is particularly prone to forming hot spots. The power plants that are in the top 10% of mercury emissions produce concentrations of mercury within a 50-kilometer (31-mile) radius of the plant that are 3.5 times higher than in the rest of the region where the plant is located.[112]

Mercury is not a criteria pollutant, but its impact on the environment is strongly influenced by the criteria pollutant SO_2. Once mercury has been released into an ecosystem, SO_4-reducing bacteria engage in mercury methylation. This is important because methylmercury is the form of mercury most likely to be taken up by microorganisms, zooplankton, and macroinvertebrates.[113] Methylmercury is also the form of mercury most toxic to organisms.[114] The amount of SO_2 deposited into the ecosystem is crucial because increases in sulfur result in SO_4-reducing bacteria greatly increasing their production of methylmercury.

Once methylmercury has been taken up by zooplankton and macroinvertebrates, it often bioaccumulates and biomagnifies as it makes its way up to higher trophic levels. As a consequence, birds and mammals that eat fish are at the greatest risk from methylmercury in the environment.[115] Organisms exposed to methylmercury may experience death, impaired growth, reduced reproductive success, and neurological impairment.[116] In mammals, fetuses are especially sensitive to methylmercury, often experiencing severe developmental problems.[117]

Reducing Sulfur Dioxide and Nitrogen Oxide Emissions

The EPA first set primary and secondary NAAQS for SO_2 and NO_2 in 1971. These standards were retained until the primary standards for both SO_2 and NO_2 were lowered in 2010. At that time, the EPA did not change the secondary standards from those set in 1971. The EPA did, however, begin a review of the secondary standards for both SO_2 and NO_2.

The policy assessment released by the EPA in 2011 found that the scientific evidence clearly shows that acidification and nutrient enrichment from SO_2 and NO_x deposition are negatively impacting ecosystems in the United States.[118] For example, in the Shenandoah area of the United States, 85% of streams show signs of acidification. Several states have soil acidification that is reducing the growth of tree species in at least some forests within the state. Additionally, nitrogen deposition continues to promote eutrophication in aquatic ecosystems, while also potentially leading to shifts in the community structure of several forest ecosystems throughout the United States.

The EPA concluded in its policy assessment that the current secondary NAAQS for SO_2 and NO_2 are not adequate for protecting sensitive ecosystems. Part of the reason is that the standards set averaging times and levels that are "not ecologically relevant."[119] To remedy this, the EPA proposed a secondary standard that would be based specifically on limiting aquatic acidification. The EPA argued that by limiting aquatic acidification, the standard would be strict enough to also limit terrestrial acidification and aquatic and terrestrial nutrient enrichment.

While the NAAQS have helped limit the emissions of SO_2 and NO_2, other provisions of the CAA have been considerably more essential. The most important provision for reducing the emissions of SO_2 and NO_2 was added by Congress in the 1990 amendments to the CAA, specifically to reduce acid rain (the provision is often called Title IV). Title IV set a goal of reducing annual SO_2 emissions by 10 million tons below 1980 levels, and annual NO_x emissions by 2 million tons below 1980 levels.[120] The primary focus of these reductions was the SO_2 and NO_x emitted by coal-fired power plants. To accomplish the reductions, Title IV created the first national cap-and-trade program in the United States.

The EPA implemented the cap-and-trade program by issuing permits to power plants for the emission of SO_2. These permits were called allowances, and emitters of SO_2 had to have one allowance for each ton of sulfur they emitted. A finite number of allowances were issued each year, meaning that there was a cap on SO_2 emissions. Most allowances were given directly to plants, but some were put up for auction by the EPA. If a plant's SO_2 emissions were going to be above the number of allowances it owned, it could either install pollution-control devices to reduce emissions or purchase allowances from another plant. This was the trade part of cap-and-trade. By putting a price on the emission of SO_2, the cap-and-trade program encouraged power plants that could cheaply reduce their emissions to reduce those emissions and sell their extra allowances to other plants. The buying and selling of allowances could be done between plants in any part of the United States.

The SO_2 cap-and-trade program did not include NO_x. Title IV did not put a cap on NO_x emissions or allow for the creation of a trading program.[121] Instead, EPA simply set limits on the emission rate of NO_x by power plants.

The cap-and-trade program and the limit on the rate of NO_x emissions were successful in reducing acid rain in the United States. Between 1990 and 2012, SO_2 emissions were reduced 76%, while NO_x emissions were reduced 52%.[122] As expected, studies have found decreasing sulfur levels in streams in the Northeast.[123] Unexpectedly, the decrease is smaller than would be predicted given the decrease in SO_2 emissions. This result is likely because soils that had accumulated sulfur during the past several decades continue to release it into nearby water bodies. Trends in stream nitrogen are less clear, with some streams showing decreases in nitrogen and others showing increases.[124] This is most likely due to the complexity of nitrogen cycling in nearby forests, which can be affected by such factors as forest maturity and logging activity. Although the emissions of SO_2 and NO_x have been reduced, the EPA has concluded that the current rates of SO_2 and NO_x emissions are still too high to protect sensitive ecosystems.

On the whole, however, due to the decrease in sulfur emissions, the SO_2 cap-and-trade program is often praised as one of the greatest accomplishments in environmental law. While the cap-and-trade program continues to operate, it has recently faced problems that have led to a near collapse of the market in SO_2 allowances. One reason for this is that the cap-and-trade program conflicts with the NAAQS program.

Recall that the good neighbor provision of the CAA requires that a state's implementation plan must not allow emissions in amounts that will "contribute significantly to nonattainment in, or interfere with maintenance by, any other State" with respect to primary or secondary NAAQS.[125] The cap-and-trade program, though, allows trading between power plants in any state, even if that means that the power plants in an upwind state buy allowances to emit SO_2 that then blows into downwind states.

This inherent tension between the cap-and-trade program and the NAAQS increased greatly when the EPA decided there was a growing need to control PM and ozone from upwind states.[126] This is relevant for the emission of SO_2 and NO_x because those two molecules are the main components of $PM_{2.5}$, and NO_x is an important precursor for ozone. The ability of power plants from upwind states to buy allowances for the emission of SO_2 contributed to nonattainment for PM and ozone in downwind states.

To help deal with this, the EPA in 2005 implemented the Clean Air Interstate Rule (CAIR).[127] CAIR required reductions in emissions of SO_2 and NO_x by 27 upwind states. A circuit court, however, quickly vacated CAIR because it did not properly reduce emissions from the upwind states.[128] In 2011 the EPA tried again with the Cross-State Air Pollution Rule (often called the Transport Rule).[129]

The Transport Rule requires reductions in SO_2 and NO_x emissions from 25 upwind states so that downwind states will be in attainment for PM and ozone. Under the Transport Rule, an upwind state falls afoul of the good neighbor provision if the upwind state exports pollution that amounts to 1% or more of the NAAQS in a downwind state, and the pollution can be eliminated cost-effectively.[130] To implement the Transport Rule, the EPA created FIPs for the upwind states. Recall that the EPA can issue a FIP for a state if it finds the SIP developed by the state to be insufficient. In this case, the EPA determined that the SIPs from the upwind states were insufficient to comply with the good neighbor provision of the CAA.

To help upwind states cut emissions, the Transport Rule created a trading program for SO_2 and NO_x allowances separate from the Title IV cap-and-trade program. The Transport Rule program allows unlimited intrastate trading of SO_2 and NO_x allowances. In contrast with the Title IV cap-and-trade program, however, the Transport Rule program greatly limits interstate trading.

The Title IV cap-and-trade program is still in effect, but it has been rendered nearly obsolete by the Transport Rule. The reason is that the cap for SO_2 emissions under the Transport Rule is lower than under the Title IV program. As a consequence, allowances distributed under the Title IV cap-and-trade program are less constraining than under the Transport Rule program.[131] This can be seen in the price for Title IV cap-and-trade allowances. In 2005, SO_2 allowances were being traded for $1,630; after finalization of the Transport Rule, they were trading for $3.[132]

After finalization the Transport Rule faced its own troubles in court. The same federal circuit court that vacated CAIR also vacated the Transport Rule.[133] The court held in 2012 that the EPA did not give the states a reasonable opportunity to create new SIPs to comply with the emission cuts required in the Transport Rule before the EPA issued FIPs. The court also held that the Transport Rule could lead to "over-control," where the reduction in emissions in a state may be greater than necessary to allow downwind states to be in attainment. Finally, the court held that EPA could not consider cost in determining how much an upwind state must reduce its emissions.

The Supreme Court in 2014 overruled the circuit court and upheld the Transport Rule.[134] The court wrote that the Transport Rule is a "permissible, workable, and equitable interpretation of the Good Neighbor Provision" of the CAA. The court wrote that the EPA had already given the upwind states an opportunity to create their own emission budgets in their original SIPs. Those SIPs were insufficient, and the EPA did not need to give the states a second opportunity before issuing FIPs. Additionally, the court stated that the EPA needs leeway in balancing the possibility of overcontrol with the possibility of undercontrol. Finally, the court held that the EPA could consider cost in determining how much each state must reduce its emissions. As a consequence, the Transport Rule is currently in effect.

TABLE 10.1. Percentage decline in emissions of common pollutants between 1990 and 2012

Pollutant	Symbol	Percentage decline
Lead		80
Sulfur dioxide	SO_2	76
Carbon monoxide	CO	65
Fine particulate matter	$PM_{2.5}$	57
Nitrogen oxides	NO_x	52
Volatile organic compounds	VOC	45
Particulate matter	PM_{10}	35

Source: Data from "Air Quality Trends," Environmental Protection Agency, accessed April 27, 2015, http://www.epa.gov/airtrends/aqtrends.html.

Note: Ozone is not listed because it usually forms from precursor molecules of NO_x and VOC.

The reduction in the emission of SO_2 under the Transport Rule will be bolstered by the Mercury and Air Toxics Standards (MATS) finalized by the EPA in 2011.[135] MATS reduce the allowable emissions of toxic air pollutants by new and existing coal- and oil-fired power plants. The toxic pollutants covered under MATS include mercury, acid gases such as hydrochloric acid (HCl), and heavy metals such as arsenic and chromium. Power plants currently emit 50% of mercury and 77% of acid gases.[136] The EPA hopes the new standards will prevent 90% of the mercury in coal from being released into the air.[137] Reduction in the emission of these toxic pollutants will impact the emission of SO_2 because the pollution-control devices that reduce the emission of the toxic pollutants also reduce SO_2 emissions. The EPA estimates that when MATS goes into effect in 2015, it will reduce SO_2 emissions from power plants by 41% below the reduction expected under the Transport Rule.[138]

As mentioned above, emissions of SO_2 and NO_x have decreased significantly since 1990. Emissions of the other criteria pollutants have also decreased considerably since then. Table 10.1 shows the percentage decreases in emissions of the criteria pollutants since 1990. The CAA has clearly done a good job of reducing the pollutants in our atmosphere. Despite the overall decreases in the emission of pollutants, though, many people in the United States still live in areas that are in nonattainment for at least one criteria pollutant. Approximately 142 million people live in a county that is in nonattainment for the primary NAAQS for at least one criteria pollutant. Most of those people live in counties that are in nonattainment for ozone or PM. Of course, not only people are negatively affected by the high levels of pollutants in these counties. Many species and ecosystems in these nonattainment areas are also negatively affected. More work is necessary to reduce the concentration of pollutants in the atmosphere. This is especially true for the air pollutant that has the potential to affect every species on this planet: carbon dioxide. Carbon dioxide and the CAA will be discussed in the next chapter.

CHAPTER ELEVEN

Global Climate Change

As will be news for no one reading this chapter, global climate change is an enormous problem that will keep getting worse until significant action is taken. "Significant" is not how one would describe the action taken by the US government so far. The federal government has recently begun taking steps to help mitigate climate change, but the process has been slow and piecemeal. Frustrated by the lack of federal action, many US states have decided to act on their own to tackle climate change.

One state-led program is the Western Climate Initiative (WCI). Formed in 2007 by several states, the WCI has the goal of reducing regional greenhouse gas emissions 15% below 2005 levels by 2020. The WCI hopes to achieve this goal mainly through a cap-and-trade program that allows trading of carbon allowances between jurisdictions (cap-and-trade programs are discussed in general later in the chapter). By 2008 the partners in the WCI included Arizona, California, Montana, New Mexico, Oregon, Utah, and Washington, and the Canadian provinces of British Columbia, Manitoba, Ontario, and Quebec.

Just as the WCI was poised to create a large regional program to reduce greenhouse gas emissions, elections happened. New governors took office in Arizona and Utah in 2009, followed by a new governor of New Mexico in 2011. All three new governors opposed a cap-and-trade program. A communications director for the governor of New Mexico denounced the cap-and-trade program, saying that it would put the state at an economic disadvantage by subjecting New Mexico businesses "to more stringent regulations than their competitors in surrounding states."[1] Meanwhile, the legislatures of Montana, Oregon, and Washington refused to pass bills authorizing a cap-and-trade program.[2] By November 2011 the states of Arizona, Montana, New Mexico, Oregon, Utah, and Washington had all dropped out of the WCI. That left only California and the four Canadian provinces in the pact, with three of the Canadian provinces also seemingly unwilling to implement cap-and-trade programs. Currently, only California and Quebec have implemented cap-and-trade programs and begun trading carbon allowances.

The rise and almost total collapse of the WCI suggests two of the difficulties in allowing the individual states to take the lead in combating climate change. First, most states do not want to implement a program to reduce greenhouse gas emissions if it places them at an economic disadvantage to other states. Recall from chapter 2 how the states in the early 1900s refused to pass legislation limiting the hunting of migratory birds because it would place their citizens at an economic disadvantage to the citizens of other states. The environmental problem may have changed, but the reaction stayed the same. Second, regulation of greenhouse gas emissions at the state level can become subject to the whims of governors and legislatures in each of the 50 states.

For the United States to make headway in confronting global climate change, a comprehensive climate change bill will likely need to be passed at the federal level. In 2009 the House of Representatives passed a bill that would have created a national cap-and-trade program, but the bill died in the Senate.

Although there is no comprehensive climate change statute at the federal level, there are several laws at the international, federal, and state levels that deal with climate change. This chapter gives a brief synopsis of the science of climate change, and then moves to an examination of the climate change laws that exist at these three levels.

CLIMATE CHANGE SCIENCE

Global climate change is an enormous challenge that has the potential to affect almost every human and wild species on the planet. To confront the problem will require the concerted efforts of most countries, altering the practices and economies of all of them. It is certainly understandable that the governments of the world want as accurate a report on the science of climate change as possible. This is the reason for the existence of the Intergovernmental Panel on Climate Change (IPCC). The IPCC was established by the World Meteorological Organization and the United Nations Environment Programme in 1988 to assess the science of climate change, and to propose recommendations for coping with climate change. The IPCC has produced five comprehensive assessment reports on climate change since being established, the latest one released in 2014. The comprehensive assessments are the work of hundreds of scientists and other experts from around the world, and the assessments are the most influential report on the state of climate change.

The IPCC has reported that from 1951 to 2010, global mean surface temperature has increased approximately 0.6 to 0.7°C.[3] The period from 1983 to 2012 was the warmest 30 years in the Northern Hemisphere of the last 1,400 years.[4] Others have estimated that the rate of climate change is an order of magnitude faster than any other warming period in the last 65 million years.[5]

The IPCC states that it is *"extremely likely* that human influence has been the dominant cause of the observed warming since the mid-20th century."[6] The IPCC defines the term "extremely likely" as being 95%–100% likely. Humans cause warming through the release of greenhouse gases (GHGs). The main GHGs are carbon dioxide (CO_2), methane (CH_4), nitrous oxide (N_2O), and halocarbons (such as chlorofluorocarbons). Of these, CO_2 is the GHG released in the greatest quantities, and is therefore the most important contributor to climate change. GHGs in the atmosphere absorb radiation, thereby directly increasing average global temperature. They also indirectly increase temperature, for example by increasing the atmospheric lifetime of other GHGs and by influencing cloud formation.[7]

The preindustrial level of CO_2 in the atmosphere was 280 parts per million. In 2011 the concentration of CO_2 in the atmosphere was 391 parts per million—40% higher than during the preindustrial era. Some scientists argue that the worst potential effects of climate change can be avoided if average global temperature rise is limited to 2°C above preindustrial levels. This can be accomplished if cumulative CO_2 emissions from 2000 to 2050 are less than 1,000 to 1,500 billion tons.[8] Since 2000, however, a total of approximately 466 billion tons has been emitted.

The effects of climate change have already been profound. There has been considerable shrinkage in the Greenland and Antarctic ice sheets, as well as shrinkage in Arctic sea ice and glaciers around the world. The result, along with thermal expansion, is that from 1901 to 2010 global average sea level has risen 0.19 meters (0.62 feet).[9] The ocean has also acidified, with the pH of the surface water decreasing by 0.1 since the preindustrial era.[10] There has also been considerable loss of permafrost thickness and extent.

The effects of climate change on species are equally profound.[11] Warming has led to earlier timing of leaf unfolding, bird migration, and breeding.[12] For example, red squirrels in Canada breed 18 days earlier than they did 10 years ago.[13] The ranges of many species have begun shifting toward the poles or up the sides of mountains. Edith's checkerspot butterfly (*Euphydryas editha*) has increased its range 90 kilometers (56 miles) north and 120 meters (394 feet) higher in elevation.[14] Meta-analyses find that terrestrial species have been moving poleward on average 1.76 kilometers (1.09 miles) per year, while marine species have been moving poleward 7.2 kilometers (4.5 miles) per year.[15]

Climate change is also altering the structure and function of ecosystems.[16] For example, warming is changing the disturbance regimes in many forest ecosystems, causing wildfires and pest outbreaks to become more frequent. As climate change continues, the resilience of some ecosystems may eventually be exceeded. Altered ecosystems will cause changes in the community composition and ecological interactions of many species, producing communities and interactions that have never existed before. The IPCC estimates that if global average temperature exceeds 2 to 3°C above preindustrial levels, then 20 to 30% of animal and plant species will be at an increased risk of extinction.[17]

Finally, climate change will also have significant effects on ecosystem services. Climate change will reduce fresh water in some areas while increasing flooding in others, reduce timber production and fishery catches, and increase vector-borne diseases. The IPCC estimates that climate change may also reduce global crop yields each decade by up to 2%.[18] Reduced yields would occur because longer and more frequent heat waves caused by climate change would kill many crops. To make up for the lost crop yields, new land would likely need to be put into production; this means plowing natural prairies and cutting down forests. Both of these actions would in turn release large amounts of carbon into the atmosphere, increasing the effects of climate change.

The IPCC assessment reports conclude that to manage climate change, both mitigation and adaptation are necessary. Mitigation means reducing the magnitude of future climate change. Adaptation means preparing for the changes that climate change will bring. As the IPCC reports are quick to acknowledge, even if GHG emissions are reduced today, many of the current effects of climate change will continue for centuries because of the inertia of climate processes and feedbacks.[19] This means that mitigation is desirable, but adaptation is required.

Laws and treaties that confront climate change focus mainly on mitigation. The primary means of mitigation is simply reducing GHG emissions. Mitigation, though, may also take the form of increasing carbon sinks, such as reforestation programs (discussed below), or carbon capture and storage underground.

To reduce GHG emissions, governments have three main options: carbon taxes, direct regulation, and cap-and-trade.

A carbon tax is a direct fee on the emission of GHGs. A government sets the amount facilities must pay for emitting GHGs, but there is no cap on the total GHGs emitted within a region or country. The goal is to set the amount of the tax just high enough to reduce GHG emissions to the desired total level. Carbon taxes are the economically most efficient method of reducing GHG emissions, and the easiest for governments to administer. However, new taxes are often politically very unpopular.

Direct regulation occurs when a government requires a facility or an entire industry to take a particular action to reduce emissions. An example of direct regulation is building codes that require certain construction materials that reduce the need for energy in operating the building and thereby reduce GHG emissions. The problem with direct regulations is that producing regulations that cover the GHG emissions of an entire economy would be extremely difficult for a government to undertake.[20]

The final option for governments to reduce emissions is cap-and-trade. Cap-and-trade schemes begin with a government placing a cap on the quantity of GHG emissions in a region or country. The government then issues or auctions allowances to facilities that permit them to emit GHGs. The total number of allowances issued equals the GHG cap. Facilities that lower their emissions below their number of allowances may sell their unused allowances to other facilities that want to emit more GHGs than they have allowances. As allowances may be traded between facilities, the market sets the price for GHG emissions. As the price for allowances rises, facilities may decide that it is less expensive to reduce emissions than to purchase allowances. Additionally, the government usually lowers the cap over time, issuing fewer total allowances.

A cap-and-trade scheme may also permit facilities to purchase carbon offsets. Carbon offsets are reductions in GHG emissions that an individual or company not under the cap-and-trade program creates. Offset projects often involve creating sinks for carbon. For example, a person who reforests her lands may be able to sell carbon offsets because the new trees will act as a carbon sink. A facility whose emissions are constrained under the cap-and-trade program may purchase carbon offsets so that it can then emit more GHGs.

If a cap-and-trade scheme does allow the purchase of offsets, there is often a requirement that the offset project be "additional." The California regulations implementing the California cap-and-trade program (discussed later in the chapter) define "additional" as an offset that reduces GHG emissions beyond any reductions required by law, or that otherwise would have occurred in a "business-as-usual scenario."[21] A reforestation project that was not required by statute or court order, and was not part of the common practice of a person or business, would likely qualify as additional.

While offset projects may ultimately help reduce the impacts of climate change, they may also have unintended ecological consequences. As already alluded to, many offset projects are based on increasing the density of trees in forests that have been harvested (reforestation), or planting trees on lands that were forested in the past but are no longer forested (afforestation). While reforestation and afforestation can be great for forest species, such projects may be terrible for grassland species.[22] A project that converts grassland to forestland to sell offsets may result in the loss of considerable habitat for grassland species. Additionally, management of forests may differ if they are part of an offset project. For example, forest managers may harvest older trees in favor of faster-growing young trees, thereby attempting to maximize the carbon sequestration of the forest. The greater the carbon sequestration, the greater the number of offsets that can be sold. Offset projects may actually not be so great for forest species that do not prosper under a management regime that maximizes carbon sequestration.

INTERNATIONAL LAW

The chapter now turns to the laws that are being used to confront climate change.[23] As climate change is a global problem, it makes sense that international law is the best means for tackling the problem. The umbrella under which most international negotiations on climate change have been conducted is the United Nations Framework Convention on Climate Change (UNFCCC). (See chapter 4 for an introduction to the basics of international law.) The objective of the convention is to stabilize GHG concentrations at a level that will "prevent dangerous anthropogenic interference with the climate system." The convention continues that this level should be achieved to ensure that food production and sustainable economic development are not threatened, and "within a time-frame sufficient to allow ecosystems to adapt naturally to climate change."[24] The UNFCCC has 195 parties, including the United States, and entered into force in 1994. As part of the framework, a conference of the parties meets every year to negotiate commitments.

The UNFCCC states as a basic principle that because developed countries are most responsible for current and past GHG emissions, they should take the lead and bear the greatest cost in mitigating climate change.[25] The UNFCCC was not written to create specific commitments, though. As its name suggests, the convention creates a framework that is meant to encourage negotiations that lead to specific commitments to reduce GHG emissions.

Negotiations under the UNFCCC have indeed led to specific commitments in the form of the Kyoto Protocol. Entering into force in 2005, the Kyoto Protocol created an initial commitment period running from 2008 to 2012. During this period, 37 industrialized countries and the European Community agreed to reduce their GHG emissions on average 5% below their 1990 levels. Developing countries were not bound under the Kyoto Protocol to reduce their emissions. The protocol permitted emissions trading between countries, so that countries having difficulty meeting their reduction commitment could buy allowances from other countries.[26] A second commitment period has been negotiated, and runs from 2013 to 2020. During this period, most industrialized countries agreed to reduce their GHG emissions 20% below 1990 levels.[27]

The Kyoto Protocol has faced challenges. While many countries met or exceeded their emission-reduction commitments during the first commitment period, several countries did not. Furthermore, Canada, Japan, New Zealand, and Russia—all countries that agreed to the first commitment period—have not agreed to be bound for the second period. Perhaps most importantly, the second largest emitter of GHGs, the United States, is not a party to the protocol. The US Congress did not ratify the Kyoto Protocol because several members of Congress objected to developing countries, especially China, not having to make binding emission reductions under the protocol. Similar to Arizona's not wanting to bear the economic costs of a cap-and-trade program if other states were not part of the program, the United States did not want to bear the economic costs of ratifying the Kyoto Protocol if other countries did not have to reduce their emissions. The result is that the Kyoto Protocol has not been nearly as effective at reducing GHG emissions as had been hoped.

The United States and China did announce a joint agreement in 2014 to reduce CO_2 emissions.[28] The United States agreed to emit 26 to 28% less CO_2 in 2025 than it did in 2005. China, meanwhile, agreed to hit its peak CO_2 emissions by 2030. While it is not a formal treaty and not legally binding, the pledge by the two largest emitters of CO_2 may help increase the likelihood of a new global climate treaty being negotiated.[29]

Incredibly, the treaty that has so far done the most to mitigate climate change is not a climate change treaty. The Montreal Protocol on Substances That Deplete the Ozone Layer (Montreal Protocol) was written to limit the use of chemicals that cause depletion of the ozone layer. Entering into force in 1989 and ratified by 197 parties, including the United States, the treaty phases out ozone-depleting chemicals such as chlorofluorocarbons and hy-drochlorofluorocarbons. These chemicals and other ozone-depleting gases are also GHGs, and actually have a much larger warming effect than CO_2. By reducing the emission of ozone-depleting gases, the Montreal Protocol has done more to mitigate climate change than the first commitment period of the Kyoto Protocol.[30] The parties to the Montreal Protocol have realized the importance of the protocol in mitigating climate change, which is one of the reasons they agreed in 2007 to adjust the protocol to accelerate the phaseout of hydrochlorofluorocarbons.

FEDERAL LAWS

As mentioned earlier, the United States is the second largest emitter of GHGs, with China in first place and the European Union in third. In fact, these three emitters release 55% of global CO_2 emissions.[31] CO_2 emissions in the United States actually decreased by 4% from 2011 to 2012. The primary reason was that a large increase in shale gas production made natural gas prices lower, leading power plants to switch from burning coal to burning natural gas.[32] Burning natural gas releases less CO_2 than burning coal, thereby leading to reduced emissions. From 2012 to 2013, however, CO_2 emissions increased by 2.5%.

Figure 11.1 shows the primary sources of GHG emissions in the United States. Generating electricity is the largest source, with transportation a close second.[33] Emissions from industry include burning fossil fuels for energy along with certain chemical reactions used in manufacturing goods. Commercial and residential emissions include burning fossil fuels for heat and using products that contain GHGs. Emissions from agriculture come primarily from livestock, agricultural soils, and rice cultivation. Of the GHGs emitted by the United States, CO_2 makes up 83.7%, with the burning of fossil fuels accounting for 94% of those CO_2 emissions.[34]

Forests in the United States act as sinks for CO_2 from the atmosphere. According to the Environmental Protection Agency (EPA), forests and other

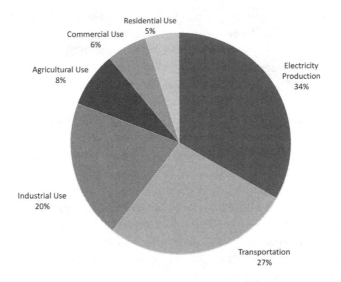

FIGURE 11.1. Percentage of greenhouse-gas emissions in the United States by economic sector. Data from "Inventory of U.S. Greenhouse Gas Emissions and Sinks: 1990–2011," Environmental Protection Agency, 2013, http://www.epa.gov/climatechange/Downloads /ghgemissions/US-GHG-Inventory-2013-Main-Text.pdf.

lands offset approximately 14% of US GHG emissions in 2011. Of the carbon sequestered, forests accounted for roughly 92%, with urban trees contributing 8%.[35] Additional sequestration comes from yard trimmings and food scraps sent to landfills, along with sequestration in nonforest soils, each accounting for roughly 1% of sequestration. As these numbers suggest, forests are a significant sink for GHGs. This means that decisions by the federal government on how to manage federal forests (discussed in chapter 5), along with decisions by local governments about where to protect forests from development (discussed in chapter 6), have a huge influence on the ability of forests in the United States to act as sinks for carbon.

As there is no comprehensive federal climate law, there is no single law laying out a national policy on climate change. That does not mean the federal government has no legal recourse to confront climate change. There has been significant regulation of GHG emissions through the Clean Air Act (CAA). Congress passed the CAA before there was much concern about climate change, so the act does not directly address GHG emissions. Through court decisions and EPA rule making, however, the CAA has become the main federal statute regulating GHGs.

Clean Air Act

Chapter 10 describes several provisions of the CAA. This section gives a brief reminder of some of those provisions.

Any stationary source that is a major emitter of an air pollutant must receive an operating permit from the state in which it is located. These Title V operating permits set limits on the amount of pollutants that sources may emit, and prescribe the pollution-control devices they must use.

New stationary sources that are major emitters of air pollutants are regulated under the new source performance standards.[36] The standards also apply to existing sources that undergo modifications. The new source performance standards set emission limits based on the best pollution-control technology available for a given industry.

The Prevention of Significant Deterioration (PSD) program protects the air quality of areas that are already in compliance with the National Ambient Air Quality Standards.[37] New or modified stationary sources in the PSD program must obtain a preconstruction permit. After receiving a permit, new major emitting facilities are required to use the "best available control technology" for each pollutant.

The CAA requires the EPA to set emission standards for new motor vehicles or new motor-vehicle engines.[38] These are called mobile source standards. The National Highway Traffic Safety Administration may also set a corporate average fuel economy. This governs the average miles per gallon that cars and light trucks must achieve.

Clean Air Act and Greenhouse-Gas Emissions

How do all of the complex provisions of the CAA relate to GHG emissions? Initially, the EPA argued that the CAA did not apply to GHGs. Under the administration of George W. Bush, the EPA stated that CO_2 is not a pollutant as defined by the CAA. The EPA then refused to regulate the emission of GHGs from mobile sources such as cars and trucks. In response, 12 states, along with several other groups, challenged the EPA's interpretation of the CAA. The Supreme Court took up the case in 2007 in *Massachusetts v. Environmental Protection Agency*.[39]

The primary question in *Massachusetts* was whether GHGs are air pollutants as defined in the CAA. The Supreme Court in *Massachusetts* began by looking at how the CAA defines its terms. The CAA defines "air pollutant" to include "any air pollution agent or combination of such agents" that enters the ambient air.[40] The section of the CAA relating to motor vehicles states that the EPA "shall" create standards for emission of "any air pollutant" from any class of new motor vehicles that "cause, or contribute to, air pollution which may reasonably be anticipated to endanger public health or welfare."[41] The act then defines "welfare" to include effects on weather and climate.[42]

By following the definition of the terms, the Supreme Court in *Massachusetts* held that GHGs, including CO_2, can be classified as air pollutants under the CAA. The court reasoned that the CAA gives a very broad definition to "air pollutant," writing that the definition "embraces all airborne compounds of whatever stripe." The court further reasoned that although Congress may not have considered GHGs as air pollutants when drafting the CAA, the broad definition of air pollutant in the statute was meant to allow the CAA to retain flexibility over time so that it does not become an obsolete statute.

Once the court had decided that GHGs could be classified as pollutants under the CAA, the next question was whether they should be classified as pollutants. The Supreme Court wrote that the EPA had not offered a "reasoned explanation for its refusal to decide whether greenhouse gases cause or contribute to climate change." The court concluded in *Massachusetts* that the EPA must make a reasoned decision whether GHG emissions from mobile sources endanger public health or welfare.

In 2009 the EPA determined that GHG emissions do indeed endanger public health or welfare. This "endangerment finding" allowed the EPA to regulate motor-vehicle GHG emissions under the CAA. The next year, the EPA and National Highway Traffic Safety Administration issued joint regulations creating standards for GHGs emitted from light-duty vehicles, with standards for heavy-duty vehicles following quickly.[43] As part of the standards, by 2025 average fuel economy for cars and light trucks must be the equivalent of 54.5 miles to the gallon.[44]

The EPA next turned to the task of regulating stationary sources of GHG emissions. The EPA began by trying to regulate facilities under the PSD program. The EPA wanted new or modified sources that emitted GHGs above a particular threshold to obtain a PSD permit. The attempt by the EPA, though, clearly shows why the CAA is not the ideal tool for regulating GHG emissions. The CAA specifically states that under the PSD program, stationary sources need permits if they pass a threshold of 100 or 250 tons of pollutants emitted per year.[45] A huge number of sources in the United States emit that quantity of GHGs every year, including many malls, hospitals, and office buildings. Issuing so many permits would be nearly impossible for the EPA.

To deal with this permitting problem, in 2010 EPA issued a "Tailoring Rule." The Tailoring Rule stated that PSD permitting for GHGs would apply only to new sources emitting at least 100,000 tons per year of CO_2-equivalent GHGs.[46] Additionally, an existing stationary source that emitted 100,000 tons per year and underwent a modification that increased emissions by 75,000 tons per year or more would also need a PSD permit. The Tailoring Rule also determined how GHG emissions would be dealt with for sources required to obtain Title V operating permits. Existing sources that needed a Title V permit for another pollutant would have to address GHG emissions in their Title V permits. More importantly, sources that would not normally need a Title V permit but that emitted 100,000 tons per year of CO_2-equivalent GHGs would now need to obtain a Title V permit.

The Supreme Court took up the legality of the EPA's requiring Title V and PSD permits for GHGs, and the subsequent Tailoring Rule, in 2014. The court held in *Utility Air Regulatory Group v. EPA* that the EPA lacked the authority to require Title V and PSD permits for stationary sources emitting GHGs.[47] Requiring permits for every facility that emitted 100 or 250 tons of GHGs per year, the court wrote, "would bring about an enormous and transformative expansion in EPA's regulatory authority without clear congressional authorization."[48] The court then examined whether the EPA could still require those permits by changing the threshold to 100,000 tons in the Tailoring Rule. The court held that the EPA also lacked the authority to issue the Tailoring Rule. The court declared that the EPA may not decide to simply change a threshold that Congress specifically wrote into the CAA. Only Congress gets to change the requirements of a statute. The Supreme Court, however, did not completely foreclose the ability of the EPA to regulate GHG emissions from stationary sources. The court held that facilities that need permits under the PSD program for other pollutants could also be required by the EPA to limit their GHG emissions using the best available control technology. These are so-called anyway sources because they have to get a PSD permit anyway. Since the EPA can require anyway sources to limit their GHG emissions, the decision by the Supreme Court will not have much practical effect. The anyway sources make up 83% of GHG emissions by stationary sources.[49] The Tailoring Rule would have added only an additional 3% of sources.

The EPA is proposing to use different provisions of the CAA to regulate the emission of GHGs from power plants. In 2013 the EPA proposed a rule under the new source performance standards to limit CO_2 emissions from new coal- and natural gas–fired power plants. Then in 2014 the EPA proposed a rule to limit emissions of CO_2 from existing fossil-fuel-fired power

plants.[50] Under the 2014 proposal, the EPA would set state-specific requirements for reducing CO_2 emissions from power plants, but would allow each
state to decide how best to meet those requirements. If implemented, by
2030 the rule would reduce CO_2 emissions from power plants by 32% below
2005 levels.

The Supreme Court took up a different aspect of climate change juris-
prudence in 2010. In *American Electric Power Co. v. Connecticut,* the court had
to decide whether states or other groups could directly sue major emitters
of GHGs to force them to reduce their emissions.[51] The case began when
several states and environmental groups, along with the city of New York,
sued several private power plants and the Tennessee Valley Authority to
force them to cap their GHG emissions.

The groups sued under the theory of common law federal public nuisance. Common law is described in box 2.2; briefly, it is law created through
court decisions that acts to fill in holes left by statutes. The Supreme Court
ruled in 1938 that for the most part, federal judges may not create common
law.[52] Federal courts, however, may create common law for issues within
national legislative power. In *American Electric Power Co.,* the Supreme Court
stated that environmental protection is an area of national legislative power,
and therefore federal courts may create common law. Public nuisance, then,
is a common law tort. (A tort is a civil wrong, such as negligence, battery, or
trespass to land.) Under tort law, a public nuisance is an unreasonable interference with a right held by the general public, such as health or use of
public property.

The Supreme Court held in *American Electric Power Co.* that with regard
to regulating GHG emissions, the CAA displaces common law federal public
nuisance. The court reasoned that the EPA may use the authority of the CAA
to set GHG emissions limits for power plants; therefore the federal courts
may not make federal common law rules that attempt to do the same thing.
The Supreme Court, however, left open the possibility of claims under state
public nuisance law. How such claims will ultimately shape climate change
law remains to be seen.

With the Supreme Court decision in *American Electric Power Co.,* the
CAA remains the primary means for regulating GHG emissions at the federal level. The EPA has certainly used the CAA to begin regulating GHG
emissions, but the process has been piecemeal, tackling one sector of the
economy at a time. While this should lead to significant emission reductions
over time, it will almost certainly be slower than if there were a comprehensive federal climate change statute.

STATE LAWS

Several states, frustrated by the slow pace of the federal government in addressing climate change, have attempted to reduce GHG emissions within their own borders. The attempts have taken several forms.

A majority of states have passed laws requiring a certain percentage of their electricity to come from renewable resources. For instance, Michigan requires 10% of its electricity to come from renewables by 2015, California requires 33% by 2020, and Hawaii requires 40% by 2030.[53]

The state doing the most to reduce GHG emissions within its borders is California. In 2006 California passed the Global Warming Solutions Act.[54] The act has the goal of reducing GHG emissions to 1990 levels by 2020, and contains enforceable penalties to reach those emission levels. The primary way California hopes to achieve its goal is through a cap-and-trade program within the state. The cap for electrical utilities was put in place in 2013, with other businesses entering the program in 2015.

As mentioned at the beginning of this chapter, California is also a member of the WCI. The Canadian province of Quebec is also a member of the WCI, and has implemented a cap-and-trade program. In 2014 California and Quebec linked their respective cap-and-trade programs, allowing trading of allowances between the two programs.

On the other side of the United States, a different set of states has created a regional cap-and-trade initiative. The Regional Greenhouse Gas Initiative (RGGI) consists of nine Northeast and mid-Atlantic states. The initiative began in 2005 with the goal of capping emissions of CO_2 from power plants at 2009 levels, and then reducing those emissions by 10% by 2018.[55]

The RGGI illustrates one of the problems that can occur with cap-and-trade programs, though. Overallocation of allowances quickly became a problem for the initiative. In 2012 the RGGI set a cap of 165 million tons of CO_2, but the power plants in the RGGI states emitted only 91 million tons. The main reason was that many power plants switched from coal to less expensive natural gas.[56] As mentioned in chapter 10, natural gas releases less CO_2 than coal when burned, thereby lowering emissions. Because emissions were so low, the cost of CO_2 allowances was also very low, providing little incentive for power plants to lower their emissions further. To remedy this, the RGGI in 2014 had to reset the CO_2 cap to 91 million tons, with the cap declining 2.5% per year until 2020.

Besides trying to mitigate climate change by reducing GHG emissions, many states are making plans to adapt to the effects of climate change that appear inevitable. Fifteen states have completed adaptation plans, with several other states in the process of writing such plans.[57] The adaptation plans usually call for increasing the number and size of preserved areas within a state, along with creating corridors for movement between such areas. For example, California's adaptation plan suggests a "large scale well connected, sustainable system of protected areas across the State."[58] The protected areas are meant to shield species as they adapt to a changing climate, and to facilitate the movement of species as their ranges shift. The state adaptation plans also tend to call for reducing other stressors on ecosystems within the states.[59] The Washington State plan suggests decreasing pollution levels to improve the resilience of ecosystems in the face of climate change.[60] Finally, the state plans often call for greater consideration of climate change when making management decisions regarding natural resources. Although the recommendations in the state adaptation plans are quite reasonable, it should be remembered that they are often just suggestions and do not have the backing of statutes or regulations. In order for adaptation to climate change to begin in earnest, the states must begin passing legislation that implements the suggestions in their plans.

As this chapter illustrates, the United States at the federal and state levels is undertaking several endeavors to reduce GHG emissions. To produce a more stable and permanent reduction in US GHG emissions, though, a comprehensive federal statute on climate change will almost certainly be necessary.

Appendix

Important Statutes and Treaties

Statute or treaty	Citation	Description	Chapter
Agreement for the Implementation of the Provisions of the United Nations Convention on the Law of the Sea relating to the Conservation and Management of Straddling Fish Stocks and Highly Migratory Fish Stocks (Fish Stocks Agreement)	2167 U.N.T.S. 3	Treaty for the management of fish species that occur in countries' exclusive economic zones and the adjacent high seas, and species that migrate great distances.	9
Agricultural Act (farm bill)	Public Law 113-79	Funds items such as guaranteed loans to farmers, crop insurance, agricultural research, nutrition programs, and the Conservation Reserve Program.	6
Antiquities Act	16 U.S.C. 431 et seq.	Authorizes president to declare objects of "historic or scientific interest" as national monuments.	9
Clean Air Act (CAA)	42 U.S.C. 7401 et seq.	Regulates emission of pollutants into the air.	10, 11
NAAQS	42 U.S.C. 7409	Creates ambient air standards for certain pollutants.	10
Massachusetts. v. Environmental Protection Agency	549 U.S. 497 (2007)	Greenhouse gases can be classified as pollutants under the Clean Air Act.	11

Statute or treaty	Citation	Description	Chapter
Clean Water Act (CWA)	33 U.S.C. 1251 et seq.	Regulates discharges of pollutants into waters of the US.	7, 8, 9
NPDES	33 U.S.C. 1342	Describes permit system for discharging pollutants.	7
Section 404	33 U.S.C. 1344	Describes permit system for filling wetlands.	8
Rapanos v. United States	547 U.S. 715 (2006)	Waters of the US are either relatively permanent bodies of water, or bodies of water such as wetlands that have a "significant nexus" to traditionally navigable waters.	8
Coastal Zone Management Act (CZMA)	16 U.S.C. 1451 et seq.	Promotes development of state-level plans for coastal zone management.	9
Comprehensive Environmental Response, Compensation and Liability Act (CERCLA)	42 U.S.C. 9601 et seq.	Authorizes Environmental Protection Agency to clean up a contaminated site whenever "any hazardous substance is released or there is a substantial threat of such a release into the environment."	6
Convention Concerning the Protection of the World Cultural and Natural Heritage (World Heritage Convention)	1037 U.N.T.S. 151	Recognizes natural and cultural areas of "outstanding universal value."	4
Convention on Biological Diversity	1760 U.N.T.S. 79	Provides framework for preserving biodiversity. (US is not a party.)	4
Convention on International Trade in Endangered Species of Wild Fauna and Flora (CITES)	993 U.N.T.S. 243	Creates restrictions on international trade of endangered plant and animal species.	4
Convention on the Conservation of Migratory Species of Wild Animals (Bonn Convention)	1651 U.N.T.S. 333	Prohibits taking of endangered migratory species. (US is not a party.)	4

Statute or treaty	Citation	Description	Chapter
Convention on Wetlands of International Importance Especially as Waterfowl Habitat (Ramsar Convention)	996 U.N.T.S. 245	Creates the List of Wetlands of International Importance.	8
Endangered Species Act (ESA)	16 U.S.C. 1531 et seq.	Authorizes the listing of threatened or endangered species, and the protection of those species from being killed or having their critical habitats modified.	3
Section 7	16 U.S.C. 1536	Prohibits federal agency actions that jeopardize the existence of a listed species or adversely modify critical habitat.	3
Section 9	16 U.S.C. 1538	Prohibits the take of listed species.	3
Tennessee Valley Authority v. Hill	437 U.S. 153 (1978)	Intent of the Endangered Species Act is to "halt and reverse the trend toward species extinction, whatever the cost."	3
Federal Insecticide, Fungicide, and Rodenticide Act (FIFRA)	7 U.S.C. 136 et seq.	Regulates chemicals used as pesticides.	6
Federal Land Policy and Management Act (FLPMA)	43 U.S.C. 1701 et seq.	Requires Bureau of Land Management to manage its lands under the principles of multiple use and sustained yield.	5
Federal Power Act	16 U.S.C. 791a et seq.	Regulates, among other things, hydroelectric dams.	7
General Agreement on Tariffs and Trade (GATT)	1867 U.N.T.S. 187	Reduces tariffs and resolves trade disputes.	4
Healthy Forests Restoration Act	16 U.S.C. 6501 et seq.	Attempts to reduce wildfires by reducing fuel on federal lands.	5
International Convention for the Regulation of Whaling (Whaling Convention)	161 U.N.T.S. 72	Currently creates a moratorium on all commercial whaling.	4
Kyoto Protocol	2303 U.N.T.S. 162	Requires parties to reduce their greenhouse gas emissions 20% below their 1990 levels. (US is not a party.)	11

Statute or treaty	Citation	Description	Chapter
Lacey Act	16 U.S.C. 3371 et seq.	Prohibits the import into the US of any animal species listed as "injurious," and prohibits the transport of any injurious species between the states.	2
Magnuson-Stevens Fishery Conservation and Management Act (Magnuson-Stevens Act)	16 U.S.C. 1801 et seq.	Creates regional councils for managing fish stocks in US waters.	9
Marine Mammal Protection Act (MMPA)	16 U.S.C. 1361 et seq.	Prohibits taking and importing of marine mammals.	9
Migratory Bird Treaty Act	16 U.S.C. 703 et seq.	Makes it unlawful to pursue, hunt, kill, or capture any migratory bird protected by one of four treaties.	2
Montreal Protocol on Substances That Deplete the Ozone Layer (Montreal Protocol)	1522 U.N.T.S. 3	Phases out use of ozone-depleting chemicals.	11
Multiple-Use, Sustained-Yield Act	16 U.S.C. 528 et seq.	Requires US Forest Service to manage forests to protect outdoor recreation, range, timber, watersheds, wildlife, and fish.	5
National Environmental Policy Act (NEPA)	42 U.S.C. 4321 et seq.	Requires federal agencies to write environmental impact statements for "major Federal actions significantly affecting the quality of the human environment."	5
National Forest Management Act (NFMA)	16 U.S.C. 1600 et seq.	Requires US Forest Service to create forest management plans that provide for multiple use and sustained yield, and that protect the "diversity of plant and animal communities."	5
National Marine Sanctuaries Act	16 U.S.C. 1431 et seq.	Prohibits injury or loss of any national marine sanctuary resource.	9
National Park Service Organic Act (Organic Act)	16 U.S.C. 1 et seq.	Requires National Park Service to protect wildlife in national parks, and to provide for recreation.	5

Statute or treaty	Citation	Description	Chapter
National Wildlife Refuge System Administration Act (NWRSAA)	16 U.S.C. 668dd, 668ee	Creates a hierarchy of uses for the Wildlife Refuge System: conservation, wildlife-dependent recreation, and all other uses.	5
Nonindigenous Aquatic Nuisance Prevention and Control Act	16 U.S.C. 4701 et seq.	Regulates ballast discharges in navigable inland waters of the US, and three nautical miles out into the ocean.	2
Ocean Dumping Act	33 U.S.C. 1411 et seq.	Prohibits dumping any material into ocean waters.	9
Plant Protection Act	7 U.S.C. 7701 et seq.	Prohibits the import of "noxious weeds" and "plant pests" into the US, and the movement of such species in interstate commerce.	2
Resource Conservation and Recovery Act (RCRA)	42 U.S.C. 6901 et seq.	Regulates the generation and disposal of solid waste.	6
Toxic Substances Control Act (TSCA)	15 U.S.C. 2601 et seq.	Regulates chemicals not used as pesticides.	6
UN Convention on the Law of the Sea (UNCLOS)	1833 U.N.T.S. 3	Acts as an international "constitution of the oceans."	9
UN Framework Convention on Climate Change (UNFCCC)	1771 U.N.T.S. 107	Provides framework for conducting international negotiations on climate change.	11
Wild and Scenic Rivers Act	16 U.S.C. 1271 et seq.	Creates a system for designating rivers to be protected from development.	7
Wilderness Act	16 U.S.C. 1131 et seq.	Prohibits all commercial enterprise and permanent roads in the National Wilderness Preservation System.	5

Acknowledgments

I would like to thank Megan Kimbrell, JD, and Cedric Worman, PhD, for reading and commenting on several chapters in the book. Cedric Worman's recent death is truly ecology's loss. I am grateful to two anonymous reviewers who made many excellent suggestions for improving this project. The book is better because of their efforts. I would also like to thank Evan White, editorial associate, and Katherine Faydash, senior manuscript editor, for their help in managing the manuscript. Joann Hoy did a great job copyediting the book, and I thank her for helping make it more readable. I would particularly like to thank Dr. Christopher Chung, the assistant editor for this book, for his help and guidance during the entire publishing process. Finally, this book would not have been possible without the love and support of Hillary, Nathan, and Hunter.

Notes

CHAPTER ONE

1. 16 U.S.C. § 1604(g)(3)(B).
2. Sierra Club v. Marita, 46 F.3d 606 (7th Cir. 1995).
3. *Marita*, 46 F.3d at 610.
4. *Marita*, 46 F.3d at 618.
5. *Marita*, 46 F.3d at 619.
6. *Marita*, 46 F.3d at 621.
7. Statutes that contain such a requirement include the Endangered Species Act, 16 U.S.C. § 1533(b)(1)(A); Magnuson-Stevens Fishery Conservation and Management Act, 16 U.S.C. § 1851(a)(2); Marine Mammal Protection Act, 16 U.S.C. § 1362(19); and the Nonindigenous Aquatic Nuisance Prevention and Control Act, 16 U.S.C. § 4711(a)(2)(D).
8. E.g., when setting National Ambient Air Quality Standards under the Clean Air Act, 42 U.S.C § 7409(b).
9. Holly Doremus, "Listing Decisions under the Endangered Species Act: Why Better Science Isn't Always Better Policy," *Washington University Law Quarterly* 75 (1997): 1038-40.
10. Kristin Carden, "Bridging the Divide: The Role of Science in Species Conservation Law," *Harvard Environmental Law Review* 30 (2006): 185.
11. Doremus, "Listing Decisions," 1038-39.
12. 16 U.S.C. § 1533(b)(1)(A).
13. Doremus, "Listing Decisions," 1112-29.
14. Abbe R. Gluck and Lisa S. Bressman, "Statutory Interpretation from the Inside—An Empirical Study of Congressional Drafting, Delegation, and the Canons," part 1, *Stanford Law Review* 65 (2013): 908.
15. Lisa S. Bressman and Abbe R. Gluck, "Statutory Interpretation from the Inside—An Empirical Study of Congressional Drafting, Delegation, and the Canons," part 2, *Stanford Law Review* 66 (2014): 739.
16. Bressman and Gluck, "Statutory Interpretation from the Inside," part 2, 742.
17. Eric Biber, "Which Science? Whose Science? How Scientific Disciplines Can Shape Environmental Law," *University of Chicago Law Review* 79 (2012): 521-25.
18. Biber, "Which Science?," 521-25.
19. Biber, "Which Science?," 550.
20. Biber, "Which Science?," 551.

21. "Rulemaking: OMB's Role in Reviews of Agencies' Draft Rules and the Transparency of Those Reviews," US General Accounting Office, 2003, http://www.gao.gov/new.items/d03929.pdf.

22. Daniel A. Farber and Anne J. O'Connell, "The Lost World of Administrative Law," *Texas Law Review* 92 (2014): 1137–89.

23. Eric Biber, "Too Many Things to Do: How to Deal with the Dysfunctions of Multiple-Goal Agencies," *Harvard Environmental Law Review* 33 (2009): 49.

24. "Rulemaking."

25. Courtney Schultz, "Responding to Scientific Uncertainty in U.S. Forest Policy," *Environmental Science and Policy* 11 (2008): 254–55.

26. Eric Biber, "Adaptive Management and the Future of Environmental Law," *Akron Law Review* 46 (2013): 934.

27. Biber, "Adaptive Management," 933–62.

28. George H. Stankey et al., "Adaptive Management and the Northwest Forest Plan: Rhetoric and Reality," *Journal of Forestry* 101 (2003): 44–45.

29. Biber, "Adaptive Management," 956–62.

30. Biber, "Adaptive Management," 956–62.

31. Ryan P. Kelly and Margaret R. Caldwell, "'Not Supported by Current Science': The National Forest Management Act and the Lessons of Environmental Monitoring for the Future of Public Resources Management," *Stanford Environmental Law Review* 32 (2013): 170–74.

32. Kelly and Caldwell, "Not Supported by Current Science," 170–74.

33. National Research Council, *Ecological Indicators for the Nation* (Washington, DC: National Academy Press, 2000), 1–180.

34. Eric Biber, "The Problem of Environmental Monitoring," *University of Colorado Law Review* 83 (2011): 35–53.

35. Biber, "Environmental Monitoring," 53–54, 76–79.

36. Frye v. U.S., 293 F. 1013 (D.C. Cir. 1923).

37. 509 U.S. 579 (1993).

38. 526 U.S. 137 (1999).

39. Lloyd Dixon and Brian Gill, *Changes in the Standards for Admitting Expert Evidence in Federal Civil Cases since the* Daubert *Decision* (Santa Monica, CA: Rand, 2001), xiv–xvi.

40. 467 U.S. 837 (1984).

41. 533 U.S. 218 (2001).

42. Peter A. Appel, "Wilderness and the Courts," *Stanford Environmental Law Journal* 29 (2010): 114.

43. 5 U.S.C. § 706(2)(A).

44. Citizens to Preserve Overton Park, Inc. v. Volpe, 401 U.S. 402 (1971).

45. 463 U.S. 29 (1983).

46. 462 U.S. 87 (1983).

47. Emily H. Meazell, "Super Deference, the Science Obsession, and Judicial Review as Translation of Agency Science," *Michigan Law Review* 109 (2011): 751–52.

48. Meazell, "Super Deference," 772–78.

49. Richard O. Brooks, Ross Jones, and Ross A. Virginia, *Law and Ecology: The Rise of the Ecosystem Regime* (Burlington, VT: Ashgate Publishing, 2002), 7–8.

50. Brooks, Jones, and Virginia, *Law and Ecology*, 365.

51. Robin K. Craig, "Learning to Think about Complex Environmental Systems in Environmental and Natural Resource Law and Legal Scholarship: A Twenty-Year Retrospective," *Fordham Environmental Law Review* 24 (2012–13): 89–90.

CHAPTER TWO

1. Michael C. Blumm and Lucus Ritchie, "The Pioneer Spirit and the Public Trust: The American Rule of Capture and State Ownership of Wildlife," *Environmental Law* 35 (2005): 691.

2. Meredith B. Lilley and Jeremy Firestone, "Wind Power, Wildlife, and the Migratory Bird Treaty Act: A Way Forward," *Environmental Law* 38 (2008): 1178.

3. Weeks-McLean Law of 1913, ch. 145, 37 Stat. 828, 847.

4. U.S. v. McCullagh, 221 F. 288 (D. Kan. 1915); U.S. v. Shauver, 214 F. 154 (E.D. Ark. 1914).

5. 252 U.S. 416 (1920).

6. For further reading on how private property rights help overcome the tragedy of the commons, see Kirsten Engel and Dean Lueck, "Symposium Introduction: Property Rights and the Environment," *Arizona Law Review* 50 (2008): 373–77; Gary D. Libecap, "Open-Access Losses and Delay in the Assignment of Property Rights," *Arizona Law Review* 50 (2008): 379–408; and Gary D. Libecap, "The Tragedy of the Commons: Property Rights and Markets as Solutions to Resource and Environmental Problems," *Australian Journal of Agricultural and Resource Economics* 53 (2009): 129–44.

7. See Wickard v. Filburn, 317 U.S. 111 (1942).

8. NLRB v. Jones & Laughlin Steel Corp., 301 U.S. 1 (1937).

9. 514 U.S. 549 (1995).

10. Douglas v. Seacoast Products, Inc., 431 U.S. 265 (1977).

11. 431 U.S. 265, 282 (1977).

12. US Constitution, Article II, Section 2.

13. US Constitution, Article I, Section 8.

14. US Constitution, Article IV, Section 3.

15. Eric T. Freyfogle and Dale D. Goble, *Wildlife Law: A Primer* (Washington, DC: Island Press, 2009), 111–13.

16. 426 U.S. 529 (1976).

17. Kleppe, 426 U.S. at 539, quoting U.S. v. San Francisco, 310 U.S. 16 (1940).

18. *McCullagh*, 221 F. 288; *Shauver*, 214 F. 154.

19. 16 U.S.C. § 703.

20. 78 Fed. Reg. 65844; 50 C.F.R. § 10.13.

21. 16 U.S.C. § 703.

22. 70 Fed. Reg. 12710.

23. 16 U.S.C. § 704; 50 C.F.R. parts 20–21.

24. 50 C.F.R. §§ 21.11, 21.23.

25. 50 C.F.R. §§ 21.43, 21.46.

26. 16 U.S.C. § 711.

27. U.S. v. FMC Corp., 572 F.2d 902 (2nd Cir. 1978).

28. U.S. v. Moon Lake Electric Association, Inc., 45 F. Supp. 2d 1070 (D. Colo. 1999).

29. Apollo Energies, Inc. v. U.S., 611 F.3d 679 (10th Cir. 2010); *Moon Lake Electric*, 45 F. Supp. 2d 1070.

30. Albert M. Manville II, "Towers, Turbines, Power Lines, and Buildings: Steps Being Taken by the U.S. Fish and Wildlife Service to Avoid or Minimize Take of Migratory Birds at These Structures," in *Proceedings of the Fourth International Partners in Flight Conference: Tundra to Tropics*, ed. Terrell D. Rich et al. (McAllen, TX: Partners in Flight, 2009), 268.

31. *Associated Press*, "Duke Energy to Pay $1 Million after Wyoming Wind Farms' Blades Kill 14 Eagles, Other Birds," November 22, 2013, http://www.nola.com/news/index.ssf/2013/11/duke _energy_to_pay_1_million_a.html.

32. "Land-Based Wind Energy Guidelines," US Fish and Wildlife Service, 2012, http://www.fws .gov/windenergy/docs/WEG_final.pdf.

33. City of Sausalito v. O'Neill, 386 F.3d 1186 (9th Cir. 2004).

34. "The State of the Birds," US Department of Interior, 2009, http://www.stateofthebirds.org /2009/pdf_files/State_of_the_Birds_2009.pdf.

35. 16 U.S.C. §§ 3371-3377; 18 U.S.C. § 42.

36. 16 U.S.C. § 3372(a).

37. 16 U.S.C. § 3371(g).

38. The Department of Agriculture maintains a website, http://www.invasivespeciesinfo.gov /laws/publiclaws.shtml, which provides the names and brief descriptions of all the federal laws that directly or indirectly touch on controlling the spread of invasive species in the United States.

39. David Pimentel, Rodolfo Zuniga, and Doug Morrison, "Update on the Environmental and Economic Costs Associated with Alien-Invasive Species in the United States," *Ecological Economics* 52 (2005): 282.

40. Cynthia S. Kolar and David M. Lodge, "Progress in Invasion Biology: Predicting Invaders," *Trends in Ecology and Evolution* 16 (2001): 201.

41. Jane Cynthia Graham, "Snakes on a Plane, or in a Wetland: Fighting Back Invasive Non-native Animals—Proposing a Federal Comprehensive Invasive Nonnative Animal Species Statute," *Tulane Environmental Law Journal* 25 (2011): 35.

42. 50 C.F.R. § 16.22.

43. 18 U.S.C. § 42(a)(1).

44. "Questions and Answers: Listing of Four Non-native Snake Species as Injurious under the Lacey Act," US Fish and Wildlife Service, 2012, http://www.fws.gov/fisheries/ANS/pdf_files/Four _snakes._QsAs.final.pdf.

45. 50 C.F.R. part 16.

46. "Injurious Wildlife: A Summary of the Injurious Provisions of the Lacey Act," US Fish and Wildlife Service, 2010, http://www.fws.gov/injuriouswildlife/pdf_files/InjuriousWildlife FactSheet2010.pdf.

47. 18 U.S.C. § 42(a)(1).

48. Peter T. Jenkins, "Invasive Animals and Wildlife Pathogens in the United States: The Economic Case for More Risk Assessments and Regulations," *Biological Invasions* 15 (2013): 244.

49. Flynn Boonstra, "Leading by Example: A Comparison of New Zealand's and the United State's Invasive Species Policies," *Connecticut Law Review* 43 (2011): 1207-14.

50. Andrea J. Fowler, David M. Lodge, and Jennifer F. Hsia, "Failure of the Lacey Act to Protect US Ecosystems against Animal Invasions," *Frontiers in Ecology and the Environment* 5 (2007): 354.

51. Jenkins, "Invasive Animals and Wildlife Pathogens," 243.

52. Fowler, Lodge, and Hsia, "Failure of the Lacey Act," 357.

53. Nonnative Wildlife Invasion Prevention Act, H.R. 669 (2009).

54. 7 U.S.C. §§ 7711(a), 7712(a).

55. 7 U.S.C. § 7702(10).

56. 7 U.S.C. § 7702(14).

57. 7 U.S.C. § 7702(14).

58. 7 U.S.C. § 7781 *et seq.*

59. 7 U.S.C. § 7782.

60. 7 U.S.C. § 7715.

61. But see James S. Neal McCubbins et al., "Frayed Seams in the 'Patchwork Quilt' of American Federalism: An Empirical Analysis of Invasive Plant Species Regulation," *Environmental Law* 43 (2013): 35-81.

62. David A. Strifling, "An Ecosystem-Based Approach to Slowing the Synergistic Effects of Invasive Species and Climate Change," *Duke Environmental Law and Policy Forum* 22 (2011): 165-67.

63. Daniel Strain, "Researchers Set Course to Blockade Ballast Invaders," *Science* 336 (2012): 664-65.

64. 16 U.S.C. §§ 4701-4751.

65. 16 U.S.C. § 4711(a).

66. 16 U.S.C. § 4711(c).

67. Northwest Environmental Advocates v. U.S. EPA, 537 F.3d 1006 (9th Cir. 2008).

68. 7 U.S.C. § 8501 *et seq.*

69. Graham, "Snakes on a Plane," 44.

70. 3 Cai. 175 (N.Y. Sup. Ct. 1805).

71. 146 U.S. 387.

72. See Joseph L. Sax, "The Public Trust Doctrine in Natural Resource Law: Effective Judicial Intervention," *Michigan Law Review* 68 (1970): 471–566.

73. 41 U.S. 367.

74. 161 U.S. 519 (1896).

75. 252 U.S. at 434.

76. 441 U.S. 322.

77. Pullen v. Ulmer, 923 P.2d 54, 60 (Alaska 1996).

78. See, e.g., Montana v. Fertterer, 841 P.2d 467 (Mont. 1992), overruled on other grounds, State v. Gatts, 928 P.2d 114 (Mont. 1996).

79. Georgia Code § 27-1-3.

80. Wildlife Society, *The Public Trust Doctrine: Implications for Wildlife Management and Conservation in the United States and Canada*, Technical Review 10-01 (Bethesda, MD: Wildlife Society, 2010), 22.

81. Raphael D. Sagarin and Mary Turnipseed, "The Public Trust Doctrine: Where Ecology Meets Natural Resources Management," *Annual Review of Environment and Resources* 37 (2012): 486–89.

82. Kentucky Revised Statutes §150.015.

83. National Audubon Society v. Superior Court of Alpine County, 658 P.2d 709 (Cal. 1983).

84. But see Jedidiah Brewer and Gary D. Libecap, "Property Rights and the Public Trust Doctrine in Environmental Protection and Natural Resource Conservation," *Australian Journal of Agricultural and Resource Economics* 53 (2009): 11–16.

85. Gregory J. Wright et al., "Selection of Northern Yellowstone Elk by Gray Wolves and Hunters," *Journal of Wildlife Management* 70 (2006): 1074.

86. Chris T. Darimont et al., "Human Predators Outpace Other Agents of Trait Change in the Wild," *Proceedings of the National Academy of Sciences* 106 (2009): 952.

87. Simone Ciuti et al., "Human Selection of Elk Behavioral Traits in a Landscape of Fear," *Proceedings of the Royal Society B* 279 (2012): 4410–12.

88. L. Palazy et al., "Rarity, Trophy Hunting and Ungulates," *Animal Conservation* 15 (2012): 5.

89. David W. Coltman et al., "Undesirable Evolutionary Consequences of Trophy Hunting," *Nature* 426 (2003): 656.

90. See, e.g., "Trends in Iowa Wildlife Populations and Harvest 2012," Iowa Department of Natural Resources, 2013, http://www.iowadnr.gov/Portals/idnr/uploads/Hunting/2012_logbook.pdf.

91. James D. Nichols et al., "Adaptive Harvest Management of North American Waterfowl Populations: A Brief History and Future Prospects," *Journal of Ornithology* 148 (2007): S343–49.

CHAPTER THREE

1. Wendell R. Haag and Melvin L. Warren Jr., "Seasonal Feeding Specialization on Snails by River Darters (*Percina shumardi*) with a Review of Snail Feeding by Other Darter Species," *Copeia* 4 (2006): 605–6.

2. 16 U.S.C. § 1536(a)(2).

3. 437 U.S. 153 (1978).

4. In *Tennessee Valley Authority v. Hill*, the Supreme Court wrote that the ESA "represented the most comprehensive legislation for the preservation of endangered species ever enacted by any nation," 180.

5. Robert V. Percival et al., *Environmental Regulation: Law, Science, and Policy* (New York: Aspen Publishers, 2006), 875–76.

6. M. J. Ashton and J. B. Layzer, "Summer Microhabitat Use by Adult and Young-of-Year Snail Darters (*Percina tanasi*) in Two Rivers," *Ecology of Freshwater Fish* 19 (2010): 610.

7. Robert M. May, "Ecological Science and Tomorrow's World," *Philosophical Transactions of the Royal Society B* 365 (2010): 42.

8. May, "Ecological Science," 42.

9. May, "Ecological Science," 43.

10. Stephen P. Ellner et al., "Precision of Population Viability Analysis," *Conservation Biology* 16 (2002): 258.

11. "The IUCN Red List of Threatened Species," International Union for Conservation of Nature, accessed January 29, 2015, http://www.iucnredlist.org.

12. "IUCN Red List Categories and Criteria, Version 3.1," International Union for Conservation of Nature, 2012, http://www.iucnredlist.org/technical-documents/categories-and-criteria/2001-categories-criteria.

13. 16 U.S.C. § 1532(6).

14. 16 U.S.C. § 1532(20).

15. 16 U.S.C. § 1533(a)(1).

16. 50 C.F.R. § 402.01(b).

17. 16 U.S.C. § 1533(b).

18. 16 U.S.C. § 1533(b)(3)(A).

19. 50 C.F.R. § 424.14(b).

20. 16 U.S.C. 1533(b)(1)(A).

21. 59 Fed. Reg. 34271.

22. 59 Fed. Reg. 34270.

23. 16 U.S.C. § 1533(b)(3)(B).

24. Kalyani Robbins, "Strength in Numbers: Setting Quantitative Criteria for Listing Species under the Endangered Species Act," *UCLA Journal of Environmental Law and Policy* 27 (2009): 20–22.

25. Robbins, "Strength in Numbers," 20–22.

26. Andrea Easter-Pilcher, "Implementing the Endangered Species Act," *BioScience* 46 (1996): 355.

27. David S. Wilcove and Lawrence L. Master, "How Many Endangered Species Are There in the United States?," *Frontiers in Ecology and the Environment* 3 (2005): 417.

28. J. Berton C. Harris et al., "Conserving Imperiled Species: A Comparison of the IUCN Red List and U.S. Endangered Species Act," *Conservation Letters* 5 (2012): 67.

29. 78 Fed. Reg. 63574.

30. Michael Wines, "Endangered or Not, but at Least No Longer Waiting," *New York Times*, March 6, 2013, http://www.nytimes.com/2013/03/07/science/earth/long-delayed-rulings-on-endangered-species-are-coming.html.

31. Robbins, "Strength in Numbers," 22–31, 34–35.

32. 16 U.S.C. § 1532(16).

33. 61 Fed. Reg. 4725.

34. 70 Fed. Reg. 46366.

35. "Endangered Species Act Authorizations," Senate Report 151, 96th Congress, 1st Session (1979).

36. 16 U.S.C. § 1532(16).

37. 16 U.S.C. § 1533(e).

38. 75 Fed. Reg. 53598.

39. 16 U.S.C. § 1531(b).

40. 16 U.S.C. § 1532(3).

41. Sierra Club v. U.S. Fish and Wildlife Service, 245 F.3d 434, 438 (5th Cir. 2001).

42. Wilcove and Master, "How Many Endangered Species," 416.

43. M. E. Gilpin and Michael E. Soulé, "Minimum Viable Populations: Processes of Extinction," in *Conservation Biology: The Science of Scarcity and Diversity*, ed. Michael E. Soulé (Sunderland, MA: Sinauer Associates, 1986), 19-34.

44. Daniel I. Bolnick et al., "Why Intraspecific Trait Variation Matters in Community Ecology," *Trends in Ecology and Evolution* 26 (2011): 189.

45. Brian Dennis, "Allee Effects in Stochastic Populations," *Oikos* 96 (2002): 398.

46. 16 U.S.C. § 1533(a)(3).

47. 16 U.S.C. § 1533(f).

48. 16 U.S.C. § 1536(a)(2).

49. 16 U.S.C. § 1536(a)(2).

50. 16 U.S.C. § 1538(a)(1).

51. 16 U.S.C. § 1539(a)(2).

52. 16 U.S.C. § 1533(a)(3).

53. 16 U.S.C. § 1532(5)(A).

54. 16 U.S.C. § 1532(5)(A).

55. 16 U.S.C. § 1536(a)(2).

56. 16 U.S.C. § 1533(b)(2).

57. 16 U.S.C. § 1533(b)(2).

58. 16 U.S.C. § 1532(5)(C).

59. 50 C.F.R. § 424.12.

60. Cape Hatteras Access Preservation Alliance v. U.S. Department of Interior, 344 F. Supp. 2d 108, 123 (D.D.C. 2010).

61. Gifford Pinchot Task Force v. U.S. Fish and Wildlife Service, 378 F.3d 1059, 1070 (9th Cir. 2004).

62. Forest Guardians v. Babbitt, 174 F.3d 1178, 1186 (10th Cir. 1999).

63. "Amendments to the Critical Habitat Requirements of the Endangered Species Act of 1973," Senate Report No. 106-126 at 2 (1999).

64. Percival et al., *Environmental Regulation*, 899-900.

65. Northern Spotted Owl v. Model, 716 F. Supp. 479 (W.D. Wash. 1988).

66. Northern Spotted Owl v. Lujan, 758 F. Supp. 621 (W.D. Wash. 1991).

67. 16 U.S.C. § 1532(5)(A)(i).

68. 50 C.F.R. § 424.02.

69. *Cape Hatteras*, 344 F. Supp. 2d at 123-24.

70. 16 U.S.C. § 1533(f)(1)(A).

71. 16 U.S.C. § 1533(f)(1).

72. D. Noah Greenwald, Kieran F. Suckling, and Martin Taylor, "The Listing Record," in *The Endangered Species Act at Thirty: Renewing the Conservation Promise*, volume 1, ed. Dale D. Goble, J. Michael Scott, and Frank W. Davis (Washington, DC: Island Press, 2005), 51-67.

73. "Interim Endangered and Threatened Species Recovery Planning Guidance, Version 1.3," NOAA Fisheries and US Fish and Wildlife Service, 2010, https://www.fws.gov/endangered/esa-library/pdf/NMFS-FWS_Recovery_Planning_Guidance.pdf.

74. "Endangered Species Bulletin," US Fish and Wildlife Service, 2006, http://www.fws.gov/endangered/news/pdf/ES%20Bulletin%2003-2006%20with%20Links.pdf.

75. "Recovery Planning Guidance."

76. 16 U.S.C. § 1533(f)(1)(B)(ii).

77. "Recovery Planning Guidance."

78. 16 U.S.C. § 1539(j).

79. 68 Fed. Reg. 26498.

80. 16 U.S.C. § 1536(a)(2).

81. 50 C.F.R. § 402.02.

82. *Gifford Pinchot Task Force*, 378 F.3d at 1069–71; N.M. Cattle Growers Association v. U.S. Fish and Wildlife Service, 248 F.3d 1277, 1283 (10th Cir. 2001); Sierra Club v. U.S. Fish and Wildlife Service, 245 F.3d 434, 441–42 (5th Cir. 2001).

83. "Consultations with Federal Agencies: Section 7 of the Endangered Species Act," US Fish and Wildlife Service, 2011, http://www.fws.gov/endangered/esa-library/pdf/consultations .pdf.

84. 16 U.S.C. § 1536(b)(3)(A).

85. 16 U.S.C. § 1536(e)–(h).

86. Amy Sinden, "In Defense of Absolutes: Combating the Politics of Power in Environmental Law," *Iowa Law Review* 90 (2005): 1504–5.

87. 16 U.S.C. § 1538(a).

88. Percival et al., *Environmental Regulation*, 915.

89. 16 U.S.C. § 1538(a)(1).

90. 50 C.F.R. § 17.31. See 16 U.S.C. § 1533(d), stating regulation may prohibit any act prohibited under Section 1538(a) for any threatened species.

91. 16 U.S.C. § 1532(19).

92. 50 C.F.R. § 17.3.

93. See Babbitt v. Sweet Home Chapter of Communities for a Great Oregon, 515 U.S. 687, 692, 708 (1995).

94. *Babbitt*, 515 U.S. at 691 n.2.

95. 16 U.S.C. § 1532(13).

96. Percival et al., *Environmental Regulation*, 917.

97. 16 U.S.C. § 1538(a)(2).

98. 16 U.S.C. § 1539(a)(1)(B).

99. 16 U.S.C. § 1539(a)(2)(A).

100. 16 U.S.C. § 1539(a)(2)(B)(iv).

101. 63 Fed. Reg. 8859.

102. Matthew E. Rahn, Holly Doremus, and James Diffendorfer, "Species Coverage in Multispecies Habitat Conservation Plans: Where's the Science?," *BioScience* 56 (2006): 616.

103. Rahn, Doremus, and Diffendorfer, "Species Coverage" 616.

104. Rahn, Doremus, and Diffendorfer, "Species Coverage" 616.

105. 50 C.F.R. § 17.22(c).

106. A. M. Trainor et al., "Evaluating the Effectiveness of a Safe Harbor Program for Connecting Wildlife Populations," *Animal Conservation* 16 (2013): 617.

107. 50 C.F.R. § 17.22(d).

108. J. B. Ruhl, Alan Glen, and David Hartman, "A Practical Guide to Habitat Conservation Banking Law and Policy," *Natural Resources and Environment* 20 (2005): 26.

109. Ruhl, Glen, and Hartman, "Habitat Conservation Banking," 27–28.

110. Ruhl, Glen, and Hartman, "Habitat Conservation Banking," 26.

111. "The Ohlone Preserve Conservation Bank," Fletcher Conservation Lands, accessed April 25, 2015, http://www.fclands.com/banks/ohlone-preserve-conservation-bank.

112. 68 Fed. Reg. 24753.

113. 68 Fed. Reg. 24753.

114. 68 Fed. Reg. 24753.

115. Ruhl, Glen, and Hartman, "Habitat Conservation Banking," 31.

116. "Conservation Banking: Incentives for Stewardship," US Fish and Wildlife Service, 2012, http://www.fws.gov/endangered/esa-library/pdf/conservation_banking.pdf.

117. James Salzman and J. B. Ruhl, "Currencies and the Commodification of Environmental Law," *Stanford Law Review* 53 (2000): 611–15.

118. Richard Stroup, "The Endangered Species Act: Making Innocent Species the Enemy," *PERC Policy Series* (1995).

119. See, e.g., Gordon H. Orians, "Endangered at What Level?," *Ecological Applications* 3 (1993): 206–8.

120. Orians, "Endangered at What Level?," 206–8.

121. "A Letter from 5,738 Biologists to the U.S. Senate concerning Science in the Endangered Species Act," Union of Concerned Scientists, 2006, http://www.ucsusa.org/sites/default/files /legacy/assets/documents/scientific_integrity/biologists_letter_ny_nj.pdf.

CHAPTER FOUR

1. "Atlantic Bluefin Tuna," NOAA Fisheries, accessed January 21, 2015, http://www.nmfs.noaa .gov/stories/2011/05/bluefin_tuna.html.

2. "Thunnus thynnus," IUCN Red List of Threatened Species, accessed January 21, 2015, http:// www.iucnredlist.org/details/21860/0.

3. "Summary Record of the Eighth Session of Committee I," fifteenth meeting of the conference of the parties, Doha, Qatar, United Nations Environment Programme, 2010, http://www.cites.org /sites/default/files/eng/cop/15/sum/E15-Com-I-Rec08.pdf.

4. David Jolly and John M. Broder, "U.N. Rejects Export Ban on Atlantic Bluefin Tuna," *New York Times*, March 18, 2010, http://www.nytimes.com/2010/03/19/science/earth/19species.html.

5. "Bluefin Tuna: Eaten Away," *Economist*, March 18, 2010, http://www.economist.com/node /15745509.

6. "Eighth Session of Committee I."

7. Vienna Convention on the Law of Treaties, Article 26.

8. For a more in-depth examination of the international laws affecting wildlife, see Michael Bowman, Peter Davies, and Catherine Redgwell, *Lyster's International Wildlife Law*, 2nd ed. (New York: Cambridge University Press, 2010).

9. Mark Giordano, "The Internationalization of Wildlife and Efforts towards Its Management: A Conceptual Framework and the Historic Record," *Georgetown International Environmental Law Review* 14 (2002): 608–9.

10. The total number of countries in the world can be difficult to determine as it depends on what definition of a country is used. It can also be politically charged, with some countries refusing to recognize the sovereignty of other countries. There are thus roughly 206 countries in the world.

11. CITES, Article II(1).

12. CITES, Article II(1).

13. CITES, Article III.

14. CITES, Article VII(6); "Resolution Conf. 11.15," CITES, accessed January 29, 2015, http:// www.cites.org/eng/res/11/11-15R12.php.

15. CITES, Article II(2).

16. CITES, Article IV.

17. CITES, Article II(3).

18. CITES, Article V.

19. CITES, Article II(4).

20. CITES, Article I(b).

21. CITES, Article I(a).

22. "Resolution Conf. 9.24," CITES, 2010, http://www.cites.org/eng/res/all/09/E09-24R15 .pdf.

23. "Resolution Conf. 9.24."

24. CITES, Articles XV, XVI.

25. CITES, Article X.

26. CITES, Article VII(3).

27. CITES, Article VII(3)(a).

28. CITES, Article VII(3)(b).

29. CITES, Article VII(4).

30. CITES, Article VII(5).

31. Matthew J. Smith et al., "Assessing the Impacts of International Trade on CITES-Listed Species: Current Practices and Opportunities for Scientific Research," *Biological Conservation* 144 (2011): 85–86.

32. Suresh K. Ghimire et al., "Demographic Variation and Population Viability in the Threatened Himalayan Medicinal and Aromatic Herb *Nardostachys grandiflora*: Matrix Modelling of Harvesting Effects in Two Contrasting Habitats," *Journal of Applied Ecology* 45 (2008): 41–51.

33. Matthew J. Smith et al., "Boosting CITES through Research," *Science* 331 (2011): 857.

34. 659 F.2d 168 (D.C. Cir. 1981), *cert. denied*, 454 U.S. 963 (1981).

35. 16 U.S.C. § 1537A(a).

36. 16 U.S.C. § 1537A(c)(2).

37. Whaling Convention, Article V.

38. "News: Whale Faecal Matter Increases Overall Marine Productivity," *Marine Pollution Bulletin* 60 (2010): 2165.

39. Whaling Convention, Article VIII.

40. "Whaling in the Antarctic (Australia v. Japan: New Zealand Intervening)," *International Court of Justice*, 2014, http://www.icj-cij.org/docket/files/148/18136.pdf.

41. "Whaling in the Antarctic," 65, internal quotation marks removed.

42. Jack Simpson, "Japan Kills 30 Minke Whales in First Hunt since ICJ Ruling," *Independent*, June 14, 2014, http://www.independent.co.uk/news/world/asia/japan-kill-30-minke-whales-in-first-hunt-since-icj-ruling-9537063.html.

43. Justin McCurry, "Japan Set to Wade into Diplomatic Row by Bypassing Ban on Whaling," *Guardian*, September 4, 2014, http://www.theguardian.com/world/2014/sep/04/japan-diplomatic-row-bypassing-whaling-ban-antarctic.

44. Convention on Biological Diversity, Article 1.

45. "Decision VII/11," Conference of the Parties, United Nations Environment Programme, 2004, http://www.cbd.int/doc/decisions/cop-07/cop-07-dec-11-en.pdf.

46. Convention on Biological Diversity, Article 6.

47. Convention on Biological Diversity, Article 3.

48. See also the *Corfu Channel* case, ICJ Reports 4, 22, decided by the International Court of Justice in 1949.

49. Convention on Biological Diversity, Article 14(2).

50. Bonn Convention, Article I(1)(e).

51. Bonn Convention, Resolution 5.3.

52. Bonn Convention, Article I(1)(a).

53. Bonn Convention, Article I(1)(a).

54. Bonn Convention, Article IV(1).

55. Bonn Convention, Article XI.

56. Bonn Convention, Article III(5).

57. Bonn Convention, Article III(4).

58. Bonn Convention, Article III(4).

59. "United States—Restrictions on Imports of Tuna," reprinted in *International Legal Materials* 30 (1991): 1594.

60. See, e.g., "US—Restriction on Imports of Tuna," *International Legal Materials* 33 (1994): 839.

61. Convention on Biological Diversity, Annex I.

62. Convention on Biological Diversity, Article 6.

63. Convention on Biological Diversity, Article 8.

64. World Heritage Convention, Preamble.

65. World Heritage Convention, Article 2.

66. "Operational Guidelines for the Implementation of the World Heritage Convention," United Nations Educational, Scientific and Cultural Organisation, 2012, http://whc.unesco.org/archive/opguide12-en.pdf.

67. "Operational Guidelines."

68. World Heritage Convention, Article 6(3).

69. John C. Kunich, "Fiddling Around while the Hot Spots Burn Out," *Georgetown International Environmental Law Review* 14 (2001): 191.

CHAPTER FIVE

1. "Investigative Report and Recommendations for the Kofa Bighorn Sheep Herd," Kofa National Wildlife Refuge and Arizona Game and Fish Department, 2007, http://www.azgfd.gov/pdfs/w_c/bhsheep/Investigative%20Report%2004-17-2007.pdf; Sean Kammer, "Coming to Terms with Wilderness: The Wilderness Act and the Problem of Wildlife Restoration," *Environmental Law* 43 (2013): 88-89.

2. Kammer, "Coming to Terms with Wilderness," 118.

3. "Investigative Report."

4. 16 U.S.C. § 1133(b).

5. 16 U.S.C. § 1133(c).

6. Wilderness Watch v. U.S. Fish and Wildlife Service, 629 F.3d 1024 (9th Cir. 2010).

7. Robert L. Glicksman and George C. Coggins, *Modern Public Land Law in a Nutshell*, 4th ed. (St. Paul, MN: West, 2012), 13-25.

8. Karin P. Sheldon, "Upstream of Peril: The Role of Federal Lands in Addressing the Extinction Crisis," *Pace Environmental Law Review* 24 (2007): 8.

9. Sheldon, "Upstream of Peril," 12.

10. US Constitution, Article IV, Section 3, Clause 2.

11. See, e.g., Kleppe v. New Mexico, 426 U.S. 529 (1976).

12. See, e.g., Utah Power & Light Co. v. U.S., 243 U.S. 389 (1917).

13. U.S. v. Alford, 274 U.S. 264 (1927).

14. Minnesota v. Block, 660 F.2d 1240, 1250 (8th Cir. 1981).

15. 43 U.S.C. § 1061.

16. United States ex rel. Bergen v. Lawrence, 848 F.2d 102 (10th Cir. 1988).

17. Robert B. Keiter, "The National Park System: Visions for Tomorrow," *Natural Resources Journal* 50 (2010): 78.

18. Keiter, "National Park System," 82-83.

19. 16 U.S.C. § 1.

20. 36 C.F.R. § 2.2(b).

21. 16 U.S.C. § 3113.

22. 36 C.F.R. § 2.3.

23. 36 C.F.R. § 2.5.

24. 16 U.S.C. § 1a-7(b).

25. "Yellowstone's Northern Range: Where Nature Takes Its Course," National Park Service, 1995, http://www.nps.gov/yell/planyourvisit/upload/paper.pdf.

26. "Yellowstone's Northern Range."

27. "Yellowstone's Northern Range."

28. Keiter, "National Park System," 84–87.

29. "Management Policies 2006," National Park Service, 2006, http://www.nps.gov/policy /mp2006.pdf.

30. Robert Manning, Yu-Fai Leung, and Megha Budruk, "Research to Support Management of Visitor Carrying Capacity of Boston Harbor Islands," *Northeastern Naturalist* 12 (2005): 201–20.

31. See, e.g., Christopher A. Monz et al., "Sustaining Visitor Use in Protected Areas: Future Opportunities in Recreation Ecology Research Based on the USA Experience," *Environmental Management* 45 (2010): 551–62.

32. "Management Policies 2006."

33. "The Visitor Experience and Resource Protection (VERP) Framework: A Handbook for Planners and Managers," National Park Service, 1997, http://verp.ysnp.gov.tw/sites/default/files /verphandbook.pdf.

34. "Management Policies 2006."

35. Wayde C. Morse, Troy E. Hall, and Linda E. Kruger, "Improving the Integration of Recreation Management with Management of Other Natural Resources by Applying Concepts of Scale from Ecology," *Environmental Management* 43 (2009): 370.

36. Morse, Hall, and Kruger, "Integration of Recreation Management," 378.

37. Rob Ament et al., "An Assessment of Road Impacts on Wildlife Populations in U.S. National Parks," *Environmental Management* 42 (2008): 481.

38. Ament et al., "Road Impacts," 488.

39. Ament et al., "Road Impacts," 486–89.

40. Nathan B. Piekielek and Andrew J. Hansen, "Extent of Fragmentation of Coarse-Scale Habitats in and around U.S. National Parks," *Biological Conservation* 155 (2012): 20.

41. Cory R. Davis and Andrew J. Hansen, "Trajectories in Land Use Change around U.S. National Parks and Challenges and Opportunities for Management," *Ecological Applications* 21 (2011): 3299.

42. Davis and Hansen, "Trajectories in Land Use," 3300.

43. Samuel A. Cushman, Erin L. Landguth, and Curtis H. Flather, "Evaluating the Sufficiency of Protected Lands for Maintaining Wildlife Population Connectivity in the U.S. Northern Rocky Mountains," *Diversity and Distributions* 18 (2012): 873–84.

44. Craig L. Shafer, "The Unspoken Option to Help Safeguard America's National Parks: An Examination of Expanding U.S. National Park Boundaries by Annexing Adjacent Federal Lands," *Columbia Journal of Environmental Law* 35 (2010): 58–60.

45. Davis and Hansen, "Trajectories in Land Use," 3314.

46. "Climate Change: Agencies Should Develop Guidance for Addressing the Effects on Federal Land and Water Resources," US Government Accountability Office, 2007, http://www.gao.gov /assets/270/265207.pdf.

47. Congress granted authority through the Forest Reserve Act of 1891.

48. 16 U.S.C. § 528.

49. 16 U.S.C. § 531(a).

50. 16 U.S.C. § 531(b).

51. 16 U.S.C. § 1604.

52. 16 U.S.C. § 1604(e).

53. 16 U.S.C. § 1604(f)(5).

54. 16 U.S.C. § 1604(g)(3)(B).

55. 16 U.S.C. § 1604(g)(3)(B).

56. Sierra Club v. Marita, 46 F.3d 606 (7th Cir. 1995).

57. *Marita*, 46 F.3d at 621.

58. Citizens for Better Forestry v. USDA, 481 F. Supp. 2d 1059 (N.D. Cal. 2007).

59. Citizens for Better Forestry v. USDA, 632 F. Supp. 2d 968 (N.D. Cal. 2009).

60. Katy Lum, "In One Acorn: The Fate of the 2008 NFMA Planning Rule under the Obama Administration," *Ecology Law Quarterly* 36 (2009): 605–10.

61. Native Ecosystems Council v. Tidwell, 599 F.3d 926 (9th Cir. 2010). *But see* The Lands Council v. McNair, 629 F.3d 1070 (9th Cir. 2010).

62. 77 Fed. Reg. 21162.

63. 36 C.F.R. § 219.19.

64. D. B. Lindenmayer, J. F. Franklin, and J. Fischer, "General Management Principles and a Checklist of Strategies to Guide Forest Biodiversity Conservation," *Biological Conservation* 131 (2006): 434-35.

65. Lindenmayer, Franklin, and Fischer, "General Management Principles," 440-41.

66. 77 Fed. Reg. 21162.

67. D. J. McRae et al., "Comparisons between Wildfire and Forest Harvesting and Their Implications in Forest Management," *Environmental Reviews* 9 (2001): 223-60.

68. Raul Rosenvald and Asko Lõhmus, "For What, When, and Where Is Green-Tree Retention Better Than Clear-Cutting? A Review of the Biodiversity Aspects," *Forest Ecology and Management* 255 (2008): 2.

69. Lucie Jerabkova et al., "A Meta-analysis of the Effects of Clearcut and Variable-Retention Harvesting on Soil Nitrogen Fluxes in Boreal and Temperate Forests," *Canadian Journal of Forest Research* 41 (2011): 1855-58.

70. Glicksman and Coggins, *Modern Public Land Law*, 234-36.

71. Izaak Walton League v. Butz, 522 F.2d 945 (4th Cir. 1975).

72. 16 U.S.C. § 1604(k).

73. 16 U.S.C. § 1604(g)(3)(E).

74. 16 U.S.C. § 1611.

75. 16 U.S.C. § 1604(g)(3)(E)(iv).

76. 16 U.S.C. § 1604(g)(3)(F).

77. "Record of Decision for Amendments to Forest Service and Bureau of Land Management Planning Documents within the Range of the Northern Spotted Owl," US Department of Agriculture and US Department of the Interior, 1994, http://www.reo.gov/riec/newroda.pdf.

78. "Pacific Northwest Research Station, Green-Tree Retention in Harvest Units: Boon or Bust for Biodiversity?," US Department of Agriculture, 2007, http://www.fs.fed.us/pnw/sciencef/scifi96.pdf.

79. Charles B. Halpern et al., "Level and Pattern of Overstory Retention Interact to Shape Long-Term Responses of Understories to Timber Harvest," *Ecological Applications* 22 (2012): 2050-53.

80. Lauren S. Urgenson, Charles B. Halpern, and Paul D. Anderson, "Twelve-Year Responses of Planted and Naturally Regenerated Conifers to Variable-Retention Harvest in the Pacific Northwest, USA," *Canadian Journal of Forest Research* 43 (2013): 49.

81. 66 Fed. Reg. 3244.

82. Monica Voicu, "At a Dead End: The Need for Congressional Direction in the Roadless Area Management Debate," *Ecology Law Quarterly* 37 (2010): 506.

83. Kirsten Rønholt, "Where the Wild Things Were: A Chance to Keep Alaska's Challenge of the Roadless Rule out of the Supreme Court," *Alaska Law Review* 29 (2012): 244.

84. California v. U.S. Department of Agriculture, 575 F.3d 999 (9th Cir. 2009).

85. "Fiscal Year 2014 Budget Overview," US Forest Service, 2013, http://www.fs.fed.us/aboutus/budget/2014/Overview%20%28COMPLETE%29.pdf.

86. Stephanie A. Grayzeck-Souter et al., "Interpreting Federal Policy at the Local Level: The Wildland-Urban Interface Concept in Wildfire Protection Planning in the Eastern United States," *International Journal of Wildland Fire* 18 (2009): 279; "National Forests on the Edge: Development Pressures on America's National Forests and Grasslands," US Forest Service, 2007, http://www.fs.fed.us/openspace/fote/GTR728.pdf.

87. 16 U.S.C. § 6501.

88. 16 U.S.C. § 6511(3).

89. Rutherford V. Platt, Thomas T. Veblen, and Rosemary L. Sherriff, "Are Wildfire Mitigation

and Restoration of Historic Forest Structure Compatible? A Spatial Modeling Assessment," *Annals of the Association of American Geographers* 96 (2006): 456.

90. Platt, Veblen, and Sherriff, "Wildfire Mitigation," 456.

91. Platt, Veblen, and Sherriff, "Wildfire Mitigation," 458-66.

92. Tania Schoennagel, Thomas T. Veblen, and William H. Romme, "The Interaction of Fire, Fuels, and Climate across Rocky Mountain Forests," *BioScience* 54 (2004): 663-66.

93. McRae et al., "Wildfire and Forest Harvesting," 249-51.

94. Keri Pidgen and Azim U. Mallik, "Ecology of Compounding Disturbances: The Effects of Prescribed Burning after Clearcutting," *Ecosystems* 16 (2013): 178-80.

95. "Statement of Tom Thompson, Deputy Chief, National Forest System, Before the Subcommittee on Public Lands and Forests, United States Senate: Hearing on the Forest Service Grazing Program," US Department of Agriculture, 2004, http://www.usda.gov/documents/Thompson .grazing.062304.doc.

96. 43 U.S.C. § 1732(a).

97. 43 U.S.C. § 1702(c), (h).

98. 43 U.S.C. § 1702(c).

99. 43 U.S.C. § 1712.

100. 43 U.S.C. § 1712(c)(2).

101. 43 U.S.C. § 1702(a).

102. Glicksman and Coggins, *Modern Public Land Law*, 272-78.

103. 16 U.S.C. § 1333(a).

104. 16 U.S.C. § 1333(a).

105. 16 U.S.C. § 1333(b)(2), (e).

106. 16 U.S.C. § 1332(f)(2).

107. 16 U.S.C. § 1334.

108. 16 U.S.C. § 1338.

109. Marya Torrez, "Cows, Congress, and Climate Change: Authority and Responsibility for Federal Agencies to End Grazing on Public Lands," *Vermont Journal of Environmental Law* 14 (2012): 8-9.

110. 43 U.S.C. § 1751(a).

111. 43 U.S.C. § 1752(a), (c).

112. 43 U.S.C. § 1752(e).

113. D. Terrance Booth et al., "Willow Cover as a Stream-Recovery Indicator under a Conservation Grazing Plan," *Ecological Indicators* 18 (2012): 512-19.

114. "Disturbance and Climate Change in United States/Mexico Borderland Plant Communities: A State-of-the-Knowledge Review," US Forest Service, 2000, http://www.fs.fed.us/rm/pubs /rmrs_gtr050.pdf.

115. 43 C.F.R. §4180.2.

116. 43 C.F.R. § 4180.2(c).

117. "Cheatgrass: The Invader That Won the West," Bureau of Land Management, 1996, http:// www.icbemp.gov/science/pellant.pdf.

118. Felicity Barringer, "Out West, 'Black Fingers of Death' Offer Hope against an Invader," *New York Times*, July 30, 2012, http://www.nytimes.com/2012/07/31/science/earth/a-fungus-emerges -as-weapon-against-cheatgrass.html.

119. Danielle R. Courtois, Barry L. Perryman, and Hussein S. Hussein, "Vegetation Change after 65 Years of Grazing and Grazing Exclusion," *Journal of Range Management* 57 (2004): 574-82.

120. Louis Provencher et al., "Comparing Alternative Management Strategies of Fire, Grazing, and Weed Control Using Spatial Modeling," *Ecological Modelling* 209 (2007): 259-61.

121. Legislative authority for Wildlife Services comes from the Animal Damage Control Act, 7 U.S.C. §§ 426-426c.

122. "Animals Taken by Wildlife Services—FY 2011," Wildlife Services, 2011, http://www.aphis

.usda.gov/wildlife_damage/prog_data/2011_prog_data/PDR_G/Basic_Tables_PDR_G/Table%20G
_ShortReport.pdf.

123. Laura R. Prugh et al., "The Rise of the Mesopredator," *BioScience* 59 (2009): 788–89.

124. Kim M. Berger, "Carnivore-Livestock Conflicts: Effects of Subsidized Predator Control and Economic Correlates on the Sheep Industry," *Conservation Biology* 20 (2006): 751–61.

125. Tyler B. Muhly and Marco Musiani, "Livestock Depredation by Wolves and the Ranching Economy in the Northwestern U.S.," *Ecological Economics* 68 (2009): 2445.

126. "Livestock Grazing: Federal Expenditures and Receipts Vary, Depending on the Agency and the Purpose of the Fee Charged," US Government Accountability Office, 2005, http://www .gao.gov/new.items/d05869.pdf.

127. Robert L. Fischman, "The National Wildlife Refuge System and the Hallmarks of Modern Organic Legislation," *Ecology Law Quarterly* 29 (2002): 458–59.

128. 16 U.S.C. § 668dd(a)(2).

129. 16 U.S.C. § 668dd(a)(4).

130. Fischman, "National Wildlife Refuge System," 526.

131. 16 U.S.C. § 668ee(2).

132. 16 U.S.C. § 668ee(1).

133. 16 U.S.C. § 668dd(c).

134. 50 C.F.R. § 29.1.

135. 16 U.S.C. § 668dd(c).

136. 16 U.S.C. § 668dd(e).

137. "Strategic Habitat Conservation Handbook: A Guide to Implementing the Technical Elements of Strategic Habitat Conservation," US Fish and Wildlife Service, 2008, http://www.fws.gov /landscape-conservation/pdf/SHCHandbook.pdf.

138. "Implementing the SHC Framework," US Fish and Wildlife Service, http://www.fws.gov /science/doc/shcframework.pdf.

139. Robert L. Fischman and Robert S. Adamcik, "Beyond Trust Species: The Conservation Potential of the National Wildlife Refuge System in the Wake of Climate Change," *Natural Resources Journal* 51 (2011): 7–8.

140. See, e.g., Bastian Steudel et al., "Biodiversity Effects on Ecosystem Functioning Change along Environmental Stress Gradients," *Ecology Letters* 15 (2012): 1397–1405.

141. 16 U.S.C. § 1131(c).

142. 16 U.S.C. § 1131(c).

143. 16 U.S.C. § 1131(b).

144. 16 U.S.C. § 1133(b).

145. 16 U.S.C. § 1133(b).

146. 16 U.S.C. § 1133(c).

147. 16 U.S.C. § 1133(d)(4).

148. 16 U.S.C. § 1133(d)(5).

149. 16 U.S.C. § 1134(a).

150. 16 U.S.C. § 1133(d)(1).

151. Sean Kammer, "Coming to Terms with Wilderness: The Wilderness Act and the Problem of Wildlife Restoration," *Environmental Law* 43 (2013): 84–87.

152. Kammer, "Coming to Terms with Wilderness," 87–91.

153. 42 U.S.C. § 4332(2)(c).

154. 40 C.F.R. § 1502.14.

155. 40 C.F.R. § 1502.14.

156. 40 C.F.R. § 1502.5.

157. 40 C.F.R. § 1502.22.

158. 40 C.F.R. § 1502.9(c)(iii).

159. 40 C.F.R. § 1508.8(a).

160. 40 C.F.R. § 1508.8(b).

161. 40 C.F.R. § 1508.7.

162. 40 C.F.R. § 1508.8(b).

163. Courtney A. Schultz, "History of the Cumulative Effects Analysis Requirement under NEPA and Its Interpretation in U.S. Forest Service Case Law," *Journal of Environmental Law and Litigation* 27 (2012): 134.

164. Schultz, "Cumulative Effects," 158–60.

165. "Canada Lynx Conservation Assessment and Strategy," US Forest Service, 2000, http://nctc.fws.gov/Pubs5/Lynx_consassess_2000.pdf.

166. "Draft NEPA Guidance on Consideration of the Effects of Climate Change and Greenhouse Gas Emissions," Council on Environmental Quality, 2010, http://www.whitehouse.gov/sites/default/files/microsites/ceq/20100218-nepa-consideration-effects-ghg-draft-guidance.pdf.

167. 40 C.F.R. § 1508.9.

168. See, e.g., Thomas v. Peterson, 753 F.2d 754 (9th Cir. 1985); 40 C.F.R. § 1508.25; 40 C.F.R. § 1508.27(b)(7).

169. 40 C.F.R. § 1508.4.

170. "Forest Service: Use of Categorical Exclusions for Vegetation Management Projects, Calendar Years 2003 through 2005," US Government Accountability Office, 2006, http://www.gao.gov/assets/260/252544.pdf.

171. Strycker's Bay Neighborhood Council, Inc. v. Karlen, 444 U.S. 223 (1980).

172. 40 C.F.R. § 1501.7.

173. 40 C.F.R. § 1503.1.

CHAPTER SIX

1. Seong-Hoon Cho et al., "The Impact of an Urban Growth Boundary on Land Development in Knox County, Tennessee: A Comparison of Two-Stage Probit Least Squares and Multilayer Neural Network Models," *Journal of Agricultural and Applied Economics* 39 (2007): 701.

2. Cho et al., "Impact of an Urban Growth Boundary," 715–16.

3. Seong-Hoon Cho, Neelam Poudyal, and Dayton M. Lambert, "Estimating Spatially Varying Effects of Urban Growth Boundaries on Land Development and Land Value," *Land Use Policy* 25 (2008): 328.

4. Lin Robinson, Joshua P. Newell, and John M. Marzluff, "Twenty-Five Years of Sprawl in the Seattle Region: Growth Management Responses and Implications for Conservation," *Landscape and Urban Planning* 71 (2005): 67–71.

5. John I. Carruthers, "Growth at the Fringe: The Influence of Political Fragmentation in United States Metropolitan Areas," *Papers in Regional Science* 82 (2003): 477–79.

6. "Major Uses of Land in the United States, 2007," US Department of Agriculture Economic Research Service, 2011, http://www.ers.usda.gov/media/188404/eib89_2_.pdf.

7. "Major Uses of Land."

8. "Major Uses of Land."

9. "Major Uses of Land."

10. "Major Uses of Land."

11. Kenneth M. Sylvester et al., "Land Transitions in the American Plains: Multilevel Modeling of Drivers of Grassland Conversion (1956–2006)," *Agriculture, Ecosystems and Environment* 168 (2013): 13–14.

12. "Agricultural Resources and Environmental Indicators," US Department of Agriculture Economic Research Service, 1997, http://www.ers.usda.gov/media/871227/ah712.pdf.

13. S. M. Munson, W. K. Lauenroth, and I. C. Burke, "Soil Carbon and Nitrogen Recovery on Semiarid Conservation Reserve Program Lands," *Journal of Arid Environments* 79 (2012): 25–26.

14. J. B. Ruhl, "Farms, Their Environmental Harms, and Environmental Law," *Ecology Law Quarterly* 27 (2000): 265–71.

15. 7 C.F.R. § 1410.6.

16. 7 C.F.R. § 1410.31(b).

17. Kevin C. Rigdon, "Stop the Planting! The 1985 Farm Bill, Conservation Compliance, and America's Agricultural Conservation Failure," *Drake Journal of Agricultural Law* 16 (2011): 498–99; "Agricultural Conservation: USDA Needs to Better Ensure Protection of Highly Erodible Cropland and Wetlands," US General Accounting Office, 2003, http://www.gao.gov/new.items/d03418.pdf.

18. "USDA Announces Results for 45th Conservation Reserve Program General Sign-Up," US Department of Agriculture, 2013, http://www.fsa.usda.gov/FSA/newsReleases?area=newsroom& subject=landing&topic=ner&newstype=newsrel&type=detail&item=nr_20130722_rel_0149.html.

19. James R. Herkert, "Response of Bird Populations to Farmland Set-Aside Programs," *Conservation Biology* 23 (2009): 1038–40.

20. Joseph A. Veech, "A Comparison of Landscapes Occupied by Increasing and Decreasing Populations of Grassland Birds," *Conservation Biology* 20 (2006): 1429–31.

21. Peer-reviewed studies were compiled and discussed in Jonathan B. Haufler, ed., *Fish and Wildlife Benefits of Farm Bill Conservation Programs: 2000–2005 Update, Wildlife Society Technical Review* 05-2 (Bethesda, MD: Wildlife Society, 2005).

22. Seth M. Munson and William K. Lauenroth, "Plant Community Recovery Following Restoration in Semiarid Grasslands," *Restoration Ecology* 20 (2012): 661.

23. 42 U.S.C. § 7545(o)(2)(B).

24. Dina Cappiello and Matt Apuzzo, "The Secret, Dirty Cost of Obama's Green Power Push," *Associated Press*, November 12, 2013, http://bigstory.ap.org/article/story-removed.

25. Gervasio Piñeiro et al., "Set-Asides Can Be Better Climate Investment Than Corn Ethanol," *Ecological Applications* 19 (2009): 280–81.

26. Ilya Gelfand et al., "Carbon Debt of Conservation Reserve Program (CRP) Grasslands Converted to Bioenergy Production," *Proceedings of the National Academy of Sciences* 108 (2011): 13864–69.

27. David Pimentel, "Environmental and Economic Costs of the Application of Pesticides Primarily in the United States," *Environment, Development and Sustainability* 7 (2005): 229–30.

28. Pimentel, "Application of Pesticides," 242–45.

29. 7 U.S.C. § 136a(c)(2); 40 C.F.R. part 158.

30. 7 U.S.C. § 136a(c)(5).

31. 7 U.S.C. § 136(bb).

32. Mary J. Angelo, "The Killing Fields: Reducing the Casualties in the Battle between U.S. Species Protection Law and U.S. Pesticide Law," *Harvard Environmental Law Review* 32 (2008): 110.

33. 40 C.F.R. § 158.230.

34. 7 U.S.C. §§ 136a(c)(1)(C), 136j(a)(2)(G).

35. 7 U.S.C. § 136a(d).

36. 7 U.S.C. § 136i-1.

37. 15 U.S.C. § 2607(b).

38. 15 U.S.C. § 2603(a).

39. 15 U.S.C. § 2603(b)(2)(A).

40. 15 U.S.C. § 2605(a).

41. 15 U.S.C. § 2604(b)(2)(B)(ii).

42. Christopher De Rosa, "The USEPA and the Toxic Substances Control Act (TSCA) of 1976: The Promise, the Reality, and the Reason(s) Why," *Human and Ecological Risk Assessment* 16 (2010): 1230.

43. Michael L. McKinney, "Urbanization, Biodiversity, and Conservation," *BioScience* 52 (2002): 883.

44. Ingo Kowarik, "Novel Urban Ecosystems, Biodiversity, and Conservation," *Environmental Pollution* 159 (2011): 1975–76.

45. McKinney, "Urbanization," 884.

46. Nancy B. Grimm et al., "Global Change and the Ecology of Cities," *Science* 319 (2008): 759.

47. Grimm et al., "Ecology of Cities," 759.

48. Eyal Shochat et al., "From Patterns to Emerging Processes in Mechanistic Urban Ecology," *Trends in Ecology and Evolution* 21 (2006): 186–91.

49. See, e.g., Colorado Revised Statutes 31-23-206 requiring comprehensive plans for cities and towns, and Colorado Revised Statutes 30-28-106 doing the same for counties.

50. See, e.g., "Arizona's Wildlife Linkages Assessment," Arizona Department of Transportation, 2006, http://azdot.gov/docs/planning/arizona_wildlife_linkages_assessment.pdf.

51. Jared G. Underwood, Joyce Francis, and Leah R. Gerber, "Incorporating Biodiversity Conservation and Recreational Wildlife Values into Smart Growth Land Use Planning," *Landscape and Urban Planning* 100 (2011): 139–40.

52. Jinki Kim and Xiaolu Zhou, "Landscape Structure, Zoning Ordinance, and Topography in Hillside Residential Neighborhoods: A Case Study of Morgantown, WV," *Landscape and Urban Planning* 108 (2012): 28–38.

53. James R. Miller et al., "Biodiversity Conservation in Local Planning," *Conservation Biology* 23 (2009): 56–60; David L. Stokes et al., "Local Land-Use Planning to Conserve Biodiversity: Planners' Perspectives on What Works," *Conservation Biology* 24 (2010): 454–56.

54. John R. Nolon and Patricia E. Salkin, *Land Use in a Nut Shell* (St. Paul, MN: Thomson/West, 2006), 63–64.

55. 272 U.S. 365 (1926).

56. Brookfield, WI, Municipal Code 17.96.010–17.96.070; Nolon and Salkin, *Land Use*, 218–22.

57. Abigail M. York and Darla K. Munroe, "Urban Encroachment, Forest Regrowth and Land-Use Institutions: Does Zoning Matter?," *Land Use Policy* 27 (2010): 471–79.

58. McKinney, "Urbanization," 883.

59. Timothy Carter, "Developing Conservation Subdivisions: Ecological Constraints, Regulatory Barriers, and Market Incentives," *Landscape and Urban Planning* 92 (2009): 118.

60. Z. Aslıgül Göçmen, "Barriers to Successful Implementation of Conservation Subdivision Design: A Closer Look at Land Use Regulations and Subdivision Permitting Process," *Landscape and Urban Planning* 110 (2013): 124.

61. Sarah E. Reed, Jodi A. Hilty, and David M. Theobald, "Guidelines and Incentives for Conservation Development in Local Land-Use Regulations," *Conservation Biology* 28 (2013): 258–68.

62. Charlotte E. Gonzalez-Abraham et al., "Patterns of Housing and Habitat Loss from 1937 to 1999 in Northern Wisconsin, USA," *Ecological Applications* 17 (2007): 2011–23.

63. Göçmen, "Conservation Subdivision Design," 123–33.

64. Jeffrey C. Milder, "A Framework for Understanding Conservation Development and Its Ecological Implications," *BioScience* 57 (2007): 757–68.

65. Buffy A. Lenth, Richard L. Knight, and Wendell C. Gilgert, "Conservation Value of Clustered Housing Developments," *Conservation Biology* 20 (2006): 1445–56.

66. Eric A. Odell and Richard L. Knight, "Songbird and Medium-Sized Mammal Communities Associated with Exurban Development in Pitkin County, Colorado," *Conservation Biology* 15 (2001): 1148–49; Michael J. Glennon and Heidi E. Krestser, "Size of the Ecological Effect Zone Associated with Exurban Development in the Adirondack Park, NY," *Landscape and Urban Planning* 112 (2013): 10–17.

67. Lenth, Knight, and Gilgert, "Clustered Housing Developments," 1449.

68. Göçmen, "Conservation Subdivision Design," 129–30.

69. Randall Arendt, "Linked Landscapes: Creating Greenway Corridors through Conservation Subdivision Design Strategies in the Northeastern and Central United States," *Landscape and Urban Planning* 68 (2004): 241–42.

70. Mark Hostetler, Will Allen, and Colin Meurk, "Conserving Urban Biodiversity? Creating Green Infrastructure Is Only the First Step," *Landscape and Urban Planning* 100 (2011): 369–71.

71. Robert C. Freeman and Kathleen P. Bell, "Conservation versus Cluster Subdivisions and Implications for Habitat Connectivity," *Landscape and Urban Planning* 101 (2011): 30.

72. James R. Miller et al., "Biodiversity Conservation in Local Planning," *Conservation Biology* 23 (2009): 62.

73. M. Soga and S. Koike, "Mapping the Potential Extinction Debt of Butterflies in a Modern City: Implications for Conservation Priorities in Urban Landscapes," *Animal Conservation* 16 (2013): 1–11.

74. 42 U.S.C. § 6903(5), (27).

75. 42 U.S.C. § 9604(a)(1)(A).

76. 42 U.S.C. § 9601(14)(F).

77. 42 U.S.C. § 9605.

78. 42 U.S.C. § 9605(a)(8)(A).

79. "Superfund: EPA Should Take Steps to Improve Its Management of Alternatives to Placing Sites on the National Priorities List," US Government Accountability Office, 2013, http://www.gao .gov/assets/660/653646.pdf.

80. 42 U.S.C. § 9613(h).

81. 42 U.S.C. § 9607(a).

82. Burlington Northern & Santa Fe Railway v. U.S., 556 U.S. 599 (2009).

83. 42 U.S.C. § 9607(o).

84. 42 U.S.C. § 9607(p).

85. 42 U.S.C. § 9601(35).

86. 42 U.S.C. § 9601(35).

87. Mira Kattwinkel, Robert Biedermann, and Michael Kleyer, "Temporary Conservation for Urban Biodiversity," *Biological Conservation* 144 (2011): 2339–42.

88. Keith H. Hirokawa, "Sustainability and the Urban Forest: An Ecosystem Services Perspective," *Natural Resources Journal* 51 (2011): 238–39.

89. 26 U.S.C. § 170(h); 26 C.F.R. § 1.170A-14.

90. 26 U.S.C. § 170(h)(4).

91. Amy Pocewicz et al., "Effectiveness of Conservation Easements for Reducing Development and Maintaining Biodiversity in Sagebrush Ecosystems," *Biological Conservation* 144 (2011): 567–74.

92. "National Conservation Easement Database," US Endowment for Forestry and Communities, accessed January 24, 2015, http://www.conservationeasement.us.

93. Zachary Bray, "Reconciling Development and Natural Beauty: The Promise and Dilemma of Conservation Easements," *Harvard Environmental Law Review* 34 (2010): 136–48.

94. See Walter and Otero County Land Trust, No. 05-CV-96 (Colo. Jud. Dist. Ct. June 21, 2005).

95. Ann T. Schwing, "Perpetuity Is Forever, Almost Always: Why It Is Wrong to Promote Amendment and Termination of Perpetual Conservation Easements," *Harvard Environmental Law Review* 37 (2013): 217–46; see also Salzburg v. Dowd, No. CV-2008-0079 (Wyo. Jud. Dist. Feb. 10, 2010), for a court that did not terminate a conservation easement.

96. Bray, "Reconciling Development and Natural Beauty," 140–45.

97. Adena R. Rissman et al., "Land Management Restrictions and Options for Change in Perpetual Conservation Easements," *Environmental Management* 52 (2013): 283–84.

98. Jean-David Gerber and Adena R. Rissman, "Land-Conservation Strategies: The Dynamic Relationship between Acquisition and Land-Use Planning," *Environment and Planning A* 44 (2012): 1851.

CHAPTER SEVEN

1. Jonathan H. Adler, "Fables of the Cuyahoga: Reconstructing a History of Environmental Protection," *Fordham Environmental Law Journal* 14 (2002): 99–100.

2. A more detailed introduction to the Clean Water Act may be found in Mark A. Ryan, ed., *The Clean Water Act Handbook*, 3rd ed. (Chicago: American Bar Association, 2011).

3. James Salzman and Barton H. Thompson Jr., *Environmental Law and Policy*, 3rd ed. (New York: Foundation Press, 2010), 151–52.

4. 33 U.S.C. § 1362(14).

5. 33 U.S.C. § 1311(a).

6. 33 U.S.C. § 1362(12).

7. 33 U.S.C. § 1362(5).

8. 33 U.S.C. § 1362(6).

9. See, e.g., D. E. Rice v. Harken Exploration Co., 250 F.3d 264 (5th Cir. 2001); Village of Oconomowoc Lake v. Dayton Hudson Corp., 24 F.3d 962 (7th Cir. 1994).

10. 33 U.S.C. § 1342(a).

11. 33 U.S.C. § 1365.

12. "Clean Watersheds Needs Survey 2008: Report to Congress," Environmental Protection Agency, 2008, http://water.epa.gov/scitech/datait/databases/cwns/upload/cwns2008rtc.pdf.

13. 33 U.S.C. § 1311(b)(1)(B).

14. 33 U.S.C. § 1311(b).

15. E. I. du Pont de Nemours & Co. v. Train, 430 U.S. 112 (1977).

16. 40 C.F.R. § 125.3.

17. 33 U.S.C. § 1362(13).

18. 40 C.F.R. § 401.15.

19. 33 U.S.C. § 1314(a)(4); 40 C.F.R. § 401.16.

20. 33 U.S.C. § 1316(b)(1)(B).

21. 40 C.F.R. § 403.3(i).

22. 40 C.F.R. § 403.5(a)(1).

23. "Enforcement Alert: EPA Targets Clean Water Act Violations at Livestock Feeding Operations," Environmental Protection Agency, 2009, http://nepis.epa.gov/Exe/ZyPDF.cgi/P10039VB.PDF?Dockey=P10039VB.PDF.

24. 33 U.S.C. § 1362(14).

25. 40 C.F.R. § 122.23(4).

26. See, e.g., 40 C.F.R. § 412.31.

27. Terence J. Centner, "Addressing Water Contamination from Concentrated Animal Feeding Operations," *Land Use Policy* 28 (2011): 708.

28. 40 C.F.R. § 122.23(e).

29. 33 U.S.C. § 1342(p).

30. 33 U.S.C. § 1342(p).

31. 40 C.F.R. § 122.26(b)(14).

32. 40 C.F.R. § 122.26(g).

33. 40 C.F.R. § 122.26(a)(3).

34. 40 C.F.R. § 122.26(b)(15).

35. Gabriel Eckstein and George W. Sherk, "Alternative Strategies for Addressing the Presence and Effects of Pharmaceutical and Personal Care Products in Fresh Water Resources," *University of Denver Law Review* 15 (2012): 371–73.

36. "Water-Quality Data for Pharmaceuticals, Hormones, and Other Organic Wastewater Contaminants in U.S. Streams, 1999–2000," US Geological Survey, 2002, http://toxics.usgs.gov/pubs/OFR-02-94/.

37. Eckstein and Sherk, "Pharmaceutical and Personal Care Products," 433.

38. Ryan J. Albrecht, "Pharmaceuticals in the Environment: Looking to Green Governance for a Remedy," *Journal of Energy and Environmental Law* 3 (2012): 186.

39. Albrecht, "Pharmaceuticals in the Environment," 186–87.

40. Marlo K. Sellin et al., "Estrogenic Compounds Downstream from Three Small Cities in

Eastern Nebraska: Occurrence and Biological Effect," *Journal of the American Water Resources Association* 45 (2009): 14–21.

41. Alan M. Vajda et al., "Reproductive Disruption in Fish Downstream from an Estrogenic Wastewater Effluent," *Environmental Science and Technology* 42 (2008): 3407–14.

42. 42 U.S.C. §§ 300h-300h-8.

43. 33 U.S.C. § 1362(14).

44. 33 U.S.C. § 1329.

45. 33 U.S.C. § 1329(b)(4).

46. Salzman and Thompson, *Environmental Law and Policy*, 161–63.

47. Thomas C. Brown and Pamela Froemke, "Nationwide Assessment of Nonpoint Source Threats to Water Quality," *BioScience* 62 (2012): 137–39.

48. Brown and Froemke, "Nonpoint Source Threats," 139.

49. Brown and Froemke, "Nonpoint Source Threats," 141–44.

50. Karl E. Havens and Dale E. Gawlik, "Lake Okeechobee Conceptual Ecological Model," *Wetlands* 25 (2005): 908–25.

51. See, e.g., Louis J. Guillette Jr. et al., "Alligators and Endocrine Disrupting Contaminants: A Current Perspective," *American Zoologist* 40 (2000): 438–52.

52. 33 U.S.C. § 1313.

53. 33 U.S.C. § 1251(a)(2).

54. 33 U.S.C. § 1313(c)(1).

55. 40 C.F.R. §§ 130.2(j), 130.7(b)(4).

56. 33 U.S.C. § 1313(d)

57. 33 U.S.C. § 1312(a).

58. Scott v. City of Hammond, 741 F.2d 992 (7th Cir. 1984).

59. Pronsolino v. Nastri, 291 F.3d 1123 (9th Cir. 2002).

60. "Draft Handbook for Developing Watershed TMDLs," Environmental Protection Agency, 2008, http://www.epa.gov/owow/tmdl/pdf/draft_handbook.pdf.

61. "National Water Quality Inventory: Report to Congress, 2004 Reporting Cycle," Environmental Protection Agency, 2009, http://water.epa.gov/lawsregs/guidance/cwa/305b/upload/2009_01_22_305b_2004report_2004_305Breport.pdf.

62. Sylvain Dolédec and Bernhard Statzner, "Responses of Freshwater Biota to Human Disturbances: Contribution of *J-NABS* to Developments in Ecological Integrity Assessments," *Journal of the North American Benthological Society* 29 (2010): 286–87.

63. Margaret A. Palmer and Catherine M. Febria, "The Heartbeat of Ecosystems," *Science* 336 (2012): 1393–94.

64. See, e.g., Guy Woodward et al., "Continental-Scale Effects of Nutrient Pollution on Stream Ecosystem Functioning," *Science* 336 (2012): 1438–40.

65. "Final Water Quality Trading Policy," Environmental Protection Agency, 2003, http://water.epa.gov/type/watersheds/trading/finalpolicy2003.cfm.

66. Marc O. Ribaudo and Jessica Gottlieb, "Point-Nonpoint Trading—Can It Work?," *Journal of the American Water Resources Association* 47 (2011): 5–14.

67. See, e.g., Utah v. Marsh, 740 F.2d 799 (10th Cir. 1984).

68. 33 U.S.C. § 1251(g).

69. Katherine A. Abend, "Avoiding Water-Intensive Energy Production: How to Keep the Water Running and the Lights On," *Environmental Law Reporter News and Analysis* 41 (2011): 11028–29.

70. "Estimated Use of Water in the United States in 2005," *US Geological Survey*, 2009, http://pubs.usgs.gov/circ/1344/pdf/c1344.pdf.

71. Stephen C. Newbold and Rich Iovanna, "Ecological Effects of Density-Independent Mortality: Application to Cooling-Water Withdrawals," *Ecological Applications* 17 (2007): 390–91.

72. 33 U.S.C. § 1326(b).

73. Riverkeeper Inc. v. Environmental Protection Agency, 475 F.3d 83 (2nd Cir. 2007).

74. Newbold and Iovanna, "Ecological Effects of Density-Independent Mortality," 390–406.

75. Stephen C. Newbold and Rich Iovanna, "Population Level Impacts of Cooling Water With-drawals on Harvested Fish Stocks," *Environmental Science and Technology* 41 (2007): 2111–13.

76. 33 U.S.C. § 1362(6).

77. 33 U.S.C. § 1326(a).

78. Y. Kanno and J. C. Vokoun, "Evaluating Effects of Water Withdrawals and Impoundments on Fish Assemblages in Southern New England Streams, USA," *Fisheries Management and Ecology* 17 (2010): 272–73.

79. Mary C. Freeman and Paula A. Marcinek, "Fish Assemblage Responses to Water Withdraw-als and Water Supply Reservoirs in Piedmont Streams," *Environmental Management* 38 (2006): 441–42.

80. Annika W. Walters and David M. Post, "How Low Can You Go? Impacts of a Low-Flow Dis-turbance on Aquatic Insect Communities," *Ecological Applications* 21 (2011): 163–74.

81. "Dams and Development: A New Framework for Decision-Making," World Commission on Dams, 2000, http://www.internationalrivers.org/files/attached-files/world_commission_on_dams_final_report.pdf.

82. 16 U.S.C. § 797.

83. Ann E. Drobot, "Transitioning to a Sustainable Energy Economy: The Call for National Co-operative Watershed Planning," *Environmental Law* 41 (2011): 743–46.

84. 16 U.S.C. § 797(e).

85. 18 C.F.R. § 16.13(a).

86. Paciencia S. Young, Joseph J. Cech Jr., and Lisa C. Thompson, "Hydropower-Related Pulsed-Flow Impacts on Stream Fishes: A Brief Review, Conceptual Model, Knowledge Gaps, and Research Needs," *Reviews in Fish Biology and Fisheries* 21 (2011): 714–15.

87. Young, Cech, and Thompson, "Hydropower-Related Pulsed-Flow," 727.

88. See, e.g., H. Lundqvist et al., "Upstream Passage Problems for Wild Atlantic Salmon (*Salmo salar* L.) in a Regulated River and Its Effect on the Population," *Hydrobiologia* 602 (2008): 123.

89. Gilson L. Volpato et al., "Fish Ladders Select Fish Traits on Migration—Still a Growing Problem for Natural Fish Populations," *Marine and Freshwater Behaviour and Physiology* 42 (2009): 307–13.

90. 16 U.S.C. §§ 1271–1287.

91. 16 U.S.C. § 1271.

92. 16 U.S.C. §§ 1131–1136.

93. 16 U.S.C. § 797c.

94. 33 U.S.C. § 1362(12).

95. See, e.g., National Wildlife Federation v. Gorsuch, 693 F.2d 156 (D.C. Cir. 1982).

96. 40 C.F.R. § 122.3(i).

97. 73 Fed. Reg. 33699–700.

98. 33 U.S.C. § 1341(a).

99. PUD No. 1 of Jefferson County v. Washington Department of Ecology, 511 U.S. 700 (1994).

100. 42 U.S.C. § 300h(d)(1).

101. 42 U.S.C. § 300h(d)(1)(B)(ii).

102. 80 Fed. Reg. 16128.

103. Stephen G. Osborn et al., "Methane Contamination of Drinking Water Accompanying Gas-Well Drilling and Hydraulic Fracturing," *Proceedings of the National Academy of Sciences* 108 (2011): 8173.

104. "An Evaluation of Fracture Growth and Gas/Fluid Migration as Horizontal Marcellus Shale Gas Wells Are Hydraulically Fractured in Greene County, Pennsylvania," US Department of Energy, National Energy Technology Laboratory, 2014, http://www.netl.doe.gov/File%20Library/Research/onsite%20research/publications/NETL-TRS-3-2014_Greene-County-Site_20140915_1_1.pdf.

105. Sally Entrekin et al., "Rapid Expansion of Natural Gas Development Poses a Threat to Surface Waters," *Frontiers in Ecology and the Environment* 9 (2011): 507–9.

106. Jennifer S. Harkness et al., "Iodide, Bromide, and Ammonium in Hydraulic Fracturing and Oil and Gas Wastewaters: Environmental Implications," *Environmental Science and Technology* 49 (2015): 1955–63.

CHAPTER EIGHT

1. J. B. Ruhl and James Salzman, "Gaming the Past: The Theory and Practice of Historic Baselines in the Administrative State," *Vanderbilt Law Review* 64 (2011): 2–5.

2. "Memorandum of Agreement between the Environmental Protection Agency and Department of the Army concerning the Determination of Mitigation under the Clean Water Act Section 404(b)(1) Guidelines," 55 Fed. Reg. 9210.

3. "Status and Trends of Wetlands in the Conterminous United States 2004 to 2009, Report to Congress," US Fish and Wildlife Service, 2011, http://www.fws.gov/wetlands/Documents/Status -and-Trends-of-Wetlands-in-the-Conterminous-United-States-2004-to-2009.pdf.

4. 33 U.S.C. § 1362(12).

5. 33 U.S.C. § 1362(7).

6. 33 C.F.R. § 328.3(a).

7. 33 C.F.R. § 328.3(b); 40 C.F.R. §§ 122.2, 230.3(t).

8. 474 U.S. 121 (1985).

9. 531 U.S. 159 (2001).

10. 547 U.S. 715 (2006).

11. U.S. v. Rapanos, 376 F.3d 629 (6th Cir. 2004).

12. Carabell v. U.S. Army Corps of Engineers, 391 F.3d 704 (6th Cir. 2004).

13. Scott G. Leibowitz et al., "Non-navigable Streams and Adjacent Wetlands: Addressing Science Needs Following the Supreme Court's *Rapanos* Decision," *Frontiers in Ecology and the Environment* 6 (2008): 370.

14. Royal C. Gardner, *Lawyers, Swamps, and Money: U.S. Wetland Law, Policy, and Politics,* (Washington, DC: Island Press, 2011), provides a more in-depth discussion of the statutes and regulations protecting wetlands.

15. 33 U.S.C. § 1344(a).

16. 33 C.F.R. § 323.2; 40 C.F.R. § 232.2.

17. 33 C.F.R. § 323.2; 40 C.F.R. § 232.2.

18. National Mining Association v. U.S. Army Corps of Engineers, 145 F.3d 1399 (D.C. Cir. 1998).

19. 33 C.F.R. § 323.6.

20. 40 C.F.R. § 230.10.

21. 40 C.F.R. § 230.10(a)(3).

22. Bonnie Malloy, "Symbolic Gestures or Our Saving Grace: The Relevance of Compensatory Mitigation for Florida's Wetlands in the Climate Change Era," *Journal of Land Use and Environmental Law* 27 (2011): 132–33.

23. 40 C.F.R. § 230.10(b).

24. 40 C.F.R. § 230.10(c).

25. 40 C.F.R. § 230.10(d).

26. 40 C.F.R. § 230.75.

27. 33 C.F.R. § 320.4(a).

28. See, e.g., Wetlands Action Network v. U.S. Army Corps of Engineers, 222 F.3d 1105 (9th Cir. 2000).

29. See, e.g., U.S. v. Mango, 199 F.3d 85 (2d Cir. 1999).

30. Gardner, *Lawyers, Swamps, and Money*, 75–76.

31. 33 U.S.C. § 1344(c).

32. "Nationwide Permit Reissuance," US Army Corps of Engineers, 2012, http://www.usace.army.mil/Portals/2/docs/civilworks/nwp/NWP2012_factsheet_15feb2012.pdf.

33. 33 U.S.C. § 1344(f).

34. 33 U.S.C. § 1344(f)(1)(E).

35. 33 U.S.C. § 1344(f)(2).

36. 129 S. Ct. 2458 (2009).

37. 40 C.F.R. § 232.2.

38. Michael B. Griffith et al., "The Effects of Mountaintop Mines and Valley Fills on the Physicochemical Quality of Stream Ecosystems in the Central Appalachians: A Review," *Science of the Total Environment* 417–18 (2012): 5.

39. Emily S. Bernhardt and Margaret A. Palmer, "The Environmental Costs of Mountaintop Mining Valley Fill Operations for Aquatic Ecosystems of the Central Appalachians," *Annals of the New York Academy of Sciences* 1223 (2011): 48; Gregory J. Pond et al., "Downstream Effects of Mountaintop Coal Mining: Comparing Biological Conditions Using Family- and Genus-Level Macroinvertebrate Bioassessment Tools," *Journal of the North American Benthological Society* 27 (2008): 724.

40. T. Ty Lindberg et al., "Cumulative Impacts of Mountaintop Mining on an Appalachian Watershed," *Proceedings of the National Academy of Sciences* 108 (2011): 20930–33.

41. Lindberg et al., "Cumulative Impacts of Mountaintop Mining," 20930–33.

42. J. D. Wickham et al., "The Effect of Appalachian Mountaintop Mining on Interior Forest," *Landscape Ecology* 22 (2007): 182–84.

43. 33 U.S.C. § 1344(e).

44. National Mining Association v. Jackson, 880 F. Supp. 2d 119 (D.D.C., 2012).

45. 40 C.F.R. § 230.10.

46. 33 C.F.R. § 332.1; 40 C.F.R. § 230.91.

47. "Memorandum of Agreement," 55 Fed. Reg. 9210.

48. 33 C.F.R. § 332.1(c)(2); 40 C.F.R. § 230.91(c)(2).

49. 40 C.F.R. § 230.91(c)(3).

50. Philip Womble and Martin Doyle, "The Geography of Trading Ecosystem Services: A Case Study of Wetland and Stream Compensatory Mitigation Markets," *Harvard Environmental Law Review* 36 (2012): 247.

51. 40 C.F.R. § 230.93(a)(2).

52. "Wetlands Protection," Environmental Protection Agency, accessed January 26, 2015, http://water.epa.gov/type/wetlands/protection.cfm.

53. U.S. v. Riverside Bayview Homes, 474 U.S. 121 (1985).

54. Nollan v. California Coastal Commission, 483 U.S. 825 (1987); Dolan v. City of Tigard, 512 U.S. 374 (1994).

55. 133 S. Ct. 2586 (2013).

56. 33 C.F.R. part 332; 40 C.F.R. part 230, subpart J.

57. 40 C.F.R. § 230.93(b)(1).

58. 40 C.F.R. § 230.93(e)(1).

59. 40 C.F.R. § 230.95.

60. 40 C.F.R. § 230.96.

61. 40 C.F.R. § 230.97.

62. 33 C.F.R. § 332.1; 40 C.F.R. § 230.91.

63. Womble and Doyle, "Geography of Trading Ecosystem Services," 248–49.

64. 33 C.F.R. § 332.2; 40 C.F.R. § 230.92; 65 Fed. Reg. 66914 (Nov. 7, 2000).

65. Womble and Doyle, "Geography of Trading Ecosystem Services," 251–52.

66. 40 C.F.R. § 230.98(b).

67. Womble and Doyle, "Geography of Trading Ecosystem Services," 249–51.

68. 40 C.F.R. § 230.98(b).

69. 40 C.F.R. § 230.93(f)(2).

70. 40 C.F.R. § 230.93(b)(2)–(3).

71. 33 C.F.R. § 332.2; 40 C.F.R. § 230.92.

72. 33 C.F.R. §§ 332.2, 332.2(d); 40 C.F.R. §§ 230.92, 230.03(d).

73. 40 C.F.R. § 230.98(d)(6)(ii)(A).

74. James M. Omernik and Robert G. Bailey, "Distinguishing between Watersheds and Ecoregions," *Journal of the American Water Resources Association* 33 (1997): 935–49.

75. Phil Roni, Karrie Hanson, and Tim Beechie, "Global Review of the Physical and Biological Effectiveness of Stream Habitat Rehabilitation Techniques," *North American Journal of Fisheries Management* 28 (2008): 858.

76. Martin W. Doyle and F. Douglas Shields, "Compensatory Mitigation for Streams under the Clean Water Act: Reassessing Science and Redirecting Policy," *Journal of the American Water Resources Association* 48 (2012): 506.

77. Doyle and Shields, "Compensatory Mitigation for Streams," 496–97.

78. Todd BenDor and Audrey Stewart, "Land Use Planning and Social Equity in North Carolina's Compensatory Wetland and Stream Mitigation Programs," *Environmental Management* 47 (2011): 245.

79. Todd BenDor, Joel Sholtes, and Martin W. Doyle, "Landscape Characteristics of a Stream and Wetland Mitigation Banking Program," *Ecological Applications* 19 (2009): 2083–85.

80. Roni, Hanson, and Beechie, "Stream Habitat Rehabilitation Techniques," 879.

81. Martin W. Doyle and Todd BenDor, "Evolving Law and Policy for Freshwater Ecosystem Service Markets," *William and Mary Environmental Law and Policy Review* 36 (2011): 166–72.

82. Jeffrey W. Matthews et al., "Relative Influence of Landscape vs. Local Factors on Plant Community Assembly in Restored Wetlands," *Ecological Applications* 19 (2009): 2108–9.

83. J. B. Ruhl and James Salzman, "The Effect of Wetland Mitigation Banking on People," *National Wetlands Newsletter* 28 (2006): 10–11; J. B. Ruhl, James Salzman, and Iris Goodman, "Implementing the New Ecosystem Services Mandate of the Section 404 Compensatory Mitigation Program—A Catalyst for Advancing Science and Policy," *Stetson Law Review* 38 (2009): 256.

84. BenDor and Stewart, "Land Use Planning and Social Equity," 245.

85. Womble and Doyle, "Geography of Trading Ecosystem Services," 289–95.

86. David Moreno-Mateos et al., "Structural and Functional Loss in Restored Wetland Ecosystems," *PLOS Biology* 10 (2012): e1001247.

87. Jeffrey W. Matthews and Greg Spyreas, "Convergence and Divergence in Plant Community Trajectories as a Framework for Monitoring Wetland Restoration Progress," *Journal of Applied Ecology* 47 (2010): 1132–34.

88. Joy B. Zedler and Suzanne Kercher, "Wetland Resources: Status, Trends, Ecosystem Services, and Restorability," *Annual Review of Environment and Resources* 30 (2005): 41.

89. Ramsar Convention, Article 6(1).

90. Ramsar Convention, Article 1(1).

91. Michael Bowman, Peter Davies, and Catherine Redgwell, *Lyster's International Wildlife Law*, 2nd ed. (New York: Cambridge University Press, 2010), 406–14.

92. Ramsar Convention, Article 2(1).

93. Ramsar Convention, Article 2(3).

94. Ramsar Convention, Article 2(2).

95. "Strategic Framework and Guidelines for the Future Development of the List of Wetlands of International Importance of the Convention on Wetlands," Ramsar Convention Secretariat, 2009, http://archive.ramsar.org/pdf/key_guide_list2009_e.pdf.

96. Ramsar Convention, Article 3(1).

97. Ramsar Convention, Resolution IX.1, Annex A.

98. Bowman, Davies, and Redgwell, *Lyster's International Wildlife Law*, 414–19.

99. Ramsar Convention, Article 3(2).

100. Ramsar Convention, Article 4(1).

101. Royal C. Gardner and Kim D. Connolly, "The Ramsar Convention on Wetlands: Assessment of International Designations within the United States," *National Wetlands Newsletter* 29 (2007): 6–9.

CHAPTER NINE

1. 16 U.S.C. § 1389(b).

2. Jessica Cheng, "One Sea Lion's Worth: Evaluating the Role of Values in Section 120," *UCLA Journal of Environmental Law and Policy* 29 (2011): 166–69.

3. Humane Society of U.S. v. Locke, 626 F.3d 1040 (9th Cir. 2010).

4. Brian D. Keller et al., "Climate Change, Coral Reef Ecosystems, and Management Options for Marine Protected Areas," *Environmental Management* 44 (2009): 1069–88.

5. UNCLOS, Article 3.

6. UNCLOS, Article 57.

7. UNCLOS, Article 58.

8. UNCLOS, Article 61(4).

9. "An Ocean Blueprint for the 21st Century," US Commission on Ocean Policy, 2004, http://govinfo.library.unt.edu/oceancommission/documents/full_color_rpt/000_ocean_full_report.pdf.

10. UNCLOS, Article 61(1)–(2).

11. UNCLOS, Article 61(3).

12. UNCLOS, Article 62.

13. UNCLOS, Article 62.

14. Rainer Froese and Alexander Proelß, "Rebuilding Fish Stocks No Later Than 2015: Will Europe Meet the Deadline?," *Fish and Fisheries* 11 (2010): 199.

15. UNCLOS, Article 76.

16. UNCLOS, Article 86.

17. UNCLOS, Articles 87, 116.

18. UNCLOS, Articles 117–118.

19. UNCLOS, Article 119(1)(a).

20. UNCLOS, Article 119(1)(b).

21. 16 U.S.C. § 1372(a)(1).

22. For a more comprehensive introduction to US laws affecting the oceans, see Donald C. Baur, Tim Eichenberg, and Michael Sutton, eds., *Ocean and Coastal Law and Policy* (Chicago: American Bar Association, 2008).

23. 16 U.S.C. § 1361(6).

24. 16 U.S.C. § 1371(a).

25. 16 U.S.C. § 1362(13).

26. 16 U.S.C. § 1362(18).

27. 50 C.F.R. § 18.3.

28. 50 C.F.R. § 216.3.

29. 16 U.S.C. § 1362(6).

30. 16 U.S.C. § 1362(6)(B).

31. 16 U.S.C. § 1362(12).

32. 16 U.S.C. § 1371(a)(1), (a)(5).

33. 16 U.S.C. § 1371(a)(5).

34. 16 U.S.C. § 1371(a)(5)(D).

35. 16 U.S.C. § 1371(b).

36. 16 U.S.C. § 1361(2).

37. 16 U.S.C. § 1362(9).

38. 50 C.F.R. § 216.3.

39. 16 U.S.C. § 1362(1).

40. 16 U.S.C. § 1383b(b).

41. "Conservation Plan for the Eastern Pacific Stock of Northern Fur Seals (*Callorhinus ursinus*)," NOAA Fisheries, 2007, http://www.nmfs.noaa.gov/pr/pdfs/conservation/plan_nfs_dec2007.pdf.

42. 16 U.S.C. § 1401.

43. 16 U.S.C. § 1403.

44. 16 U.S.C. § 1402.

45. 16 U.S.C. § 1402(d).

46. 16 U.S.C. § 1403(c).

47. Richard L. Wallace and Kathryn A. Semmens, "Social and Institutional Challenges in Species and Ecosystem Conservation: An Appraisal of the U.S. Marine Mammal Commission," *Policy Sciences* 43 (2010): 225–26.

48. "Ocean Blueprint."

49. 16 U.S.C. § 1387.

50. 16 U.S.C. § 1387(f).

51. 16 U.S.C. § 1362(20).

52. 16 U.S.C. § 1362(20).

53. 16 U.S.C. § 1387(b).

54. 50 C.F.R. § 229.2.

55. 16 U.S.C. § 1362(19).

56. 16 U.S.C. § 1386(c).

57. 16 U.S.C. § 1387(f).

58. 16 U.S.C. § 1387(f)(9).

59. 16 U.S.C. § 1387(f)(2).

60. Martin D. Robards et al., "Limitations of an Optimum Sustainable Population or Potential Biological Removal Approach for Conserving Marine Mammals: Pacific Walrus Case Study," *Journal of Environmental Management* 91 (2009): 57–58.

61. Jeffrey E. Moore et al., "A Review of Marine Mammal, Sea Turtle and Seabird Bycatch in USA Fisheries and the Role of Policy in Shaping Management," *Marine Policy* 33 (2009): 447–48.

62. Barbara L. Taylor et al., "Lessons from Monitoring Trends in Abundance of Marine Mammals," *Marine Mammal Science* 23 (2007): 157–75.

63. Taylor et al., "Lessons from Monitoring Trends," 158.

64. Robards et al., "Limitations of an Optimum Sustainable Population," 59.

65. Jeremy B. C. Jackson et al., "Historical Overfishing and the Recent Collapse of Coastal Ecosystems," *Science* 293 (2001): 635.

66. 16 U.S.C. § 1802(12).

67. 16 U.S.C. § 1802(13).

68. 16 U.S.C. § 1852.

69. 16 U.S.C. § 1852(a)(3).

70. 16 U.S.C. § 1802(21).

71. 16 U.S.C. § 1852(h).

72. 16 U.S.C. § 1851.

73. 16 U.S.C. § 1853(a).

74. 16 U.S.C. § 1802(33).

75. 16 U.S.C. § 1802(10).

76. 50 C.F.R. § 600.815(a)(2)(iv).

77. 16 U.S.C. § 1855(b)(2).

78. 50 C.F.R. § 600.815(a)(8).

79. 16 U.S.C. § 1853(b).

80. 16 U.S.C. § 1852(g).

81. 16 U.S.C. § 1852(h)(6).

82. 16 U.S.C. § 1854(a).

83. Robert Holahan, "Investigating Interest Group Representation on the Pacific Fisheries Management Council," *Marine Policy* 36 (2011): 783.

84. 16 U.S.C. § 1855(f).

85. 5 U.S.C. § 706.

86. Massachusetts v. Daley, 10 F. Supp. 2d 74 (D. Mass. 1998).

87. See, e.g., Hadaja, Inc. v. Evans, 263 F.3d 710 (9th Cir. 2002).

88. Jackson et al., "Historical Overfishing," 635-36.

89. Jackson et al., "Historical Overfishing," 634.

90. Jackson et al., "Historical Overfishing," 634.

91. Michael J. Fogarty and Steven A. Murawski, "Large-Scale Disturbance and the Structure of Marine Systems: Fishery Impacts on Georges Bank," *Ecological Applications* 8 (1998): S7.

92. Fogarty and Murawski, "Fishery Impacts on Georges Bank," S6.

93. Fogarty and Murawski, "Fishery Impacts on Georges Bank," S10.

94. Fogarty and Murawski, "Fishery Impacts on Georges Bank," S16-S17.

95. "Status of Stocks: Report on the Status of U.S. Fisheries for 2011," NOAA Fisheries, 2012, http://www.nmfs.noaa.gov/stories/2012/05/docs/status_of_stocks_2011_report.pdf.

96. 50 C.F.R. § 600.310(e)(2)(i)(E).

97. 50 C.F.R. § 600.310(e)(2)(i)(B).

98. 50 C.F.R. § 600.310(e)(2)(ii)(B).

99. T. A. Branch et al., "Fisheries Conservation and Management: Finding Consensus in the Midst of Competing Paradigms," *Animal Conservation* 15 (2012): 1-2.

100. Josh Eagle, "Domestic Fishery Management," in *Ocean and Coastal Law and Policy*, ed. Donald C. Baur, Tim Eichenberg, and Michael Sutton (Chicago: American Bar Association, 2008), 275-93.

101. Eric P. Palkovacs, "The Overfishing Debate: An Eco-evolutionary Perspective," *Trends in Ecology and Evolution* 26 (2011): 616-17.

102. Douglas P. Swain, Alan F. Sinclair, and J. Mark Hanson, "Evolutionary Response to Size-Selective Mortality in an Exploited Fish Population," *Proceedings of the Royal Society B* 274 (2007): 1018-19.

103. Fish Stocks Agreement, Article 5.

104. Fish Stocks Agreement, Article 5(b).

105. Fish Stocks Agreement, Article 8.

106. Fish Stocks Agreement, Article 21.

107. 16 U.S.C. § 1433(a).

108. 16 U.S.C. § 1433(b).

109. 16 U.S.C. §§ 1433(b)(2), 1434.

110. 16 U.S.C. § 1434(a)(2).

111. 16 U.S.C. § 1434(b)(1).

112. 16 U.S.C. § 1436.

113. 15 C.F.R. § 922.42.

114. 15 C.F.R. § 922.30.

115. 15 C.F.R. §§ 922.61, 922.72.

116. 16 U.S.C. § 1441.

117. 16 U.S.C. § 1441(c)(1).

118. 16 U.S.C. § 431.

119. 71 Fed. Reg. 36443; 50 C.F.R. § 404.6.

120. See, e.g., Joseph Briggett, "An Ocean of Executive Authority: Courts Should Limit the

President's Antiquities Act Power to Designate Monuments in the Outer Continental Shelf," *Tulane Environmental Law Journal* 22 (2009): 403–22.

121. Brent J. Hartman, "Extending the Scope of the Antiquities Act," *Public Land and Resources Law Review* 32 (2011): 173–74.

122. Arnaud Grüss et al., "Consequences of Adult and Juvenile Movement for Marine Protected Areas," *Biological Conservation* 144 (2011): 699–700.

123. Grüss et al., "Marine Protected Areas," 693.

124. Grüss et al., "Marine Protected Areas," 695.

125. Erin S. Dunlop et al., "Propensity of Marine Reserves to Reduce the Evolutionary Effects of Fishing in a Migratory Species," *Evolutionary Applications* 2 (2009): 372.

126. "The National Action Plan to Conserve Coral Reefs," US Coral Reef Task Force, 2000, http://www.coralreef.gov/about/CRTFAxnPlan9.pdf.

127. Peter J. Mumby and Robert S. Steneck, "Coral Reef Management and Conservation in Light of Rapidly Evolving Ecological Paradigms," *Trends in Ecology and Evolution* 23 (2008): 561–62.

128. Brian D. Keller et al., "Climate Change, Coral Reef Ecosystems, and Management Options for Marine Protected Areas," *Environmental Management* 44 (2009): 1077–78.

129. Robin K. Craig, "Ocean Governance for the 21st Century: Making Marine Zoning Climate Change Adaptable," *Harvard Environmental Law Review* 36 (2012): 344–46.

130. Randall S. Abate, "Marine Protected Areas as a Mechanism to Promote Marine Mammal Conservation: International and Comparative Law Lessons for the United States," *Oregon Law Review* 88 (2009): 308.

131. 33 U.S.C. § 1362(12).

132. 33 U.S.C. § 1362(7).

133. 33 U.S.C. § 1362(8).

134. 33 U.S.C. § 1362(9).

135. But see Proclamation No. 7219, which extended the US contiguous zone to 44 kilometers (24 nautical miles).

136. 33 U.S.C. § 1362(10).

137. 33 U.S.C. § 1344(a).

138. 33 U.S.C. § 1342(a)(1).

139. 33 U.S.C. §§ 1312(a), 1313(c)(2)(A).

140. 33 U.S.C. § 1343.

141. 33 U.S.C. § 1343(c).

142. 40 C.F.R. § 125.120–124.

143. 40 C.F.R. § 125.122.

144. 40 C.F.R. § 125.121(e).

145. 40 C.F.R. § 125.122.

146. 33 U.S.C. § 1362(12).

147. 40 C.F.R. § 122.3(a).

148. Northwest Environmental Advocates v. U.S. Environmental Protection Agency, 537 F.3d 1006 (9th Cir. 2008).

149. In 2008 Congress passed the Clean Boating Act as an amendment to the CWA; it exempts recreational vessels from NPDES permitting for discharges.

150. 33 U.S.C. § 1362(6)(A).

151. 33 U.S.C. § 1322(a)(6).

152. 33 U.S.C. § 1322.

153. 33 U.S.C. § 1322(b)(1).

154. 33 U.S.C. § 1322(f).

155. Robert J. Diaz and Rutger Rosenberg, "Spreading Dead Zones and Consequences for Marine Ecosystems," *Science* 321 (2008): 926–27.

156. Diaz and Rosenberg, "Dead Zones," 927.

157. The Ocean Dumping Act implements the Convention on the Prevention of Marine Pollution by Dumping of Wastes and Other Matter 1972 (London Convention), of which the United States is a party.

158. 33 U.S.C. § 1411(a).

159. 33 U.S.C. § 1411(b).

160. 33 U.S.C. § 1402(b); 40 C.F.R. § 220.2(c).

161. 33 U.S.C. § 1402(c).

162. 33 U.S.C. §§ 1412, 1413.

163. 40 C.F.R. §§ 227.2, 227.4.

164. 40 C.F.R. § 228.4.

165. 40 C.F.R. § 228.5.

166. 33 U.S.C. § 1414b(a).

167. 33 U.S.C. § 1412(d).

168. Stacie E. Crowe et al., "Impact of the Charleston Ocean Dredged Material Disposal Site on Nearby Hard Bottom Reef Habitats," *Marine Pollution Bulletin* 60 (2010): 679.

169. 33 U.S.C. § 1413.

170. 33 U.S.C. § 1413(c).

171. See, e.g., Crowe et al., "Dredged Material," 679–91; James A. Blake et al., "Long-Term Benthic Infaunal Monitoring at a Deep-Ocean Dredged Material Disposal Site off Northern California," *Deep-Sea Research II* 56 (2009): 1775–1803.

172. "National Coastal Condition Report IV," Environmental Protection Agency, 2012, http://water.epa.gov/type/oceb/assessmonitor/nccr/upload/0_NCCR_4_Report_508_bookmarks.pdf.

173. "Coastal Condition Report."

174. Barton H. Thompson Jr., "A Federal Act to Promote Integrated Water Management: Is the CZMA a Useful Model?," *Environmental Law* 42 (2012): 218–21.

175. 16 U.S.C. § 1453(1).

176. 16 U.S.C. § 1455(b).

177. 16 U.S.C. § 1455(d).

178. Kristen M. Fletcher, "Managing Coastal Development," in *Ocean and Coastal Law and Policy*, ed. Donald C. Baur, Tim Eichenberg, and Michael Sutton (Chicago: American Bar Association, 2008), 149.

179. 15 C.F.R. § 923.21.

180. 15 C.F.R. § 923.20.

181. 15 C.F.R. § 923.133.

182. 16 U.S.C. § 1455b.

183. 16 U.S.C. § 1455b(g)(5).

184. 16 U.S.C. § 1455b(c)(3), (4).

185. 16 U.S.C. § 1455b(c)(2).

186. 16 U.S.C. § 1455b(f).

187. 16 U.S.C. § 1456.

188. 16 U.S.C. § 1456(c)(1)(A).

189. 16 U.S.C. § 1456(c)(3)(A).

190. 16 U.S.C. § 1461.

191. 16 U.S.C. § 1461(c).

192. 15 C.F.R. § 921.1(d).

193. 16 U.S.C. § 1461(f)(3).

CHAPTER TEN

1. "Effects of Acid Rain—Forests," Environmental Protection Agency, 2012, http://www.epa.gov/acidrain/effects/forests.html.

2. "Effects of Acid Rain—Surface Waters and Aquatic Animals," Environmental Protection Agency, 2012, http://www.epa.gov/acidrain/effects/surface_water.html.

3. Vickie L. Patton, "The New Air Quality Standards, Regional Haze, and Interstate Air Pollution Transport," *Environmental Law Reporter* 28 (1998): 10160–61.

4. Patton, "New Air Quality Standards," 10160–61.

5. "The Acidic Deposition Phenomenon and Its Effects, Critical Assessment Review Papers, volume 1, Atmospheric Science," Environmental Protection Agency, 1983, http://nepis.epa.gov/Exe/ZyPDF.cgi/2000G3OU.PDF?Dockey=2000G3OU.PDF.

6. "Acidic Deposition Phenomenon."

7. Richard L. Revesz, "Federalism and Interstate Environmental Externalities," *University of Pennsylvania Law Review* 144 (1996): 2353.

8. Revesz, "Federalism and Interstate Environmental Externalities," 2351–52.

9. "Acidic Deposition Phenomenon."

10. "Acidic Deposition Phenomenon."

11. "Acidic Deposition Phenomenon."

12. Patton, "New Air Quality Standards," 10161–63.

13. Patton, "New Air Quality Standards," 10161–63.

14. 42 U.S.C. § 7410(a)(2)(E), amended in 1990 by 42 U.S.C. § 7410(a)(2)(D), where it was changed to "contribute significantly to nonattainment."

15. 42 U.S.C. § 7426(b).

16. 42 U.S.C. § 7423.

17. 42 U.S.C. § 7423(c).

18. North Carolina v. Environmental Protection Agency, 531 F.3d 896 (D.C. Cir. 2008), modified on rehearing, 550 F.3d 1176.

19. 42 U.S.C. § 7408(a).

20. 42 U.S.C. § 7408(a).

21. 42 U.S.C. § 7409.

22. 42 U.S.C. § 7409(b)(1).

23. 42 U.S.C. § 7409(b)(2).

24. 42 U.S.C. § 7602(h).

25. 531 U.S. 457 (2001).

26. 42 U.S.C. § 7409(d).

27. 42 U.S.C. § 7409(d)(2).

28. 42 U.S.C. § 7410.

29. 42 U.S.C. § 7410(a)(2)(D).

30. 42 U.S.C. § 7410(c).

31. 42 U.S.C. § 7661a.

32. 42 U.S.C. §§ 7412(a)(1), 7602; 40 C.F.R. § 70.2.

33. 42 U.S.C. § 7511.

34. 42 U.S.C. § 7502(c)(1).

35. 42 U.S.C. § 7502(c)(5).

36. 42 U.S.C. § 7503(a)(2).

37. 42 U.S.C. § 7503(c).

38. 42 U.S.C. § 7411.

39. 42 U.S.C. § 7411(a)(2).

40. 42 U.S.C. § 7411(a)(4).

41. James Salzman and Barton H. Thompson Jr., *Environmental Law and Policy*, 3rd ed. (New York: Foundation Press, 2010), 97–98.

42. 42 U.S.C. § 7470.

43. 42 U.S.C. §§ 7472, 7474.

44. 42 U.S.C. § 7473.

45. 42 U.S.C. § 7473(b)(1).

46. 42 U.S.C. § 7473(b)(2).

47. 42 U.S.C. § 7475(a)(4).

48. 42 U.S.C. § 7412.

49. 42 U.S.C. § 7412(b)(2).

50. 42 U.S.C. § 7412(b)(1).

51. 42 U.S.C. § 7521; this section of the CAA is often referred to as Title II.

52. 42 U.S.C. § 7521(a)(1).

53. 42 U.S.C. § 7521(a)(3)(A)(i).

54. J. Nillson and P. Grennfelt, *Critical Loads for Sulfur and Nitrogen*, Miljörapport 15 (Copenhagen: Nordic Council of Ministers, 1988).

55. Tara L. Greaver et al., "Ecological Effects of Nitrogen and Sulfur Air Pollution in the U.S.: What Do We Know?," *Frontiers in Ecology and the Environment* 10 (2012): 371.

56. An De Schrijver et al., "Cumulative Nitrogen Input Drives Species Loss in Terrestrial Ecosystems," *Global Ecology and Biogeography* 20 (2011): 812.

57. De Schrijver et al., "Cumulative Nitrogen Input," 812.

58. "Policy Assessment for the Review of the Particulate Matter National Ambient Air Quality Standards," Environmental Protection Agency, 2011, 5-2, http://www.epa.gov/ttn/naaqs/standards /pm/data/20110419pmpafinal.pdf.

59. "Integrated Science Assessment for Particulate Matter," Environmental Protection Agency, 2009, 3-4, http://www.epa.gov/ncea/pdfs/partmatt/Dec2009/PM_ISA_full.pdf.

60. "Assessment for Particulate Matter," 3-4.

61. "Assessment for Particulate Matter," 3-6.

62. "Assessment for Particulate Matter," 9-193.

63. "Assessment for Particulate Matter," 9-193.

64. "Assessment for Particulate Matter," 3-18.

65. "Assessment for Particulate Matter," 9-175.

66. 78 Fed. Reg. 3086.

67. "Integrated Science Assessment for Carbon Monoxide," Environmental Protection Agency, 2010, 2-2, http://ofmpub.epa.gov/eims/eimscomm.getfile?p_download_id=494432.

68. "Policy Assessment for the Review of the Carbon Monoxide National Ambient Air Quality Standards," Environmental Protection Agency, 2010, 3-4, http://www.epa.gov/ttn/naaqs/standards /co/data/20101022copafinal.pdf.

69. "Integrated Science Assessment for Ozone and Related Phytochemical Oxidants," Environmental Protection Agency, 2013, 1-2, http://ofmpub.epa.gov/eims/eimscomm.getfile?p_download _id=511347.

70. "Assessment for Ozone," 1-2.

71. "Assessment for Ozone," 3-6.

72. "Assessment for Ozone," 1-2.

73. "Assessment for Ozone," 2-35.

74. "Policy Assessment for the Review of the Ozone National Ambient Air Quality Standards: Second External Review Draft," Environmental Protection Agency, 2014, 5-54, http://www.epa .gov/ttn/naaqs/standards/ozone/data/20140131pa.pdf.

75. "Assessment for Ozone," 2-39.

76. "Assessment for Ozone," 9-81.

77. "Policy Assessment for the Review of the Ozone National Ambient Air Quality Standards," 5-55.

78. "Policy Assessment for the Review of the Ozone National Ambient Air Quality Standards," 6-47.

79. "Assessment for Ozone and Related Phytochemical Oxidants," 9-118.

80. Richard G. Newell and Kristian Rogers, "The U.S. Experience with the Phasedown of Lead in Gasoline," *Resources for the Future* (2003): 1.

81. Newell and Rogers, "Phasedown of Lead in Gasoline," 2.

82. "Integrated Science Assessment for Lead," Environmental Protection Agency, 2013, 1-6, http://ofmpub.epa.gov/eims/eimscomm.getfile?p_download_id=518908.

83. "Assessment for Lead," 2-2.

84. "Assessment for Lead," 2-5.

85. "Assessment for Lead," 1-60.

86. "Assessment for Lead," 6-255.

87. "Assessment for Lead," 6-191.

88. "Assessment for Lead," 6-222.

89. "Assessment for Lead," 1-46.

90. "Policy Assessment for the Review of the Lead National Ambient Air Quality Standards," Environmental Protection Agency, 2014, ES-5, http://www.epa.gov/ttn/naaqs/standards/pb/data/140501_pa_pb_fin.pdf.

91. "Policy Assessment for the Review of the Lead National Ambient Air Quality Standards," 1-14.

92. "Integrated Science Assessment for Oxides of Nitrogen and Sulfur—Ecological Criteria," Environmental Protection Agency, 2008, 2-4, http://ofmpub.epa.gov/eims/eimscomm.getfile?p_download_id=485280.

93. "Assessment for Oxides of Nitrogen and Sulfur," 2-11.

94. "Assessment for Oxides of Nitrogen and Sulfur," 3-5.

95. Greaver et al., "Nitrogen and Sulfur Air Pollution," 367-68.

96. "Assessment for Oxides of Nitrogen and Sulfur," 3-29.

97. "Assessment for Oxides of Nitrogen and Sulfur," 3-60.

98. "Assessment for Oxides of Nitrogen and Sulfur," 3-22.

99. Greaver et al., "Nitrogen and Sulfur Air Pollution," 367-68.

100. "Assessment for Oxides of Nitrogen and Sulfur," 3-14.

101. "Assessment for Oxides of Nitrogen and Sulfur," 3-48.

102. Daniel C. Josephson et al., "Chemical and Biological Recovery from Acid Deposition within the Honnedaga Lake Watershed, New York, USA," *Environmental Monitoring and Assessment* 186 (2014): 4397.

103. Josephson et al., "Honnedaga Lake Watershed," 4403-4.

104. Josephson et al., "Honnedaga Lake Watershed," 4404.

105. Josephson et al., "Honnedaga Lake Watershed," 4406.

106. Greaver et al., "Nitrogen and Sulfur Air Pollution," 369.

107. "Assessment for Oxides of Nitrogen and Sulfur," 3-148.

108. Sara Jovan et al., "Eutrophic Lichens Respond to Multiple Forms of N: Implications for Critical Levels and Critical Loads Research," *Ecological Applications* 22 (2012): 1910.

109. Jovan et al., "Eutrophic Lichens," 1910.

110. Greaver et al., "Nitrogen and Sulfur Air Pollution," 370.

111. 65 Fed. Reg. 79825.

112. 76 Fed. Reg. 24976.

113. "Assessment for Oxides of Nitrogen and Sulfur," 12.

114. "Mercury Study Report to Congress, volume 6, An Ecological Assessment for Anthropo-

genic Mercury Emissions in the United States," Environmental Protection Agency, 1997, ES-7, http://www.epa.gov/ttn/caaa/t3/reports/volume6.pdf.

115. "Mercury Study Report to Congress," ES-8.

116. "Mercury Study Report to Congress," 4-3.

117. "Mercury Study Report to Congress," 2-3.

118. "Policy Assessment for the Review of the Secondary National Ambient Air Quality Standards for Oxides of Nitrogen and Oxides of Sulfur," Environmental Protection Agency, 2011, ES-5, http://www.epa.gov/ttn/naaqs/standards/no2so2sec/data/20110114pamain.pdf.

119. "Policy Assessment for Oxides of Nitrogen and Oxides of Sulfur," ES-5.

120. 42 U.S.C. § 7651(b).

121. 42 U.S.C. § 7651(f).

122. "Air Quality Trends," Environmental Protection Agency, 2014, http://www.epa.gov/airtrends/aqtrends.html.

123. M. Alisa Mast, "Evaluation of Stream Chemistry Trends in U.S. Geological Survey Reference Watersheds, 1970–2010," *Environmental Monitoring and Assessment* 185 (2013): 9352.

124. Mast, "Evaluation of Stream Chemistry Trends," 9354.

125. 42 U.S.C. § 7410(a)(2)(D).

126. David A. Evans and Richard T. Woodward, "What Can We Learn from the End of the Grand Policy Experiment? The Collapse of the National SO_2 Trading Program and Implications for Tradable Permits as a Policy Instrument," *Annual Review of Resource Economics* 5 (2013): 332.

127. 70 Fed. Reg. 25162.

128. North Carolina v. Environmental Protection Agency, 531 F.3d 896 (D.C. Cir. 2008), modified on rehearing, 550 F.3d 1176.

129. 76 Fed. Reg. 48208.

130. Environmental Protection Agency v. EME Homer City Generation, No. 12-1182 (2014).

131. Evans and Woodward, "SO_2 Trading Program," 336.

132. Evans and Woodward, "SO_2 Trading Program," 334–37.

133. EME Homer City Generation v. Environmental Protection Agency, 696 F.3d 7 (D.C. Cir. 2012).

134. Environmental Protection Agency v. EME Homer City Generation.

135. Evans and Woodward, "SO_2 Trading Program," 343.

136. "Fact Sheet: Mercury and Air Toxics Standards for Power Plants," Environmental Protection Agency, http://www.epa.gov/mats/pdfs/20111221MATSsummaryfs.pdf.

137. 77 Fed. Reg. 9304.

138. As this book was going to press, the Supreme Court ruled that the EPA did not properly consider cost when creating the Mercury and Air Toxics Standards (*Michigan v. Environmental Protection Agency*, No. 14-46). The EPA will have to review the standards, taking cost of compliance into consideration before deciding whether regulation is "appropriate and necessary."

CHAPTER ELEVEN

1. Geoffrey Craig, "Six U.S. States Leave the Western Climate Initiative," *Platts*, November 18, 2011, http://www.platts.com/latest-news/electric-power/washington/six-us-states-leave-the-western-climate-initiative-6695863.

2. Craig, "Western Climate Initiative."

3. "Climate Change 2013: The Physical Science Basis," Intergovernmental Panel on Climate Change, 2013, http://www.ipcc.ch/pdf/assessment-report/ar5/wg1/WG1AR5_ALL_FINAL.pdf.

4. "Physical Science Basis."

5. National Research Council of the National Academies, *Abrupt Impacts of Climate Change: Anticipating Surprises* (Washington, DC: National Academies Press, 2013), 4.

6. "Physical Science Basis," emphasis in original.

7. "Inventory of U.S. Greenhouse Gas Emissions and Sinks: 1990–2011," Environmental Protection Agency, 2013, http://www.epa.gov/climatechange/Downloads/ghgemissions/US-GHG -Inventory-2013-Main-Text.pdf.

8. "Trends in Global CO_2 Emissions: 2013 Report," European Commission's Joint Research Centre, 2013, http://edgar.jrc.ec.europa.eu/news_docs/pbl-2013-trends-in-global-co2-emissions -2013-report-1148.pdf.

9. "Physical Science Basis."

10. "Physical Science Basis."

11. The Working Group II of the IPCC does an excellent job of reviewing and summarizing the scientific research on the effects of climate change on ecosystems in its *Impacts, Adaptation and Vulnerability* report.

12. "Climate Change 2007: Synthesis Report," Intergovernmental Panel on Climate Change, 2008, http://www.ipcc.ch/pdf/assessment-report/ar4/syr/ar4_syr_full_report.pdf.

13. "Climate Change 2007: Impacts, Adaptation and Vulnerability," Intergovernmental Panel on Climate Change, 2007, http://www.ipcc.ch/pdf/assessment-report/ar4/wg2/ar4_wg2_full_report .pdf.

14. "Impacts, Adaptation and Vulnerability."

15. National Research Council, *Climate Change*, 114.

16. "Synthesis Report."

17. "Synthesis Report."

18. Justin Gillis, "Climate Change Seen Posing Risk to Food Supplies," *New York Times*, November 1, 2013, http://www.nytimes.com/2013/11/02/science/earth/science-panel-warns-of-risks-to -food-supply-from-climate-change.html.

19. "Synthesis Report."

20. "Policy Options to Reduce Greenhouse Gas Emissions," Parliamentary Information and Research Service, Canada, 2008, http://www.parl.gc.ca/content/LOP/ResearchPublications/prb0819 -e.pdf.

21. California Code of Regulations, Title 17, § 95802(a)(4).

22. Chadwick D. Rittenhouse and Adena R. Rissman, "Forest Cover, Carbon Sequestration, and Wildlife Habitat: Policy Review and Modeling of Tradeoffs among Land-Use Change Scenarios," *Environmental Science and Policy* 21 (2012): 95.

23. A book that does a good job of pulling together disparate strands concerning the law and science of climate change is Hari M. Osofsky and Lesley K. McAllister, *Climate Change Law and Policy* (New York: Wolters Kluwer Law and Business, 2012).

24. UNFCCC, Article 2.

25. UNFCCC, Article 3.

26. Kyoto Protocol, Article 6.

27. Kyoto Protocol, Annex B.

28. Mark Landler, "U.S. and China Reach Climate Accord after Months of Talks," *New York Times*, November 11, 2014, http://www.nytimes.com/2014/11/12/world/asia/china-us-xi-obama -apec.html.

29. Landler, "Climate Accord."

30. Guus J. M. Velders et al., "The Importance of the Montreal Protocol in Protecting Climate," *Proceedings of the National Academy of Sciences* 104 (2007): 4818.

31. "Trends in Global CO_2 Emissions."

32. "Trends in Global CO_2 Emissions."

33. "Inventory of U.S. Greenhouse Gas Emissions and Sinks."

34. "Inventory of U.S. Greenhouse Gas Emissions and Sinks."

35. "Inventory of U.S. Greenhouse Gas Emissions and Sinks."

36. 42 U.S.C. § 7411.

37. 42 U.S.C. § 7470.

38. 42 U.S.C. § 7521.

39. 549 U.S. 497 (2007).

40. 42 U.S.C. § 7602(g).

41. 42 U.S.C. § 7521(a)(1).

42. 42 U.S.C. § 7602(h).

43. 75 Fed. Reg. 25324; 76 Fed. Reg. 57106.

44. 77 Fed. Reg. 62624.

45. 42 U.S.C. § 7479(1).

46. 75 Fed. Reg. 31514.

47. 134 S. Ct. 2427 (2014).

48. 134 S. Ct. at 2444.

49. Howard A. Learner, "Emerging Clarity on Climate Change Law: EPA Empowered and State Common Law Remedies Enabled" *Environmental Law Reporter News and Analysis* 44 (2014): 10745.

50. 79 Fed. Reg. 34830.

51. 131 S. Ct. 2527 (2011).

52. Erie Railroad Co. v. Tompkins, 304 U.S. 64 (1938).

53. "Renewable Portfolio Standard," US Department of Energy, accessed April 23, 2015, http://energy.gov/savings/renewable-portfolio-standard.

54. Assembly Bill No. 32 (2006).

55. "Regional Greenhouse Gas Initiative: Memorandum of Understanding," Regional Greenhouse Gas Initiative Inc., 2005, http://www.rggi.org/docs/mou_final_12_20_05.pdf.

56. Beth Daley, "Mass. to Lower Cap on Emissions by Power Plants," *Boston Globe*, February 8, 2013, http://www.bostonglobe.com/lifestyle/health-wellness/2013/02/08/massachusetts-and -other-states-cut-allowed-power-plant-emissions/lsSzXaN111BoZ4nvyY3WdJ/story.html.

57. "State and Local Climate Adaptation," Center for Climate and Energy Solutions, accessed January 29, 2015, http://www.c2es.org/us-states-regions/policy-maps/adaptation.

58. "2009 California Climate Adaptation Strategy," California Natural Resources Agency, 2009, http://resources.ca.gov/docs/climate/Statewide_Adaptation_Strategy.pdf.

59. See, e.g., "Responding to Climate Change in New York State," New York State Energy Research and Development Authority, 2011, http://www.nyserda.ny.gov/climaid.

60. "Preparing for a Changing Climate: Washington State's Integrated Climate Response Strategy," Department of Ecology, State of Washington, 2012, https://fortress.wa.gov/ecy/publications /publications/1201004.pdf.

Index